Algorithms for Intelligent Systems

This book series publishes research on the analysis and development of algorithms for intelligent systems with their applications to various real world problems. It covers research related to autonomous agents, multi-agent systems, behavioral modeling, reinforcement learning, game theory, mechanism design, machine learning, meta-heuristic search, optimization, planning and scheduling, artificial neural networks, evolutionary computation, swarm intelligence and other algorithms for intelligent systems.

The book series includes recent advancements, modification and applications of the artificial neural networks, evolutionary computation, swarm intelligence, artificial immune systems, fuzzy system, autonomous and multi agent systems, machine learning and other intelligent systems related areas. The material will be beneficial for the graduate students, post-graduate students as well as the researchers who want a broader view of advances in algorithms for intelligent systems. The contents will also be useful to the researchers from other fields who have no knowledge of the power of intelligent systems, e.g. the researchers in the field of bioinformatics, biochemists, mechanical and chemical engineers, economists, musicians and medical practitioners.

The series publishes monographs, edited volumes, advanced textbooks and selected proceedings.

More information about this series at http://www.springer.com/series/16171

Algorithms for Intelligent Systems

This book series publishes research on the analysis and development of algorithms for intelligent systems with their applications to various real world problems. It covers research related to autonomous agents, multi-agent systems, behavioral modeling, reinforcement learning, game theory, mechanism design, machine learning, meta-heuristic search, optimization, planning and scheduling, artificial neural networks, evolutionary computation, swarm intelligence and other algorithms for intelligent systems.

The book series includes recent advancements, modification and applications of the artificial neural networks, evolutionary computation, swarm intelligence, artificial immune systems, fuzzy system, autonomous and multi agent systems, machine learning and other intelligent systems related areas. The material will be beneficial for the graduate students, post-graduate students as well as the researchers who want a broader view of advances in algorithms for intelligent systems. The contents will also be useful to the researchers from other fields who have no knowledge of the power of intelligent systems, e.g. the researchers in the field of bioinformatics, biochemists, mechanical and chemical engineers, economists, musicians and medical practitioners.

The series publishes monographs, edited volumes, advanced textbooks and selected proceedings.

More information about this series at http://www.springer.com/series/16171

Jyotsna Kumar Mandal · Joyanta Kumar Roy
Editors

Proceedings of International Conference on Computational Intelligence and Computing

ICCIC 2020

Springer

Editors
Jyotsna Kumar Mandal
Department of Computer Science
and Engineering
University of Kalyani
Kalyani, West Bengal, India

Joyanta Kumar Roy
Department of Electronics
and Communication Engineering
Eureka Scientech Research Foundation
Agarpara, West Bengal, India

ISSN 2524-7565 ISSN 2524-7573 (electronic)
Algorithms for Intelligent Systems
ISBN 978-981-16-3367-6 ISBN 978-981-16-3368-3 (eBook)
https://doi.org/10.1007/978-981-16-3368-3

This Springer imprint is published by the registered company Springer Nature Singapore Pte Ltd.
The registered company address is: 152 Beach Road, #21-01/04 Gateway East, Singapore 189721, Singapore

Preface

With explosive growth of e-business, artificial intelligence and IoT-based applications, the researchers in academics and industry face more and more challenging problems in computing that require computational intelligence to cater the intelligent e-applications, high-speed wireless networks, autonomous systems and in almost every application with enhanced quality of intelligence. The International Conference on Computational Intelligence and Computing-2020 (ICCIC-2020ne) has been organized by SRGI Group of Institution, Jhansi, Eureka Scientech Research Foundation, Kolkata, IETE, Kolkata. Springer is the publication partner of this conference. The conference was held on virtual platform during 19–20 February 2021 hosted by Bundelkhand University, Jhansi. The conference received 87 papers through Easy Chair from authors from various countries on various topics such as soft computing and artificial intelligence, networking and architectures, image processing and signal analysis, safety and security, modelling, simulation and management, Internet of Things and other allied areas. The papers were processed through peer review selection of high standard, and therefore, 30 papers are successfully accepted and presented.

The inaugural session was highlighted and sparkled with the galaxy of eminent dignitaries like vice chancellor of universities, distinguish speakers of IEEE, Ex-IIT Directors, President of IETE. There were seven keynotes, and out of them, five were from different countries. The conference was a truly international, since the papers were received from Australia, Bangladesh, UK, Vietnam. The conference was truly interdisciplinary platform for scientists, researchers, engineers, practitioners and educators to present and discuss the most recent innovations, trends and concerns as well as practical challenges encountered and solutions adopted in the field of computational Intelligence and computing for various applications.

The main objective of the conference was to bring together academic and industrial experts of the research community to highlight key issues, identify trends and develop a vision of the computational intelligence for future applications from a design, deployment and operation standpoints, and the objectives were fulfilled with grand success.

This conference was technically co-sponsored by IEEE, CAS, Kolkata, and IET (UK) Kolkata Section. There was a pre-conference one-day tutorial on Internet of Things, This Tutorial was hosted by IETE, Kolkata, on Virtual platform on 7 February 2021.

On behalf of the organization committee of the ICCIC-2020ne, our gratitude to all speakers, esteem reviewers, authors and participants.

The proceedings is published in one volume. This will be a valuable document to the researchers, budding engineers and graduate and postgraduate students.

Jyotsna Kumar Mandal
Professor, Department of Computer Science
and Engineering, University of Kalyani
Kalyani, India

Joyanta Kumar Roy
Former Dean (Research and Consultancies)
and Professor, Department of Electronics
and Communication Engineering
MCKV Institute of Engineering
Howrah, India

Contents

About the Editors

Professor Jyotsna Kumar Mandal received his M.Sc. in Physics from Jadavpur University in 1986 and M.Tech. in Computer Science from the University of Calcutta and was awarded a Ph.D. in Computer Science and Engineering Jadavpur University in 2000. Currently, he is a Professor of Computer Science and Engineering and was Dean of the Faculty of Engineering, Technology, and Management, Kalyani University, West Bengal, for two consecutive terms. He started his career as Lecturer at NERIST, Arunachal Pradesh, in September 1988 and has 30 years of teaching and research experience. His research areas include coding theory, data and network security, remote sensing and GIS-based applications, data compression, error correction, visual cryptography, steganography, security in MANET, wireless networks, and unified computing. He has been a Life Member of the Computer Society of India since 1992, CRSI since 2009, ACM since 2012, IEEE since 2013, and Fellow of IETE since 2012. He has chaired more than 60 sessions at various international conferences and delivered more than 60 expert/invited lectures during the last 5 years. He has acted as Program Chair of several international conferences and edited more than 30 proceedings volumes. He is a Reviewer for various international journals and conferences and has published over 360 articles and 6 books. He is one of the editors for the Springer AISC and CCIS Series. He has published more than 450 articles, out of which 160 papers are published in the international journal. He has authored 12 books. Twenty-four scholars were awarded Ph.D. degrees under his supervision.

Prof. (Dr.) Joyanta Kumar Roy, Ph.D. (Tech.), has been associated with academics and industry in Electronics and Automation Engineering for the last 40 years. His present assignments include Visiting Professor, Narula Institute of Technology (Autonomous unit of MAKAUT), India; Company Director, System Advance Technologies, Kolkata, India, and Founder Chairman, Eureka Scientech Research Foundation, Kolkata. He is also a Freelance Consultant to many industries for providing design support toward smart technology in the water industry. Earlier, he served three engineering institutions in India with various capacities, including Principal, Dean, and Professor. He is a Senior Member of IEEE and past Chairman of IET Kolkata network, Fellow Member of IETE and IWWA, regular Reviewer of IEEE and Springer Journals, and Associate Editor of S2IS journal. He has published a

Design of Portable Integrated Sensor System for Local Monitoring of Climate Change

Bansari Deb Majumder and Joyanta Kumar Roy

1 Introduction

Due to the ever-increasing demand for developing an intelligent system that has human-like intelligence, scientists have focused mainly on emulating the sensing capacity of a human being. A sensor is a kind of physical device which has response to any kind of physical stimulus by giving an electrical output as the response of stimulus. The modern-day state-of-the-art instrumentation system is equipped with different sensors. Each sensor measures a specific parameter independently, and necessary signal processing algorithms are used to combine all the independent measurements to provide a complete measurement that takes inference from all the measurements. Such kind of system can be termed as a multi-sensor system. In a multi-sensor system, coordination of all the available sensors is essential to achieve a system objective. One of the types of multi-sensor system is the integrated sensors. This is also called as multimode sensors. Figure 1 presents a generalized framework of an integrated sensor. The multiple sensors are embedded in one board which provides accurate measurement data to the sensor fusion level where state-of-the-art signal processing application is carried out. The success and reliability of the integrated sensor depend on the design of reliable sensor technology which gives the superior sensing capability to the system.

Traditionally, a single unit of the sensor provides a single measurement, but with the advancement of technology, it is observed that many measurements can be extracted from a single unit of multi-sensor. Due to the advancement of the semiconductor industry, MEMS strain gauges came into existence in 1958 [1]. Zean

B. D. Majumder (✉)
Department of Electronics and Instrumentation Engineering, Narula Institute of Technology, Kolkata, India

J. K. Roy
Eureka Scientech Research Foundation, Kolkata, India

J. K. Mandal and J. K. Roy (eds.), *Proceedings of International Conference on Computational Intelligence and Computing*, Algorithms for Intelligent Systems,
https://doi.org/10.1007/978-981-16-3368-3_1

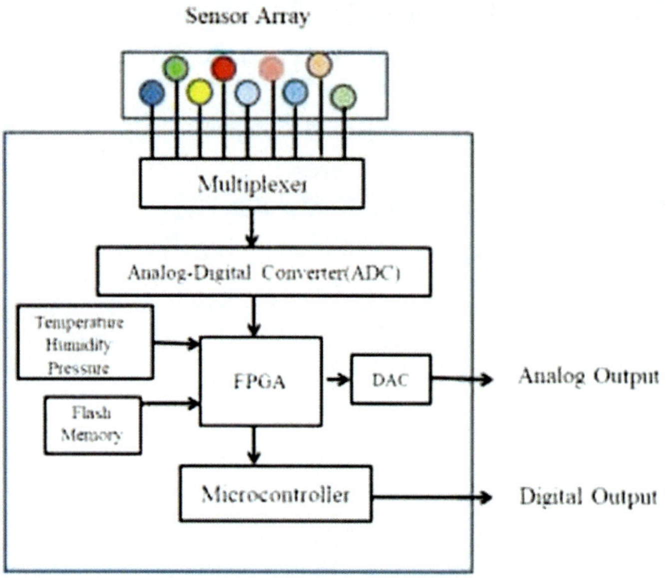

Fig. 1 General framework of integrated sensor

Fang et al. have designed a micro-integrated system that measures temperature and humidity simultaneously for weather station [2]. Vankatesh et al. developed integrated sensing model for weather station based on insect antenna [3]. Marco Crescentini et al. have designed an integrated and autonomous conductivity temperature depth (CDS) sensors for environmental monitoring [4, 5]. In [6–10], many integrated sensors have been developed for various applications like greenhouse environment, air quality monitoring, environmental monitoring, and portable weather stations.

An attempt has been made to study the static characteristic of an integrated sensor commercially available, i.e. MS8607, and have the capability to measure the temperature, humidity, and pressure simultaneously [4]. Further an experimental set-up is designed and can be used as portable local climate monitoring system. It has been associated with signal processing modules and remote logging and storage of data eventually using IoT technology. The concept of IoT has been proposed more than 15 years, and it is not just staying at the concept level but is becoming a reality with the rapid development and wide application of wireless sensor network (WSN) and cloud computing. The climate data can be monitored on the application developed in android platform.

This paper is organized as follows. Section 2 provides the methods and materials of experimental design. Section 3 provides the results and analysis with error analysis of the integrated sensor, and Sect. 4 provides the concluding remarks.

2 Methods and Materials

2.1 Overview of Integrated Sensor

The MS8607 sensor is the integrated sensor manufactured by Texas Instruments. It is the novel digital combination sensor under modernizing extension and advisory services (MEAS). The self-contained sensor measures three physical parameters all-in-one: pressure, humidity, and temperature sensor (PHT). The sensor is optimal for applications in which essential requirements such as ultralow power consumption, high PHT accuracy, and compactness are critical. The advantages of using MS8607 are high-pressure resolution combined with high PHT linearity. Therefore, MS8607 is an ideal sensor for environmental monitoring purposes [11]. It can also be used as an altimeter in smartphones and tablet PC, as well as PHT applications such as HVAC and weather stations. Figure 2 shows the MS8607 sensor.

MS8607-02BA01 is an eight-pin integrated sensor. The pin configuration of the sensor is provided in Fig. 3. The SDA and SCL pins are used for I2C communications. The MS8607 includes two sensors to measure pressure, humidity, and temperature with unique MEMS technologies. The first sensor is a piezo-resistive sensor providing pressure and temperature. The second sensor is a capacitive-type humidity sensor providing relative humidity. Each sensor is interfaced to a $\Delta\Sigma$ ADC-integrated circuit for the digital conversion. The MS8607 converts both analogue output voltages to a 24-bit digital value for the pressure and temperature measurements and a 12-bit digital value for the relative humidity measurement. An external microcontroller is connected to the sensor module which uses the input SCL and SDA to provide clocks in the data. SCL is a serial clock, and SDA is serial data. Both sensors respond on the same pin SDA, which is bidirectional for the I2C bus interface. Two different I2C addresses are used: pressure and temperature, the other for relative humidity.

Fig. 2 TE connectivity MS8607-02BA01 sensor

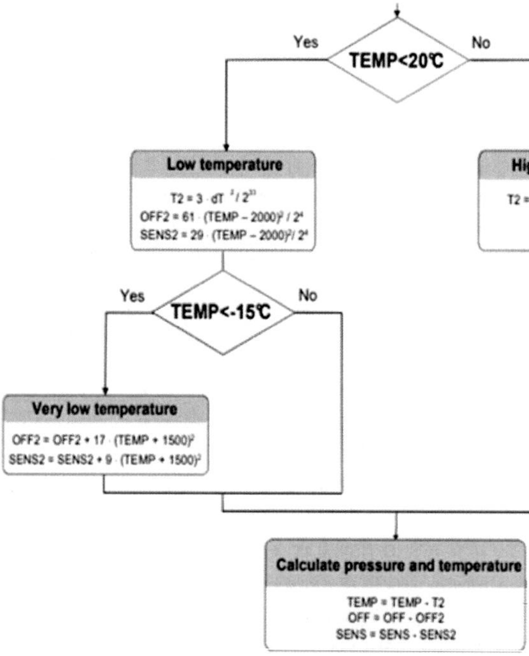

Fig. 3 Flow chart of calculation of temperature and pressure with optimal accuracy

Calculation of Pressure and Temperature

Initially, the range of each of the parameters are considered as follows:

$$P_{\text{MIN}} = 10\,\text{mbar} \quad \text{and} \quad P_{\text{MAX}} = 2000\,\text{mbar}$$
$$T_{\text{MIN}} = -40\,°\text{C}, \quad T_{\text{MAX}} = 85\,°\text{C} \quad \text{and} \quad T_{\text{REF}} = 20\,°\text{C}$$

Difference between actual and reference temperature can be calculated by Eq. 1.

$$\text{d}T = D2 - T_{\text{REF}} = D2 - C5 * 2^8 \tag{1}$$

For actual temperature,

$$\text{TEMP} = 20\,°\text{C} + \text{d}T * \text{TEMPSENS} = 2000 + \text{d}T * C6/2^{23} \tag{2}$$

where

$D2$ is digital temperature value,
$C5 = T\text{REF}$ is reference temperature,
$C6 = \text{TEMPSENS}$ is the temperature coefficient of the temperature.

Both the pressure and temperature values are calculated using standard formulae. However, there is a temperature error below 20 °C. Therefore, a software-based compensation has been incorporated and shown in the flow chart shown in Fig. 3.

Figure 3 shows, for temperature $T \geq 20$ °C, the pressure and temperature can be calculated directly, whereas the calculation method differs for the other case. In other cases, the calculation is subdivided into low and very low temperatures. Finally, the temperature and pressure are calculated for both the zones of temperature.

The measured relative humidity is also calculated and compensated using the formula given in Eq. 3.

$$\mathrm{RH} = -600 + 12500 * D3/2^{16} \tag{3}$$

where RH is the actual humidity and $D3$ is the digital relative humidity value.

$$\mathrm{RH}_{\mathrm{compensated}} = \mathrm{RH} + (20 - T)T_{\mathrm{coeff}} \tag{4}$$

where

$\mathrm{RH}_{\mathrm{compensated}}$ is the compensated relative humidity.

T is the calculated temperature calculated (°C).

Tcoeff is the temperature correction coefficient unit (%RH/°C).

Optimal relative humidity accuracy over [0… + 85 °C] temperature range is obtained with $T\mathrm{coeff} = -0.18$.

The MS8607 can pick and place using vacuum nozzles, and it will not be damaged. The sensor does not show pressure hysteresis effects. However, all the contact pads must be soldered properly. It can be used on flexible PCB with interconnections. One of the application areas will be designing smartwatches or any special devices. The MS8607 has been manufactured in cleanroom conditions.

2.2 *Experimental Design*

For experimentation, a logging system for temperature, relative humidity, and pressure has been built. The embedded board chosen is the Raspberry Pi 3 Model B. The MS8607 is connected with the Raspberry Pi board. The sensor's size is tiny and is embedded on an extended evaluation board made by Texas Instruments. Figure 4 shows the pin diagram of the Raspberry Pi 3 processing board. Raspberry Pi board has the advantage of having portability and credit card size, and it has its operating system, documents, and programs. The Python Version 3 has been used as a programming language and platform for analysing the experimental outputs.

As the experimental set-up has been developed, the following process Raspbian has been configured, and the system has been successful, as shown in Figs. 5 and 6. As the Raspbian has been configured and all the packages have been updated, the SSH, VNC, and I2C have been enabled and installing I2C eventually. Further, it has

Connection Label	From Raspberry Pi 3 Model B	To Sensor Board DPP901G000
I2C-SCL	PIN-5	PIN-1
I2C-SDL	PIN-3	PIN-2
VCC (5V)	PIN-2	PIN-3
GND	PIN-6	PIN-4

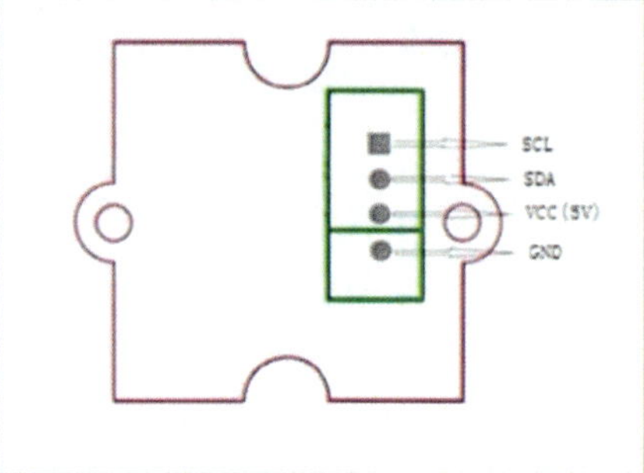

Sensor Board Pin out

Raspberry Pi Pin out

Fig. 4 Pin configuration of sensor board and Raspberry Pi

Fig. 5 Experimental set-up

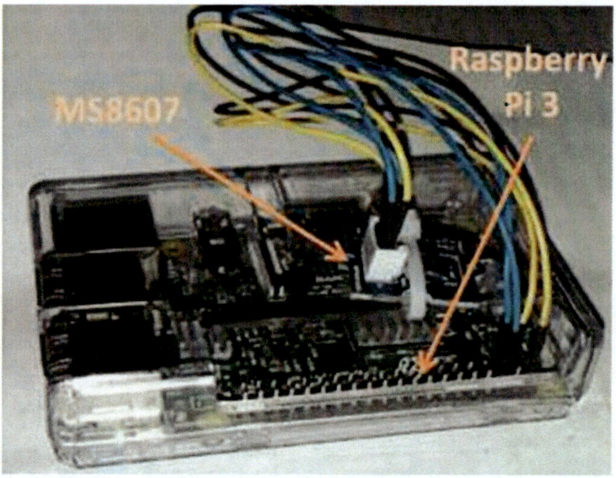

been verified that the Raspberry Pi can communicate with the sensor board over I2C. The integrated sensor board has two I2C devices. One device address is for measuring pressure and temperature and the other address for measuring the relative humidity. As the THP logger source code has been downloaded and installed, then it has been executed. Finally, the system is ready to measure the physical parameters (THP),

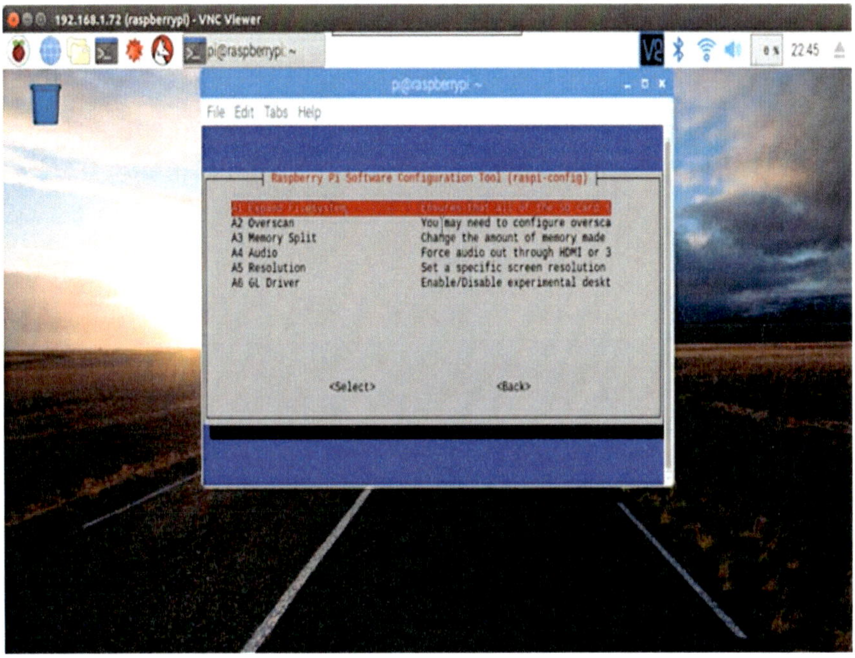

Fig. 6 Configuration of Raspbian

and the data is recorded for further analysis. Figure 7a–c shows the recorded data of relative humidity, atmospheric pressure, and room temperature. The data has been recorded from 12:00 AM 14 April 2020 to 12 AM on 17 April 2020 with an interval of 1 h. The recorded charts shown in Fig. 7 are generated with the Libre Office Cal program using the data generated over 70,000 s (19 h) and with one measurement for every 60 s.

The complete working flow chart from data acquisition to the data display at the Python environment is shown in Fig. 8.

According to the flow chart, at the initial stage, the data pins are activated, and data is read and being assigned to variables. The following data is read:

(a) Pressure sensitivity
(b) Pressure offset
(c) Temperature coefficient of pressure sensitivity
(d) Temperature coefficient of pressure offset
(e) Reference temperature
(f) Temperature coefficient of temperature.

These 12 bytes of calibration data are read and converted and assigned to again a new set of variables. After that, the initiation of temperature and pressure conversion has been done. A further difference in temperature (dT), temperature sensitivity, and offset has been calculated. According to the three conditions of temperature, as

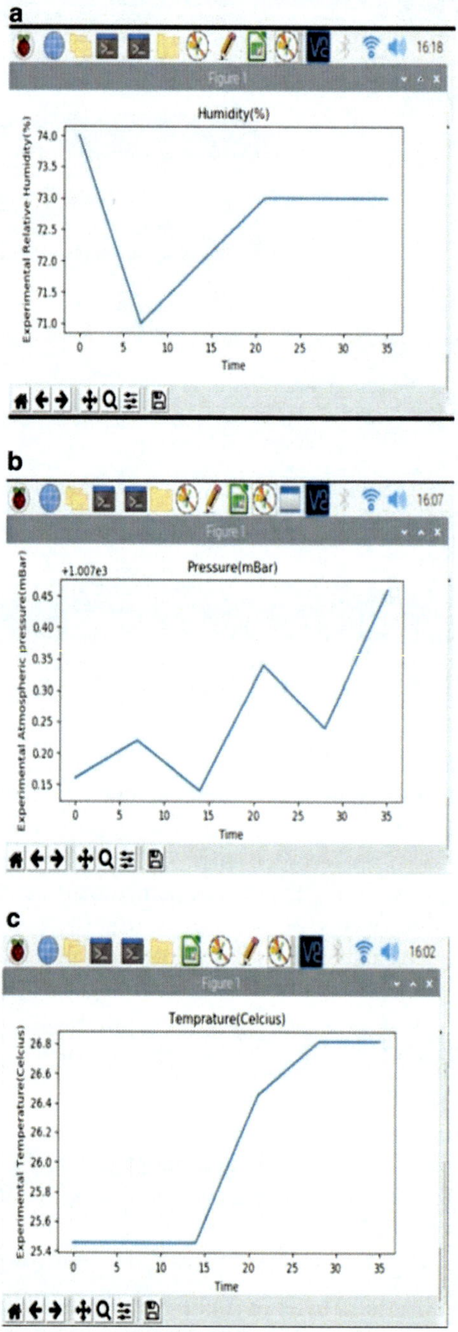

Fig. 7 **a** Real-time acquisition of relative humidity. **b** Real-time acquisition of atmospheric pressure. **c** Real-time acquisition of room temperature

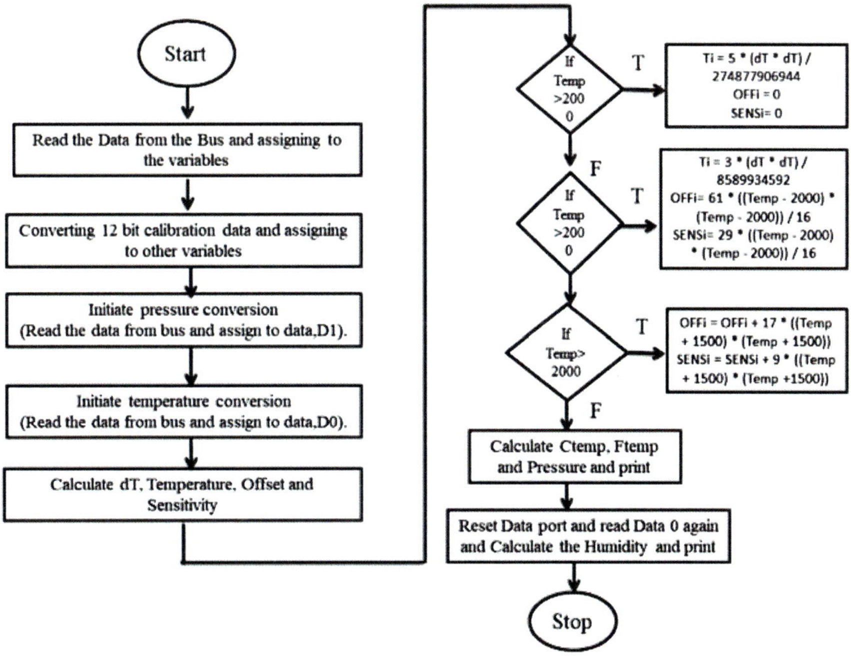

Fig. 8 Flow chart of the measurement technique

shown in the flow chart, the parameters temperature and pressure have been found. The program is reset for some time, and again data has been read from pin data0. This time the read data is used for the calculation of humidity. All the observed data has been recorded and shown in real-time graphical mode.

3 Results and Discussions

A comparative study has been carried out between the recorded data of the integrated sensor MS8607 and available standard sensors. Figures 9, 10, and 11 show the comparative study between experimental observation of change of temperature, relative humidity, atmospheric pressure, and over time versus ideal readings. The observation shows the accuracy of the sensor compared with the standard observation.

Error Analysis

The graphical views show the recorded experimental data correlates with the standard climate data of temperature, relative humidity, and atmospheric pressure. The comparison chart of atmospheric pressure has $R^2 = 1$, which shows the high correlation between the two of the observable data. The other two graphs of temperature and temperature data are not linear, but it has significant correlations.

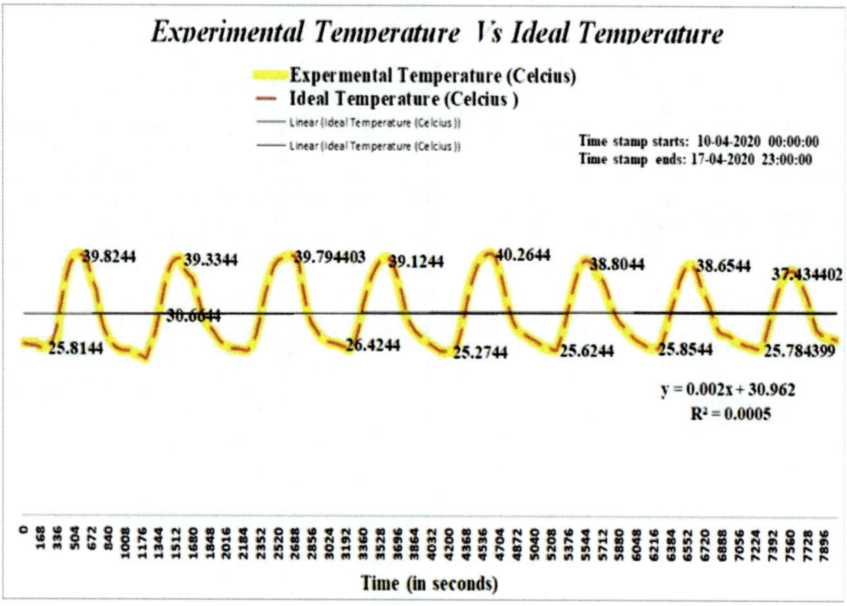

Fig. 9 Experimental temperature versus ideal temperature

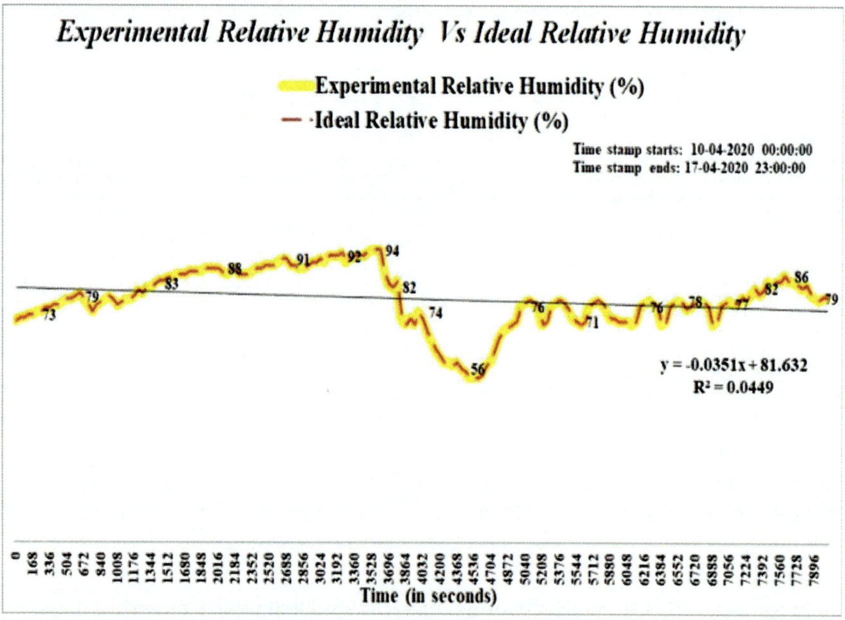

Fig. 10 Experimental relative humidity versus ideal relative humidity

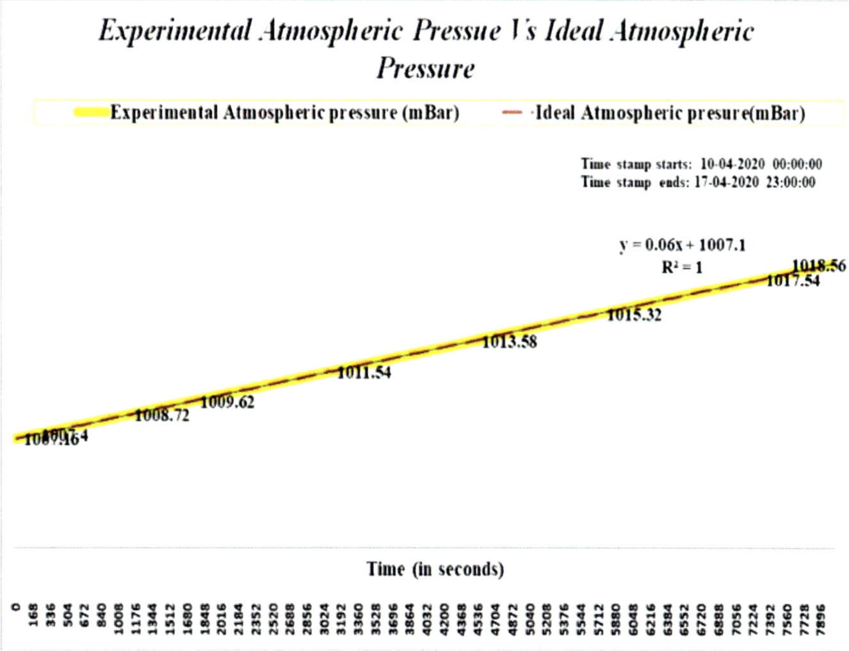

Fig. 11 Experimental atmospheric pressure versus ideal atmospheric pressure

4 Conclusions

The integrated sensors are having advantages of being compact and ease of use rather than using multiple sensors and creating a complicated circuit board. The MS8607 sensor, along with the eval board, makes a very simple system for monitoring climate change in the local area. The experiment of recording the data using MS8607 associated with Raspberry Pi board provides continuous data of temperature, relative humidity, and atmospheric temperature simultaneously. The error analysis shows the recorded data has acceptable measurement error. Further modifications in the experimental prototype can make it usable for some real-life applications where the installation of such devices is required to record and analyses the change of climate data periodically at very remote locations.

Acknowledgements The experiment has been conducted in Eureka Scientech Research Laboratory, Kolkata, West Bengal, India.

References

1. Prime Faraday Technology Watch, Loughborough University, pp. 4–5 (2002)
2. Z. Fang, et al.: A new portable micro weather station. In: Proceedings of 2010 IEEE 5th International Conference on Nano/Micro Engineered and Molecular Systems, China, 20–23 Jan 2010. https://doi.org/10.1109/NEMS.2010.5592239
3. V. Chakravartula, et al.: Integrated weather monitoring device for multi-parameter sensing modelled on Insect Antennae. In: Plastics, Proceedings of IEEE International Conference on Advanced Networks and Telecommunication Systems, Indore, 16–19 Dec 2018. https://doi.org/10.1109/ANTS.2018.8710092
4. M. Crescentini, et al.: Integrated and autonomous conductivity-temperature-depth (CDS) sensors for environmental monitoring. In: Proceedings of 2011 IEEE 54th International Midwest Symposium on Circuits and Systems (MWSCAS), South Korea, 7–10 Aug 2011. https://doi.org/10.1109/MWSCAS.2011.6026396
5. G.S. Gupta, et al.: Multi-sensor integrated systems for wireless of greenhouse environment. In: Proceedings .of. 2018 IEEE Sensors Applications Symposium (SAS), South Korea, 12–14 March 2018. https://doi.org/10.1109/SAS.2018.8336723
6. J.-Y. Kim, et al.: Designing integrated sensing systems for real-time air quality monitoring. In: Proceedings of 2014 International Conference on Information Science & Applications (ICISA), South Korea, 6–9 May 2014. https://doi.org/10.1109/ICISA.2014.6847385
7. S.H. Lin, et al.: Integrated humidity and temperature sensing circuit fabricated by inkjet printing technology. In: Proceedings of 2016 11th International Microsystems, Packaging, Assembly and Circuits Technology Conference (IMPACT), Taiwan, 26–28 Oct 2016. https://doi.org/10.1109/IMPACT.2016.7800045
8. A. Murandar, et al.: Design of real-time weather monitoring system based on a mobile application using automatic weather station. In: Proceedings of 2017 2nd International Conference on Automation, Cognitive Science, Optics, Micro Electro-Mechanical System, and Information Technology (ICACOMIT), Indonesia, 23–24 Oct 2017. https://doi.org/10.1109/ICACOMIT.2017.8253384
9. A.E. Ruano, et al.: A neural network-based intelligent weather station. In: Proceedings of 2015 IEEE 9th International Symposium on Intelligent Signal Processing (WISP), Italy, 15–17 May 2015. https://doi.org/10.1109/WISP.2015.7139169
10. Fang, S., et al.: An integrated system for regional environmental monitoring and management based on Internet of Things. IEEE Trans. Ind. Inf. **10**(2), 1596–1605 (2014). https://doi.org/10.1109/TII.2014.2302638
11. Product Manual, MEAS MS8607 Sensor for Grove System, TE Connectivity

Hexalevel Grayscale Imaging and K-Means Clustering to Identify Cloud Types in Satellite Visible Range Images

Sanjay Goswami, Barnali Goswami, and Gupinath Bhandari

1 Introduction

The problem of identifying the types of the clouds from digital satellite images has been an interest of research for quite a many years [1, 2]. With the advent of statistical classification techniques such as k-means clustering, several fruitful attempts have been made to develop methods to classify rain clouds from digital satellite images [1, 3]. In the present work, an attempt has been made to develop an indigenous approach to classify convective clouds from visible band of satellite images (VIS), using a new imaging technique and combining it with k-means clustering.

Clouds belong to several categories. Though it is difficult to classify clouds into discrete classes, broadly they can be categorized into three main genii—Cumuliform, Stratiform and Cirriform.

Cumuliform clouds are of irregular shapes and demonstrate vertical development. Their span is from medium to high altitude. The subclasses of cumuliform clouds are—cumulus, towering cumulus and cumulonimbus. Cumulus clouds are classic fair—weather puffy clouds resembling cotton balls in the sky. Towering cumulus clouds are those which grow tall and thick. Cumulonimbus clouds are an extreme case of towering cumulus genre when they grow very high and thick so that their top assumes a flat anvil like shape due to strong upper layer winds. Though towering cumulus clouds are responsible for medium to heavy rain, Cumulonimbus clouds are mainly responsible for thunderstorms. Towering cumulus and Cumulonimbus clouds appear very bright in visible satellite imagery.

S. Goswami (✉)
Center for Disaster Preparedness and Management, Jadavpur University, Kolkata, India

B. Goswami
School of Computer Science, MIT World Peace University, Pune, India

G. Bhandari
Department of Civil Engineering, Jadavpur University, Kolkata, India

J. K. Mandal and J. K. Roy (eds.), *Proceedings of International Conference on Computational Intelligence and Computing*, Algorithms for Intelligent Systems,
https://doi.org/10.1007/978-981-16-3368-3_2

Stratiform clouds are characterized by a flat and layered organization. Subclasses of Stratiform are—stratus, altostratus and cirrostratus. Stratus clouds are low-level clouds that appear to cover the entire sky, making it gray and overcast. Very low stratus clouds are called fog. Altostratus clouds are middle-level stratiforms, while Cirrostratus are high-level stratiforms. Medium- and low-level stratus clouds are responsible for moderate to low rain or drizzle. In visible imagery, stratus clouds appear as broad flat clouds with a very smooth texture. Higher stratus clouds often appear darker as they are often more thin than lower- or medium-level stratus clouds.

Cirriform clouds are very high-level ice crystal clouds. They have very low presence of water vapour. Subclasses of Cirriform include—cirrus, cirrocumulus and cirrostratus. Cirrus clouds are classic Cirriform clouds and appear as windswept wispy clouds. They are fibrous in texture. Cirrocumulus clouds are puffy high-level clouds. Cirrostratuses also fall into this category as they are high-altitude flat layers. Since Cirriform clouds are usually very thin, they appear transparent in VIS imagery.

In this study, from the VIS images, cloudy pixels have been identified. Then the type of cloud has been determined. Also the area under cloud cover is obtained from this work. The output of this study can be further utilized for rainfall prediction in the field of meteorology.

At the beginning, a VIS image in colour is taken as input to the developed model. This image is then converted to grayscale image, which is then further converted to hexalevel grayscale image. Feature vector is calculated for this image, and finally, k-means clustering is applied.

This paper has been organized into eight sections. Section 1 is the introduction to the types of clouds. In Sect. 2, difference between binary, grayscale and hexalevel images has been explained. Section 3 points out the significance of hexalevel images for the current objective. The procedure to generate hexalevel grayscale images from grayscale images has been explained in Sect. 4. K-means clustering is explained in Sect. 5. Finally, Sect. 6 represents the developed cloud classification technique. Section 7 is about experiments carried out and the results obtained. This study is concluded in Sect. 8, which is followed by references list and appendix.

2 Binary, Grayscale and Hexalevel Images

Grayscale images are distinct from black and white images. Black and white images are bi-tonal, each pixel defined by binary bits 0 or 1, representing the two colours—black and white (also called bi-level or binary images). However, grayscale images have several shades of gray, defined by values within 0–255. These shades are called levels. They are also called monochromatic images due to the absence of any chromatic variations in terms of the colour frequencies. A hexa level image that's why contains six levels. So, a hexalevel grayscale image can be obtained from a full grayscale image by reducing the 256 grayscale levels to six.

3 Significance of Hexalevel Grayscale Images in Identification of Clouds

According to the requirement of meteorology, clouds can be categorized as no rain, very low rain, low rain, medium rain, heavy rain and very heavy rain. That is, there can be six categories. If the different cloud pixels in a digital image can be represented using six different shades of gray, then they can be conveniently identified and classified into six classes. Hexalevel grayscale images are perfect for this type of problem domain. There can be a situation where in a particular region in the image, there is existence of more than one types of clouds. This can be solved by calculating the percentage of the presence of pixels of various shades and then selecting the dominating class of cloud to represent that particular area.

4 Generation of Hexalevel Grayscale Images

The steps for generating hexalevel grayscale images are as follows:

1. Input a colour image in BMP format only.
2. Convert this colour image into grayscale image:
   (i) Obtain the values of its red, green and blue (RGB) primaries in linear intensity encoding by Gamma expansion.
   (ii) Convert to grayscale using the equation [5]:

$$\text{Grey} = (\text{red} * 0.3) + (\text{green} * 0.59) + (\text{blue} * 0.11)$$

3. Reduce the 32-bit grayscale image thus obtained to hexalevel grayscale image with the following pixel value mapping scheme (Table 1).

Pixel values in 240–255 in the input grayscale image are discarded because those values are used to represent region boundaries and longitude and latitude lines.

Thus, the final image contains six intensity values of gray shades for each pixel instead of 256 shades.

Table 1 Hexalavel grayscale mapping scheme

256 level pixel range	Hexalevel pixel value	Cloud type
0–40	0	No rain cloud
41–80	50	Very low rain cloud
81–110	90	Low rain cloud
111–150	130	Medium rain cloud
151–180	180	Heavy rain cloud
181–240	255	Very heavy rain cloud

5 K-means Clustering

K-means clustering [4] is the simplest and very famous unsupervised learning algorithm that follows a simple and easy way to classify a given set of data among a certain number of predefined clusters (k say) fixed a priori. The main idea is to define k centroids, one for each cluster, representing cluster centres. These centroids should be placed carefully as different cluster initializations lead to different results. It is always an intelligent idea to place them as much far away from each other as possible that leads to more crisp classification. The next step is to assign each data point in the given set to the nearest centroid in the space. At the end of assignment of all the data points to the centroids, the first step is said to be completed, and an early grouping is assumed to be done. At this point, we need to re-calculate k new centroids resulted from this grouping. After the k new centroids have been calculated, re-assigning of the same data points from the set is to be carried out until the average change in the centroid positions is nearly zero. The algorithm is depicted as below:

Step-1: Initialize the number of clusters, k, and the k centroid points in the space.
Step-2: Calculate the distance of each data point from each cluster centre.
Step-3: Assign the data point to the cluster whose centroid's distance from the data point is smallest.
Step-4: Take the average values of the data points in a cluster to obtain the new centroid.
Step-5: Repeat Steps 2–4, until the recent and previous centroid values merge.

6 Proposed Cloud Classification Technique

In order to apply k-means clustering on the hexalevel grayscale image, feature vectors were computed. Based on the feature vector, the clustering process was completed.

The step-by-step process of the entire developed cloud classification technique, right from the beginning, is given below:

1. *Input VIS satellite image.*
2. *Convert input image to grayscale image (0–255 pixel intensity levels).*
3. *Convert grayscale image to hexalevel grayscale image (6 pixel intensity levels).*
4. *Calculate pixel ratios for each shade.*
5. *Prepare feature vectors comprising pixel ratio values for each shade, $f = (p1, p2, p3, p4, p5, p6)$*
6. *Apply k-means clustering to classify the cloud into one of six clusters.*

7 Experimental Results

To implement the proposed technique, an in-house programme was developed in C language and was tested on visible spectrum satellite images obtained from satellite KALPANA-1 (www.mosdac.gov.in). First, a colour composite image in BMP format (Fig. 1) was taken as input and was converted to a normal grayscale image (Fig. 2a). On applying the hexalevel reduction, the corresponding hexalevel grayscale image was obtained (Fig. 2b).

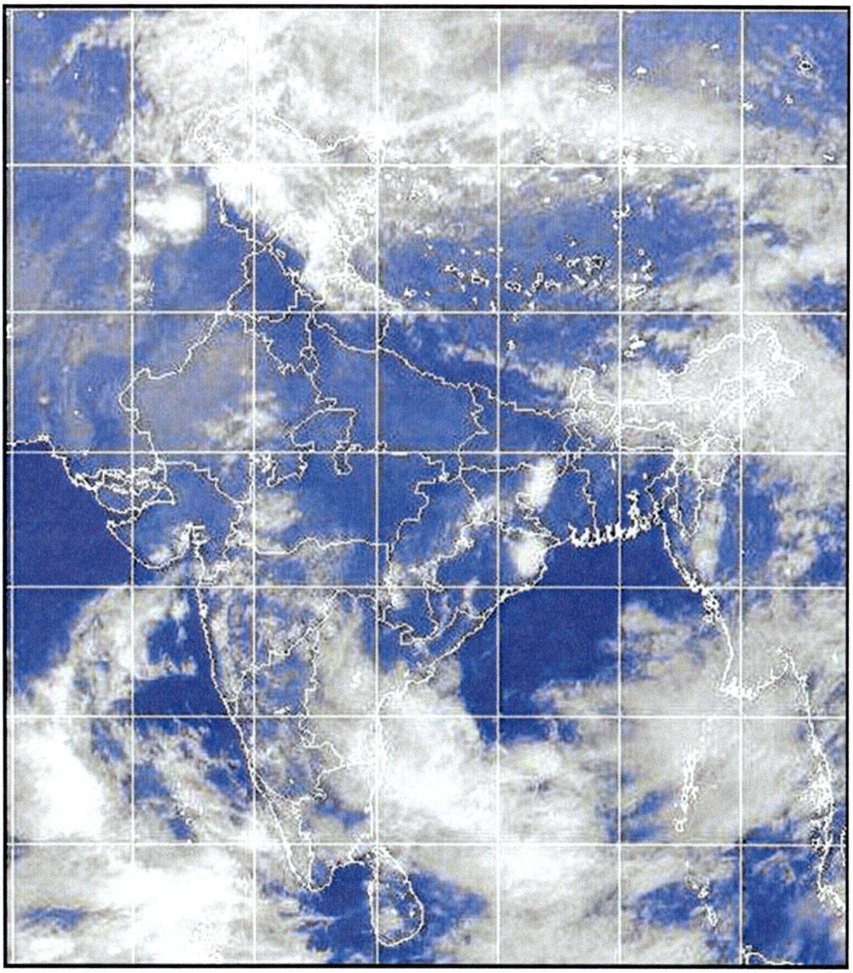

Fig. 1 Input image (VIS)

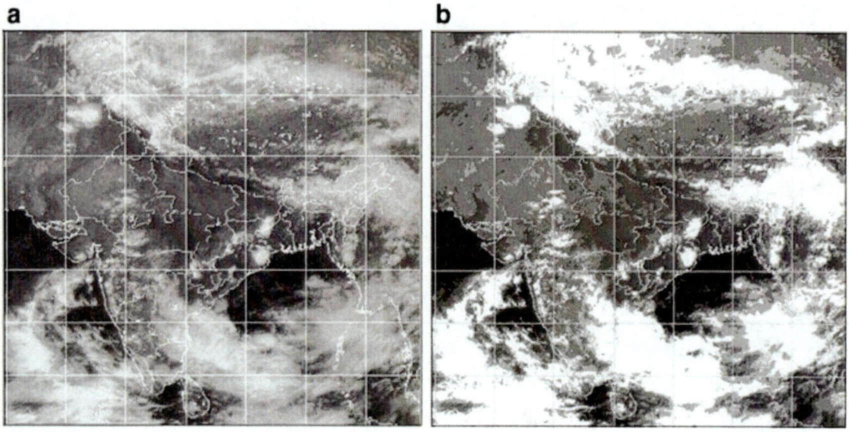

Fig. 2 **a** Grayscale image (overall). **b** Hexalevel image (overall)

The images in Figs. 1, 2a and b are representing the entire Indian subcontinent, taken from INSAT satellite KALPANA-1. As can be observed from the images, the subcontinent is covered with differential spread of medium to very heavy clouds. Moreover, there are several areas where there is no cloud cover. On applying the developed hexalevel imaging technique and extracting pixel ratio features from these images, it can be concluded from the proposed algorithm that there is a spread of very heavy rain clouds in several regions. Several separate regions of the image were analysed, and the results of the analysis are summarized in Table 2. The images of the regions analysed from the input image are provided in the Appendix.

8 Conclusions

An indigenous attempt has been made to develop a new method to classify clouds from a visible band satellite image. The developed method is unique as it represents the image in hexalevel grayscale format which reduces unnecessary calculations. K-means clustering is used for segmentation of the image, and finally, region-wise cloud type is determined. The results obtained are quite interesting and indicates its usability for further research. This technique can be further utilized in the field of meteorology for prediction of precipitation.

Table 2 Classification of cloud region-wise

	Region	Feature vector *(p1, p2, p3, p4, p5, p6)*	Predicted cloud type	Actual perception
2a, 2b	Overall	*(0.310321, 0.170751, 0.266380, 0.139809, 0.059796, 0.000011)*	Very heavy rain clouds	Medium to very heavy rain clouds
3a, 3b	North	*(0.456086, 0.234434, 0.182703, 0.072833, 0.001644, 0.000000)*	Very heavy rain clouds	Very heavy rain clouds
4a, 4b	South	*(0.466639, 0.158552, 0.145582, 0.086316, 0.095714, 0.000035)*	Very heavy rain clouds	Very heavy rain clouds
5a, 5b	East	*(0.298479, 0.189776, 0.183146, 0.148280, 0.107895, 0.000034)*	Very heavy rain clouds	Very heavy rain clouds
6a, 6b	West	*(0.069537, 0.105486, 0.432215, 0.138346, 0.191193, 0.000035)*	Medium rain clouds	Medium to low rain clouds
7a, 7b	Central	*(0.057763, 0.072212, 0.270157, 0.474309, 0.062211, 0.000000)*	Low rain clouds	Low rain clouds
8a, 8b	North-West	*(0.283756, 0.096712, 0.465974, 0.107649, 0.000304, 0.000000)*	Medium rain clouds	Medium to heavy rain clouds
9a, 9b	South-East	*(0.361615, 0.272404, 0.179978, 0.100481, 0.035874, 0.000034)*	Very heavy rain clouds	Very heavy rain clouds

Appendix

Images of the several regions, that have been analysed from the same input image, are shown in Figs. 3, 4, 5, 6, 7, 8 and 9.

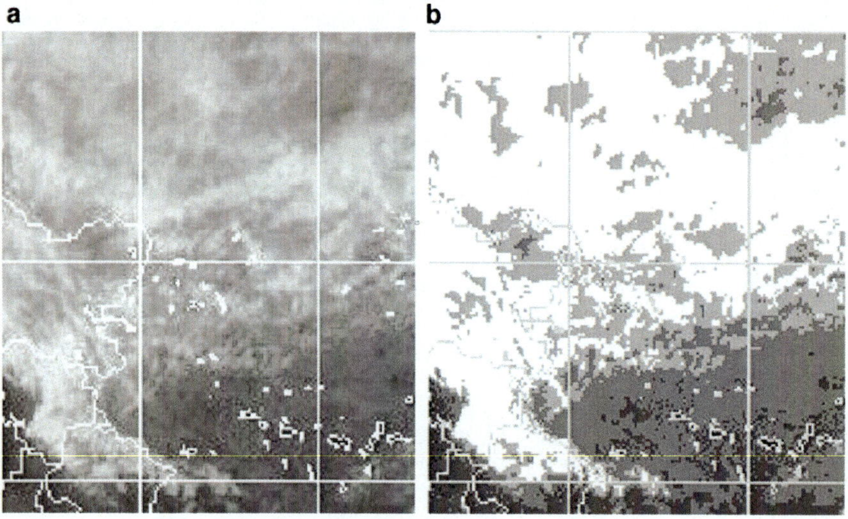

Fig. 3 **a** North grayscale. **b** North hexalevel

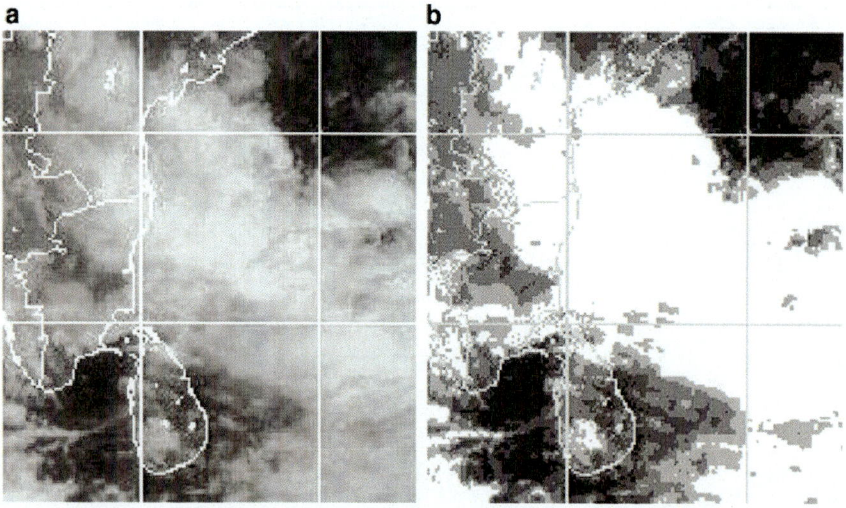

Fig. 4 **a** South grayscale. **b** South hexalevel

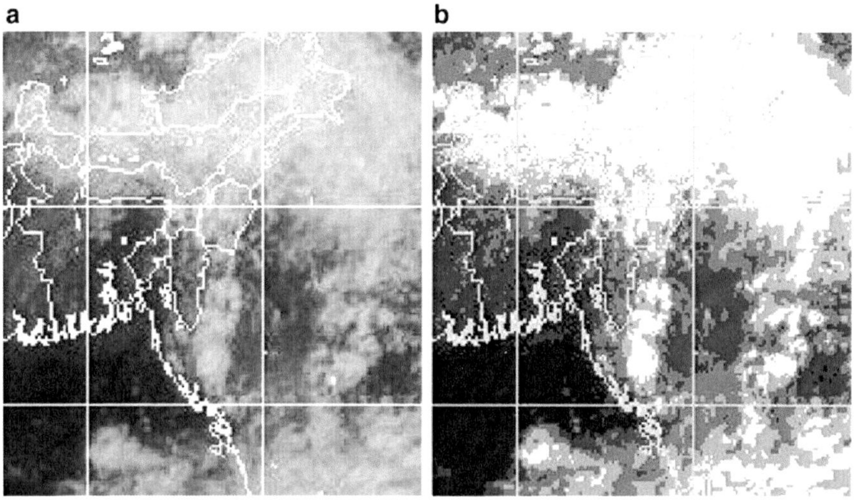

Fig. 5 **a** East grayscale. **b** East hexalevel

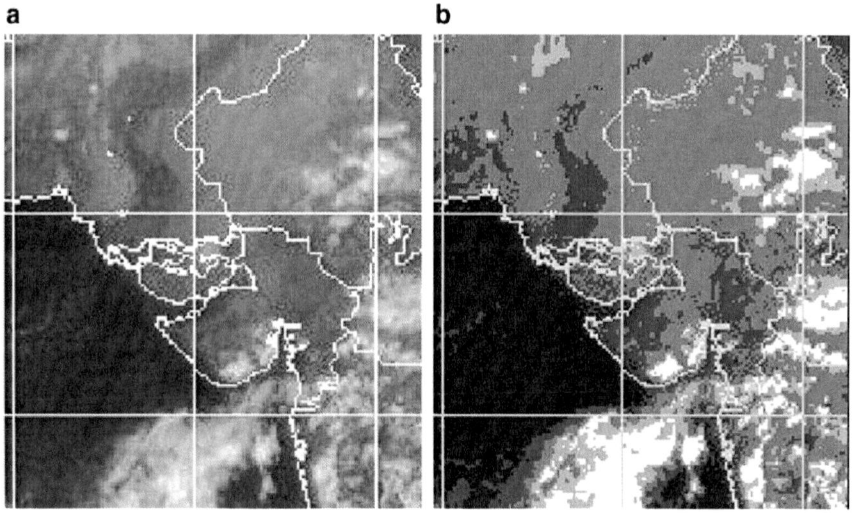

Fig. 6 **a** West grayscale. **b** West hexalevel

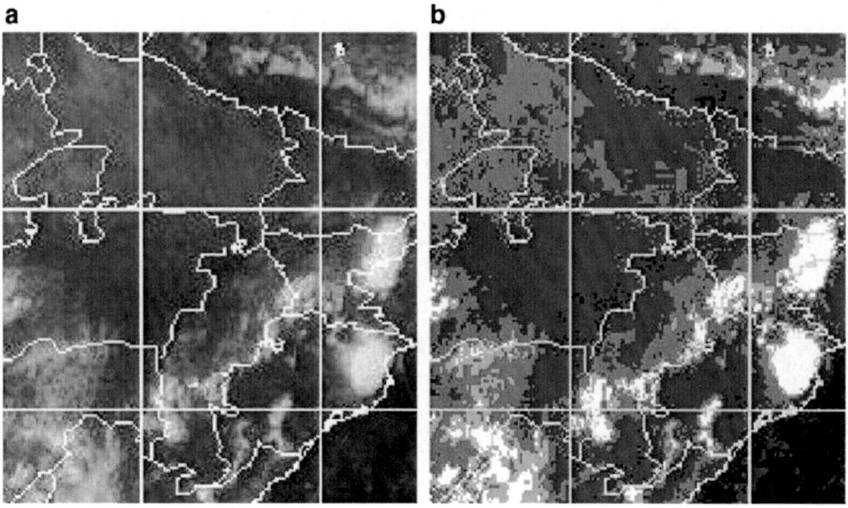

Fig. 7 **a** Central grayscale. **b** Central hexalevel

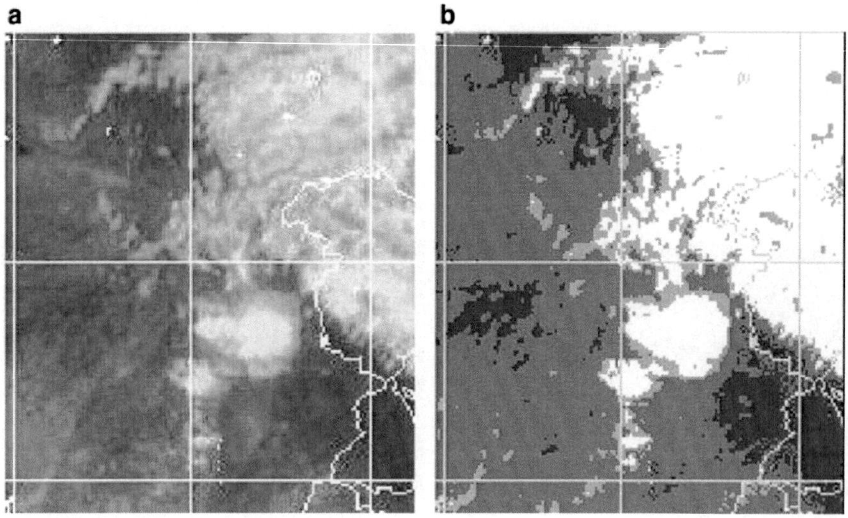

Fig. 8 **a** North-West grayscale. **b** North-West hexalevel

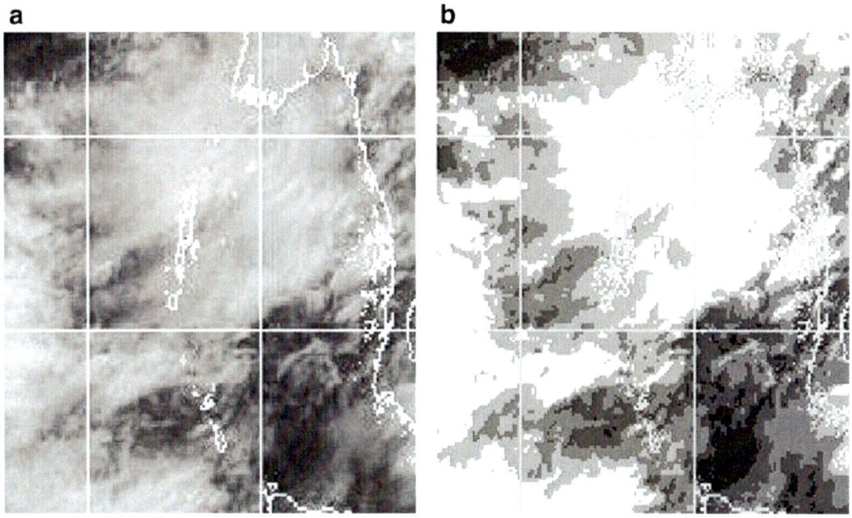

Fig. 9 a South-East grayscale. **b** South-East hexalevel

References

1. Mandal, A.K., Pal, S., De, A.K., Mitra, S.: Novel approach to identify good tracer clouds from a sequence of satellite images. IEEE Trans. Geosci. Remote Sens. **43**(4), 813–818 (2005)
2. Goswami, B., Bhandari, G.: convective cloud detection and tracking from series of infrared images. J. Indian Soc. Remote Sens. **41**(2), 291–299 (2013)
3. Goswami, B., Bhandari, G.: Near real-time detection of heavy rain clouds from IR image for estimation of precipitation. In: Proceedings of the 3rd International Conference on Water and Flood Management (ICWFM-2011), pp. 277–281, BUET, Dhaka (2011)
4. Hartigan, J.A., Wong, M.A.: Algorithm AS 136: a K-means clustering algorithm. J. Royal Stat. Soc. Ser. C Appl. Stat. **28**(1), 100–108 (1979)
5. ITU-R: Recommendation BT.601: Studio Encoding Parameters of Digital Television for Standard 4:3 and Wide-Screen 16:9 Aspect Ratios. International Telecommunication Union (2011). https://www.itu.int/rec/R-REC-BT.601/
6. Phillips, D.: Image Processing in C: Analyzing and Enhancing Digital Images. R&D Publications Inc., Lawrence, KS, USA (1994)

DC and Noise Performance of Si IMPATT Oscillator at Ka Band

S. J. Mukhopadhyay, R. Dhar, and M. Mitra

1 Introduction

The Ka band has range of 26.5–40 GHz and has multiple applications across many domains. Silicon-based IMPATTs are capable of delivering enough RF output power over various frequency bands [1–3]. The fabrication of Ka band IMPATT has been accomplished by Mitra et al. [4], displayed in Fig. 1.

2 Simulation Method

As shown in Fig. 2, given model has been taken for the simulation purpose. The source is designed for a particular frequency (fd) from formula [5] $W_{n,\,p} = 0.37$ (vsn, sp/fd). Use of technique developed previously [6], the E-field profile is received. The high-frequency simulation highlighted beforehand [6, 7] is taken to obtain high-frequency parameters [8].

2.1 *Static and Small-Signal Noise Simulation*

By solving some basic device equations, the E-field and normalized current density profiles are received. DC and high-frequency simulation techniques have been highlighted beforehand [6–8]. The flowchart of the static and small-signal simulation techniques is exhibited in Figs. 3 and 4. The details on noise simulation method have been depicted earlier [9–13].

S. J. Mukhopadhyay (✉) · R. Dhar · M. Mitra
Department of Electronics and Telecommunication, IIEST, Shibpur, India

J. K. Mandal and J. K. Roy (eds.), *Proceedings of International Conference on Computational Intelligence and Computing*, Algorithms for Intelligent Systems,
https://doi.org/10.1007/978-981-16-3368-3_3

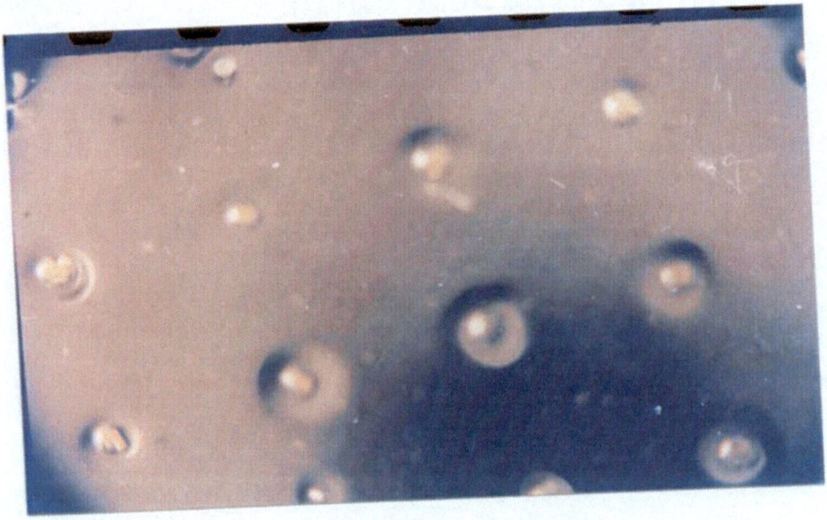

Fig. 1 Silicon p + n structure for IMPATT diode (mesa etched)

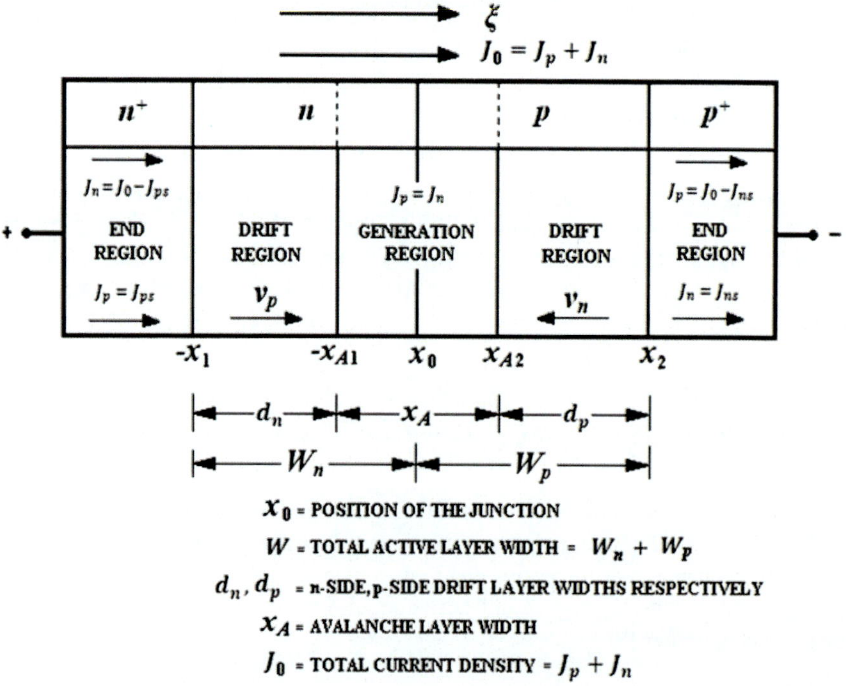

Fig. 2 1-D DDR IMPATT

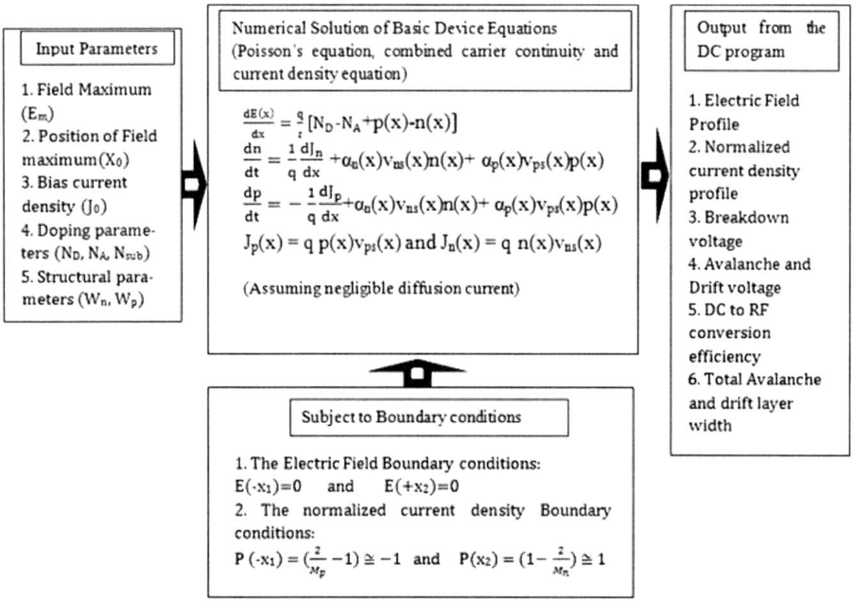

Fig. 3 DC simulation program outline

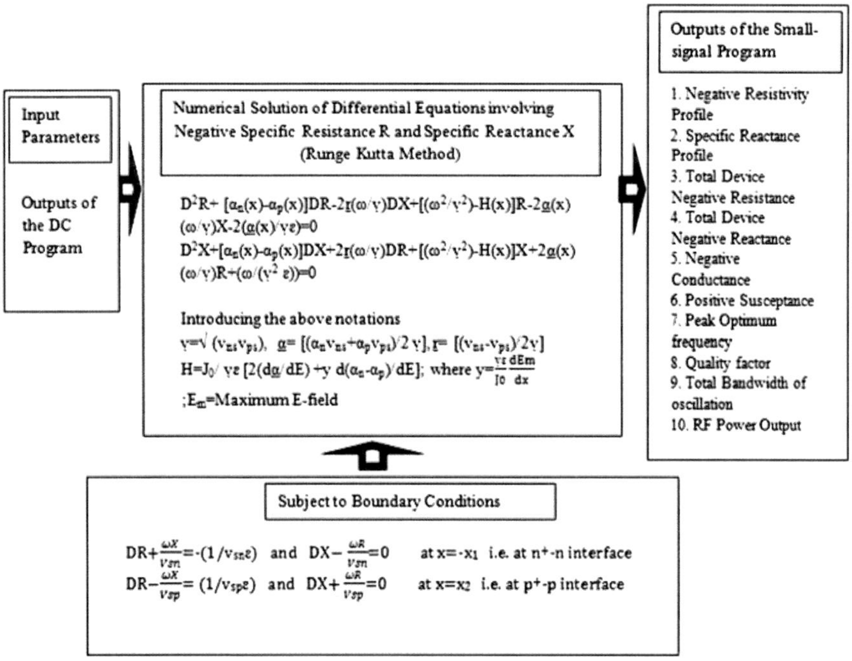

Fig. 4 Small-signal simulation program outline

3 Results and Observation

Noise performance of Ka band Si IMPATT formulated from plot and exhibited in Fig. 6. The experimental set up is exhibited in Fig. 5 [14], where the previously [4] fabricated IMPATT is used. Note from Fig. 6 that the simulated value is in good agreement to experimental value.

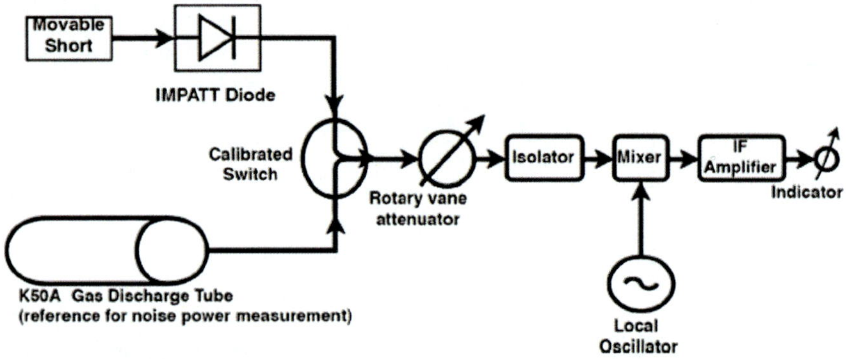

Fig. 5 Experimental block-diagram

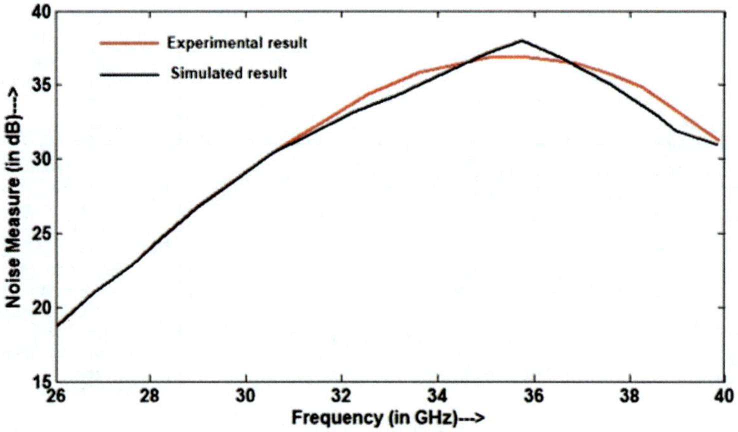

Fig. 6 Plot for validation with experimental

4 Conclusion

In this current article, DC and noise performance is examined based on Si-based IMPATTs. Simulated results are in good agreement with experimental which confirms the simulation strategy acquired by the authors.

References

1. Midford, T.A., Bernick, R.L.: Millimeter-wave CW IMPATT diodes and oscillators. IEEE Trans. Microwave Theory Tech. **27**, 483 (1979)
2. Chang, Y., Hellum, J.M., Paul, J.A., Weller, K.P.: Millimeter-wave IMPATT sources for communication applications. In: IEEE MTT-S International Microwave Symposium Digest, pp. 216–219 (1977)
3. Ghoshal, D.: Measurement of electrical resistance of Silicon Single Drift Region IMPATT diode based on the study of the device and mounting circuit at threshold condition. J. Electron Dev. **11**, 625–631 (2011)
4. Mitra, M., Ganguly, A., Roy, S.K., Banerjee, J.P.: Experimental studies on process steps for fabrication of IMPATT diodes and corresponding study of the DC breakdown voltage. IETE Tech. Rev. **10**(4), 351–354 (1993)
5. Sze, S.M., Ryder, R.M.: Microwave avalanche diodes. Proc. IEEE **59**, 1140–1154 (1971)
6. Roy, S.K., Sridharan, M., Ghosh, R., Pal, B.B.: Computer methods for the dc field and carrier current profiles in IMPATT devices starting from the field extremum in the depletion layer. In: Proceedings of NASECODE-I Conference on Numerical Analysis of Semiconductor Devices, pp. 266–274. Boole Press, Dublin (1979)
7. Roy, S.K., Banerjee, J.P., Pati, S.P.: A computer analysis of the distribution of high frequency negative resistance in the depletion layers of IMPATT diodes. In: Proceedings of NASECODE-IV Conference on Numerical Analysis of Semiconductor Devices, pp. 494–500. Boole Press, Dublin (1985)
8. Gummel, H.K., Blue, J.L.: A small-signal theory of avalanche noise in IMPATT diodes. IEEE Trans. Electron Dev. **14**, 569–580 (1967)
9. Dash, G.N., Mishra, J.K., Panda, A.K.: Noise in mixed tunneling avalanche transit time (MITATT) diodes. Solid State Electron. **39**, 1473–1479 (1996)
10. Mishra, J.K., Panda, A.K., Dash, G.N.: An extremely low-noise heterojunction IMPATT. IEEE Trans. Electron Dev. **ED-44**, 2143–2148 (1997)
11. Acharyya, A., Mukherjee, M., Banerjee, J.P.: Noise performance of millimeter-wave silicon based mixed tunneling avalanche transit time (MITATT) diode. Int. J. Electr. Electron. Eng. **4**, 577–584 (2010)
12. Acharyya, A., Mukherjee, M., Banerjee, J.P.: Noise in millimeter-wave mixed tunneling avalanche transit time diodes. Arch. Appl. Sci. Res. **3**, 250–266 (2011)
13. Haus, H.A., Statz, H., Pucel, R.A.: Optimum noise measure of IMPATT diode. IEEE Trans. MTT **19**, 801 (1971)
14. Goedbloed, J.J.: Determination of the intrinsic response time of semiconductor avalanches from microwave measurements. Solid State Electron. **15** (1972)

Inversion of Integral Equation Using Laplace Transform

Priyank Jain and Archana Lala

1 Introduction

Boundary value problem expressible in the form of integral equations involving special functions and their generalization as kernels have got tremendous place as a remarkable and significant branch of applied mathematics and pure mathematics with its great number of applications in the field of industry, economics, physics, biology, and applied sciences.

Integral equations have same importance as of differential equations and arises in the most applied areas as mentioned by Wazwaz [1]. Several boundary value problems can be transformed as a problem of solving integral equations whose kernel includes numerous well-known classical polynomials such as modified Hermite, modified Legurre, and modified Jacobi. Many researchers have made many attempts with the use of Rodrigue's formulae to unify and generalize classical polynomials of that kinds (for more details, see [2–10]). Integral Transform is a powerful tool to solve integral equation as well as differential equation and is very useful in solving various types of boundary value problems, by using various values to the kernel $k(t, s)$ and prescribing the interval (a, b) as $(0, \infty)$, $(-\infty, \infty)$ generally. Time to time, a number of integral transforms have been introduced and studied by several authors i.e., Laplace Transform, Mellin Transform, Hankel Transform and Fourier Transform. A book by Erdelyi et al. [11], which contains most of the tables of integral transform which is very useful and time saving as it contains direct results of many transforms. Here in this work we have used Laplace transform and generalised Hermite polynomials which was given by Gould Hopper [12], as

$$H_n^r(t, a, p) = (-1)^n t^{-a} e^{pt^r} D^n \left[t^a e^{-pt^r} \right] \tag{1}$$

P. Jain (✉) · A. Lala
SR Group of Institutions, Jhansi, Uttar Pradesh, India

J. K. Mandal and J. K. Roy (eds.), *Proceedings of International Conference on Computational Intelligence and Computing*, Algorithms for Intelligent Systems,
https://doi.org/10.1007/978-981-16-3368-3_4

and we have used

$$H_n^1(t, a, p) = (-1)^n t^{-a} e^{pt} D^n\left[t^a e^{pt}\right]; a \geq n, p > 0 \tag{2}$$

where $D \equiv \frac{\mathrm{d}}{\mathrm{d}t}$ and r, a, and p are parameters, using appropriate values of r, a, and p; Eq. (1) can be transformed to modified Hermite, modified Laguerre, and modified Bessel polynomials. Taking into account these generalizations, it is worth considering integral equations involving $H_n{}^1(t, a, p)$ as kernel and such we prove the following theorem in the next section.

2 Theorem

Statement: If $f(t)$ is an unknown function which satisfies the convolution integral equation

$$g(t) = \int_0^t k(t-y) f(y) \mathrm{d}y, \quad t > 0 \tag{3}$$

where

$$k(t) = t^a e^{-pt} \cdot H_n^1(t, a, p); a \geq n, p > 0 \tag{4}$$

and $g(t)$ is a prescribed function, then $f(t)$ is given by

$$f(t) = e^{pt} \int_0^t e^{-p(t-y)} \frac{(t-y)^{n-1}}{(n-1)!} \left\{e^{-py} g(y)\right\} \mathrm{d}y$$

Proof We shall use of Laplace transform as tool to prove this theorem and shell discuss case $p > 0$.

By the convolution theorem for Laplace transform (3) reduces to

$$K(s)F(s) = G(s) \tag{5}$$

where $K(s)$, $F(s)$, $G(s)$ are respective Laplace transform of $k(t)$, $f(t)$, $g(t)$ and defined by

$$F(s) = \int_0^\infty f(t) \cdot e^{-st} \mathrm{d}t = L[f(t), s] \tag{6}$$

When $p > 0$, applying Laplace transform of Eq. (4) and using the result of Erdelyi [11], we get

$$\begin{aligned} K(s) &= (-1)^n L\left[D^n\left(t^a \cdot e^{-pt}\right); s\right] \\ &= (-1)^n (-1)^a s^n \frac{1}{(s+p)^a}; \quad a \geq n \\ &= (-1)^{n+a} s^n \frac{1}{(s+p)^a}; \quad a \geq n \end{aligned} \tag{7}$$

We write Eq. (5) in the form

$$F(s) = \frac{G(s)}{K(s)}$$

replacing s by $s + p$

$$F(s+p) = G(s+p)L(s) \tag{8}$$

where

$$L(s) = \frac{(-1)^{a+n}}{(s+2p)^a K(s+p)} = \frac{1}{(s+p)^n} \tag{9}$$

Taking inverse Laplace transform of

$$l(t) = e^{-pt} \frac{t^{n-1}}{(n-1)!} \tag{10}$$

Now, from Eq. (8) and with the use of inverse Laplace transform

$$e^{-pt} f(t) = \int_0^t e^{-p(t-y)} \frac{(t-y)^{n-1}}{(n-1)!} \left\{e^{-py} g(y)\right\} \mathrm{d}y$$

$$f(t) = e^{pt} \int_0^t e^{-p(t-y)} \frac{(t-y)^{n-1}}{(n-1)!} \left\{e^{-py} g(y)\right\} \mathrm{d}y$$

This proves the theorem.

3 Conclusion

The purpose of the present work is to further develop and extend the work done so far on the inversion of integral equations involving special functions and their generalization as kernels. We have proposed a new theorem to find the unknown function satisfying the given integral equation of Volterra type. The solution has the advantage that most quantity of physical interest is easily evaluated in terms of known function. We hope that this work can motivate and encourage further scientific works, and the proposed theorem will be greatly useful to the researchers involved in applied mathematics.

References

1. Wazwaz, A.M.: A first course in integral equations. 2nd edn., World Scientific Publishing Co., Inc., NJ, United States (2015)
2. Habibullah, G.M., Shakoor, A.: A generalization of hermite polynomials. Int. Math. Forum **8**(15), 701–706 (2013)
3. Goyal, S.P., Salim, T.O.: A class of convolution integral equations involving a generalized polynomial set. Proc. Indian Acad. Sci. (Math. Sci.) **108**(1), 55–62 (1998)
4. Srivastava, R.: The inversion of an integral equation involving a general class of polynomials. J. Math. Anal. Appl. **186**, 11–20 (1994)
5. Lala, A., Shrivastava, P.N.: Inversion of an integral involving a generalized function. Bull. Calcutta Math. Soc. **82**, 115–118 (1990)
6. Lala, A., Shrivastava, P.N.: Inversion of an integral involving a generalized Hermite polynomial. Indian J. Pure App/. Math. **21**, 163–166 (1990)
7. Chatterjea, S.K.: Some operational formulas connected with a function defined by a generalized Rodrigues' formula. Acta Math. Acad. Sci. Hungar. **17**, 379–385 (1996)
8. Srivastava, H.M., Singhal, J.P.: A class of polynomials defined by generalized Rodrigues formula. Ann. Mat. Pura Appl. **90**, 75–85 (1971)
9. Jain, P., Lala, A., Singh, C.: Inversion of integral equation associated with Leguerre polynomial obtained from hermite polynomial. Int. J. Eng. Res. Technol. **8**(4) (2019)
10. Jain, P., Lala, A., Singh, C.: Inversion of integral equation involving certain class of generalized polynomial. J. Emerg. Technol. Innov. Res. **6**(6), (2019)
11. Erdelyi, A., Magnus, W., Oberhettinger, F., Triconi, F.G.: Tables of Integral Transforms, vol. I. McGraw-Hili, New York (1954)
12. Gould, H.W., Hopper, A.T.: Operational formulas connected with two generalizations of Hermite polynomials. Duke Math. J. **29**, 51–63 (1962)

Entropy-Based Intelligent Computation for Decision-Making Models of Pandemic Analytics

Debabrata Datta and Joyanta Kumar Roy

1 Introduction

Coronavirus disease (COVID-19) outbreak was first identified by World Health Organization (WHO) and reported from Wuhan city, Hubei province of China on December 31, 2019 [1]. Looking at the rapid transmission of coronavirus and increasing trend of the mortality worldwide, WHO declared COVID-19 as a pandemic [2, 3]. As of December 2, 2020 as per report received from WHO, we have 63,136,866 confirmed cases including deaths of 1,469,402. With this information, the researchers are still looking into the existence of the virus and trying to investigate the trend of the disease including its long- and short-term effects. Accordingly, COVID-19 caused a public health emergency of international concern posing a higher risk to vulnerable healthcare system [4]. COVID-19 is the public name given to the scientific name SARS-CoV-2 which has been studied by coronavirus study group of the International Committee on Taxonomy of Viruses3 and has found its symptoms in accordance with severe acute respiratory syndrome coronavirus (SARS-CoV) which first came into light in 2002 and this re-emerged with a different name ten years later as the Middle East respiratory syndrome coronavirus (MERS-CoV). With the research done on the impact of this virus and its origin, the preliminary details trace its origin to a family of single-stranded RNA viruses which are also known as Coronaviridae. On January 30, 2020, the overall

D. Datta (✉)
Radiological Physics and Advisory Division, Bhabha Atomic Research Centre, Mumbai 400085, India

SRM Institute of Science & Technology, Kattankulathur, Chennai 603203, India

J. K. Roy
Eureka Scientech Research Foundation, Kolkata 700051, India

Narula Institute of Technology, Kolkata 700109, India

J. K. Mandal and J. K. Roy (eds.), *Proceedings of International Conference on Computational Intelligence and Computing*, Algorithms for Intelligent Systems,
https://doi.org/10.1007/978-981-16-3368-3_5

mortality rate was 3.4%. In order to prevent the abrupt rise of COVID-19 infections, most of the countries of the world have adopted various protective measures such as wearing face mask, social distancing, and lockdown. In effect of the lockdown, there exists a landslide in economy and possibility of policy change in education. One of the adverse impacts of lockdown measures takes place on the disruption of supply chain. Literature study [6] has evidenced the substantial reduction of air pollution due to industrial shutdown and temporary cuts in air emissions worldwide [7]. It is worth to mention some specific cases related to environmental disturbance through air pollution. In this context, the lockdown has reduced the level of pollution in China leading the lives of 77,000 people indirectly [8]. In the domain of air pollution lockdown initiation at the starting phase of COVID-19 pandemic, the average particulate matter-2.5 (PM2.5) level in New Delhi, India, was reduced by 71% from 91 $\mu g/m^3$ on March 20 to 26 $\mu g/m^3$ on March 27 [9]. Health analytics resulted an improvement of air quality by 44% in Wuhan. Many of the health issues are linked directly or indirectly with COVID-19 pandemic. The concentration of PM2.5, PM10, CO_2, NO_2, Ozone, and SO_2 over 22 cities of India in March and April 2020 has been reported proving a substantial breakthrough of lockdown measures along with social distancing [10]. A huge amount of data related to public as confirmed, recovered, and death cases is available during COVID-19 pandemic and also a large amount of information related to other social events such as supply chain in conventional medicine, nuclear medicine, and other products. Overall, results of the rate of increasing the information proved that COVID-19 pandemic situation is a domain of big data. Basically big data is defined as a dataset in which we consider five properties such as Volume, Velocity, Varity, Variety, and Value. Variety provides the uncertainty of the data. Hence, data analytics plays a major role including decision making problems under uncertainty. Since data is linked with COVID-19 pandemic, we developed a new branch of data analytics named as Pandemic Analytics (PA), wherein we have developed three categories of model namely (1) descriptive model providing the data description with little bit of statistical analysis, (2) predictive model providing the information on forecasting of the spread of pandemic with respect to various epidemiological models, and (3) prescriptive model wherein we have demonstrated optimization techniques for making decisions under uncertainty. Finally, the status of the pandemic situation is measured by entropy as it points out the time arrow. The present paper discusses the details about the Pandemic Analytics. Paper mostly focuses the entropy-oriented intelligent decision-making process including forecasting of the PANDEMIC situation. The remaining part of the paper is organized in the following way. Section 2 presents a small review of research publications in the domain of COVID-19. Section 3 presents the concept of Pandemic Analytics. Section 4 presents various computational methodologies related to forecasting of the PANDEMIC situation subject to the SIRD and RD_COVID19 epidemic models. Section 5 presents the results of Pandemic Analytics. The paper is concluded in Sect. 6.

2 Literature Review

Multiple efforts have sought to characterize COVID-19 social distancing policy responses. These efforts have used various coding frameworks to strengthen policy coding methodologies to improve COVID-19 modeling [11]. Kriston has studied a methodology to have a predictive accuracy of a hierarchical logistic model of cumulative SARS-COV-2 case growth [12]. A substantial amount of data mining research on COVID-19 data has been carried out [13]. These reviews motivate the evolution of entropy-oriented intelligent decision-making reasoning to develop Pandemic Analytics.

3 Concept of Pandemic Analytics

Big data has been accumulated in various domains such as healthcare, public administration, retail, biochemistry, and many other interdisciplinary sectors. Generally, big data is defined as five V's which are (1) **V**olume of the data, (2) **V**elocity of the data, the rate at which the data generates, (3) **V**ariety of the data means the different types of data present in the dataset, (4) **V**eracity of the data which provides the uncertainty of the data, and (5) **V**alue of the data (importance of data). Analytics pertaining to big data has a substantial advantage, and it provides new opportunities in the knowledge processing tasks for forthcoming research. A schematic diagram representing the characteristics of big data is shown in Fig. 1.

In order to handle large complexities associated with big data, we need to apply appropriate statistical methods that have the learning abilities and evolution techniques. Present circumstances of worldwide COVID-19 generates large

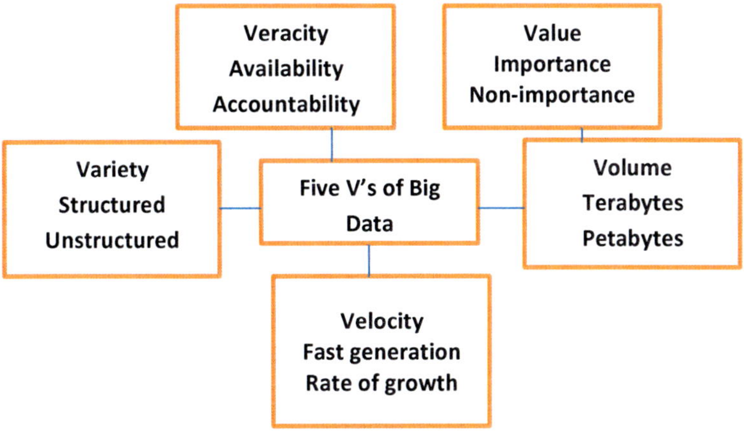

Fig. 1 Block diagram of Big Data

amount of data which are basically grouped in the form of a time series of global confirmed cases, recovered cases, and death cases. Data analytics pertaining to the COVID-19 data is named as Pandemic Analytics (PA) as we need to analyze and develop forecasting models with machine-learning techniques. Similar to usual branch of data analytics, in PA, we have descriptive models, predictive models, and prescriptive models. Descriptive model presents the traditional statistics of time series of COVID-19 data, trend analysis, and categorization of the time series whether stationary or non-stationary. Predictive model provides the knowledge of forecasting on the basis of data-driven model and traditional epidemiological models such as Susceptible-Infected-Recovered (SIR), Susceptible-Infected-Recovered-Dead (SIRD), and Susceptible-Exposed-Infected-Quarantine-Recovered-Dead (SEIQRD). However, in this paper, our analysis is based on SIRD and RD_COVID19 epidemiological model. RD_COVID19 epidemiological model is a networked model that has been developed based on graph theory. Decisions such as case fatality rate (CFR), active cases, prevalence of disease, and relative risks are taken into account in this paper. In the prescriptive model of PA, we have carried out an optimization technique based on Levenberg–Marquardt algorithm to estimate the parameters of the epidemiological model as mentioned. As epidemiological models are compartmental and mathematically described as a system of coupled first-order ordinary differential equation, estimation of parameters is classified as inverse problem which must be routed through optimization technique justifying the necessity of switching on the optimization method. Overall PA is a subset of data analytics and health analytics. Severity of model is judged on the basis of basic reproduction number, R_0 which presents the ratio of secondary infection to the primary infection. Ranking of model based decision is computed in PA using the concept of entropy. More the entropy, higher is the stochasticity of the model, and more is the risk from the point of transmissibility and chaotic nature of the disease.

4 Computational Methodologies of Pandemic Analytics

In the event of pandemic, data and science together plays major role in awareness of health of the public, environment, and economy of the world. Due to lockdown, various countries have adopted their lockdown protective measures which disrupt worldwide and nationwide economy of the various countries. In this paper, Pandemic Analytics is designed with respect to descriptive, predictive, and prescriptive models. Input dataset used in Pandemic Analytics is collected from John Hopkins University and WHO dashboard. The computation used in Pandemic Analytics will be presented into various subsections.

4.1 Input Dataset

Input data used in Pandemic Analytics is basically a time series and related to public. Data is grouped into confirmed cases, recovered cases, and death cases in addition to the latitude, longitude, and population of the country. Table 1 presents the segment of a dataset recorded for country India from March 24, 2020 to April 10, 2020. Definition of Pandemic Analytics is as follows:

- Pandemic Analytics is a modern way to combat a problem as old as humanity itself: the proliferation of disease.
- Pandemic Analytics can address health-related issues (health analytics) as well as economics-related issues (business analytics).
- Exploratory data analysis (EDA) pertaining to data related to COVID-19 pandemic situation in a quantified manner is named as Pandemic Analytics.

In fact, our analysis is based on the complete data ranged from January 22, 2020 to November 30, 2020. Each class of dataset like confirmed, deaths, and recovered are a time series. Our descriptive model is based on traditional statistical analysis. As we need to compute statistical property of time series, we need to be familiar of basic terminologies such as active cases, close contact, and outbreak. We define active cases, close contact, and outbreak in the following way:

Table 1 Pandemic dataset

Lat	Long	Date	Confirmed	Deaths	Recovered
20.6	78.96	03-24-2020	536	10	40
20.6	78.96	03-25-2020	657	12	43
20.6	78.96	03-26-2020	727	20	45
20.6	78.96	03-27-2020	887	20	73
20.6	78.96	03-28-2020	987	24	84
20.6	78.96	03-29-2020	1024	27	95
20.6	78.96	03-30-2020	1251	32	102
20.6	78.96	03-31-2020	1397	35	123
20.6	78.96	04-01-2020	1998	58	148
20.6	78.96	04-02-2020	2543	72	191
20.6	78.96	04-03-2020	2567	72	192
20.6	78.96	04-04-2020	3082	86	229
20.6	78.96	04-05-2020	3588	99	229
20.6	78.96	04-06-2020	4778	136	375
20.6	78.96	04-07-2020	5311	150	421
20.6	78.96	04-08-2020	5916	178	506
20.6	78.96	04-09-2020	6725	226	620
20.6	78.96	04-10-2020	7598	246	774

- Active cases: The number of confirmed cases minus the number of recovered cases and deaths. It is the number of cases still considered to be infectious
- Close contact: A person who has had face-to-face contact for more than 15 min or shared a closed space for more than 2 h with someone who has been diagnosed with COVID-19 (while they were considered infectious)
- Outbreak: An outbreak is the occurrence of more cases of disease than would normally be expected in a specific place or group of people over a given period of time.

4.2 Descriptive Model

Computation of simple monthly average of confirmed cases for months ranging from March to November is carried out and shown in Fig. 2. Other statistical parameters such as median, skewness, minimum, maximum, 25th percentile, 75th percentile, and 95th percentile of confirmed cases are computed and given in Table 2.

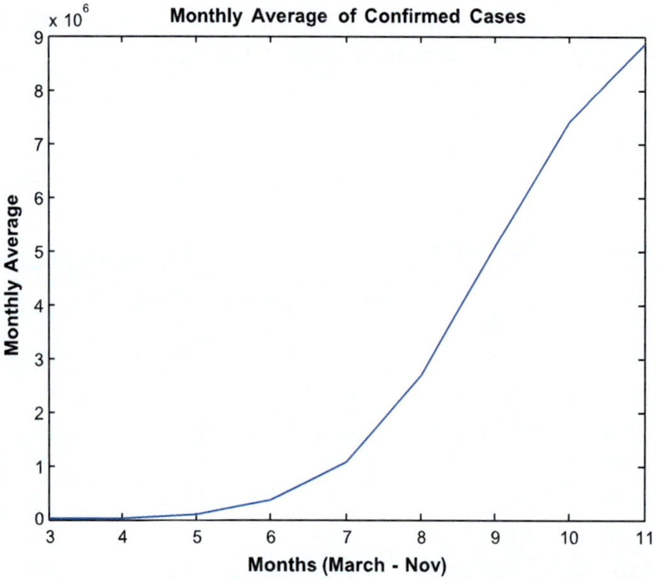

Fig. 2 Monthly average of confirmed cases

Table 2 Statistical parameters of confirmed cases

Month	Median	Std deviation	Min	First Q	Third Q	Max
March	119	404.27	3.0	41.0	517.5	1397
April	12,876	10,181.9	1998	6118	22,650	34,863
May	90,648	45,933.57	37,257	61,252	134,980	190,609
June	348,578	115,940.4	198,370	268,483	452,191	585,481
July	1,003,832	329,123.5	604,641	780,549	1,312,566	1,695,988
Aug	2,647,663	586,590.8	1,750,723	2,184,042	3,136,836	3,136,836
Sep	5,069,306	783,783.3	3,769,523	4,394,062	5,710,891	6,312,584
Oct	7,432,680	542,272.2	6,394,068	6,942,787	7,839,746	8,184,082
Nov	8,859,709	374,594.1	8,229,313	8,563,175	9,168,346	9,462,809

4.3 Predictive Model

Predictive model is one of the components of predictive analytics which is the branch of the advanced analytics used to make predictions of unknown future events. Predictive analytics uses techniques such as data mining, modeling, machine learning, and artificial intelligence to analyze current data to predict future events. In this case, we have carried out forecasting of pandemic situation. Since COVID-19 dataset is a time series, forecasting can be computed in various methods. We have carried out an autoregressive integrated moving average model of order p and q (ARIMA(p, q)). Here p denotes the order of autoregressive and q denotes the order of moving average. p and q are determined by partial autocorrelation and autocorrelation coefficients. On the other hand, we have carried out forecasting using data-driven model where in directly data is fitted using nonlinear least square regression technique. Parameters of the data-driven model are synchronized with the parameters (infection rate, recovery rate, and mortality rate) of dynamic characteristics of the epidemiological models (SIRD and RD-COVID19). Finally data-driven model, SIRD, and RD_COVID19 are networked as per graph theory. Success of a networked model is assessed by R_0, a parameter defined as the average number of people infected by a single infectious person. If $R_0 > 1$, infection spreads till herd immunity reached, and if $R_0 < 1$, infection dies out.

4.4 Prescriptive Model

Here in this case, optimization techniques have been applied to compute pandemic severity index shown in Fig. 3.

Prescriptive modeling is termed as advanced analytics that consists of machine learning techniques, neural networks, and regression analysis. Prescriptive analytics of Pandemic Analytics use optimization techniques. Entropy computation is carried

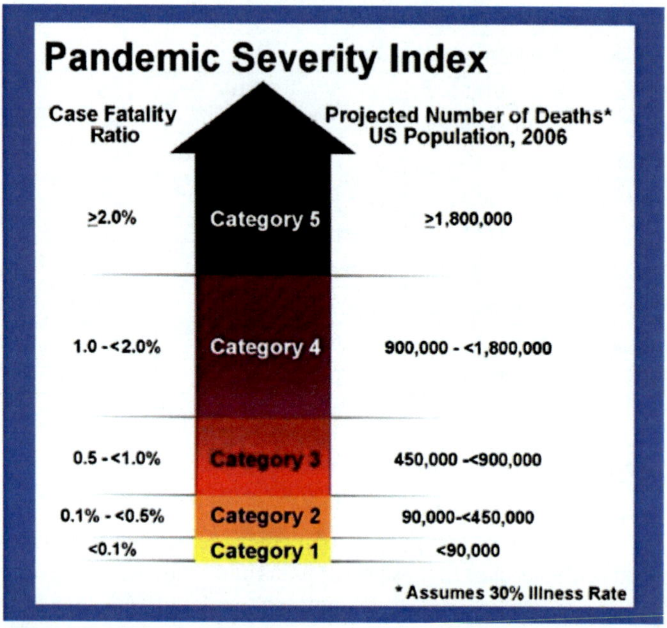

Fig. 3 Pandemic severity index

out in this section of Pandemic Analytics. Entropy being a probabilistic parameter, it ranks the decision taken in pandemic situation. Entropy also directs the uncertainty analysis of epidemiological model. Uncertainty analysis of basic reproduction number is computed.

5 Results of Pandemic Analytics

Pandemic Analytics basically focus various decision models used for econometric analysis, especially in the field of supply chain and management. Figure 4 presents

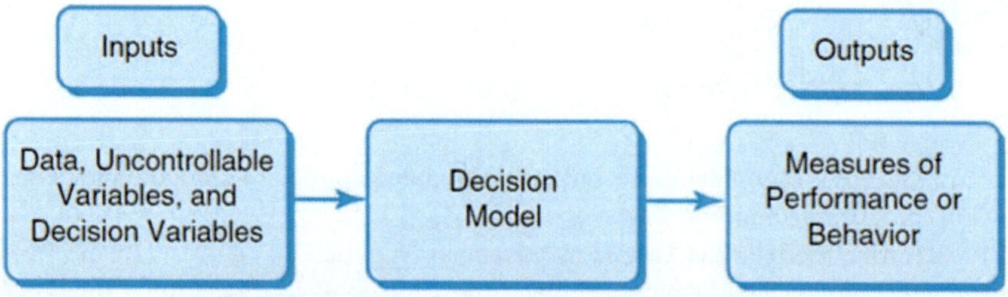

Fig. 4 Schematic diagram of decision model

the schematic diagram of decision models. A mathematical model of the S-curve has been applied as $S = ae^{bect}$, where S is sales, t is time, e is the base of natural logarithms, and a, b, and c are constants. Figure 5 represents the sales curve.

Table 3 presents a typical analysis of sales model, where sales is represented by Eq. (1) as

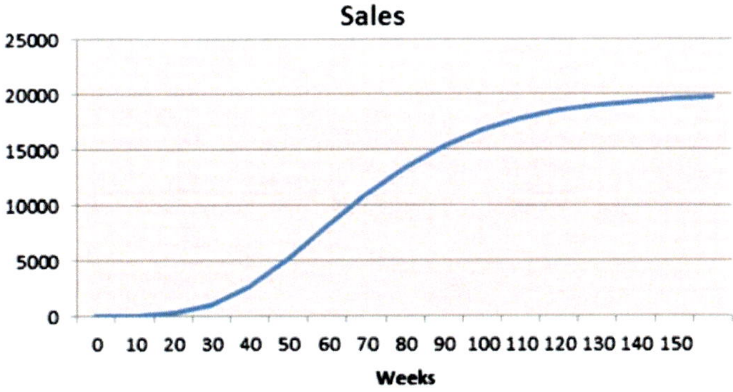

Fig. 5 Sales curve

Table 3 Sales model

Week	Price ($)	Coupon (0, 1)	Advertising ($)	Store 1 Sales (units)	Store 2 Sales (units)	Store 3 Sales (units)
1	$6.99	0	$0	501	510	481
2	S6.99	0	$150	772	748	775
3	$6.99	1	$0	554	528	506
4	$6.99	1	$150	838	785	834
5	S6.49	0	$0	521	519	500
6	$6.49	0	S150	723	790	723
7	$6.49	1	$0	510	556	520
8	S6.49	1	S150	818	773	800
9	$7.59	0	$0	479	491	486
10	$7.59	0	S150	825	822	757
11	$7.59	1	$0	533	513	540
12	$7.59	1	S150	839	791	832
13	$5.49	0	$0	484	480	508
14	$5.49	0	$150	686	683	708
15	$5.49	1	$0	543	531	530
16	$5.49	1	S150	767	743	779

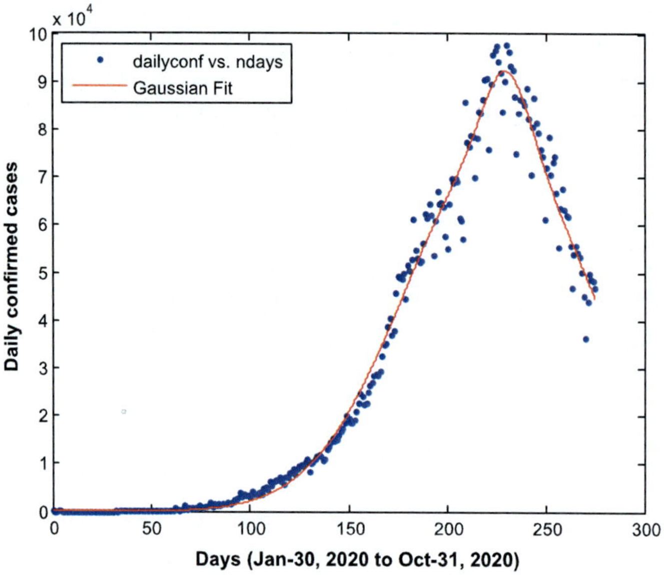

Fig. 6 Double Gaussian model of daily confirmed cases of India

$$\text{Sales} = 500 - 0.05(\text{price}) + 30(\text{coupons}) + 0.08(\text{advertising}) \\ + 0.25(\text{price})(\text{advertising}) \quad (1)$$

Nonlinear least square regression analysis including Levenberg–Marquardt (LM) algorithm as optimization technique results the status of COVID-19 peak and case fatality rate (CFR) in India. LM algorithm allows for fitting a double Gaussian model of daily confirmed cases and single Gaussian model of daily death cases in India (Figs. 6 and 7). Double Gaussian and single Gaussian models are given in Eqs. (2) and (3).

$$f(x) = a_1 e^{-\left(\frac{x-b_1}{c_1}\right)^2} + a_2 e^{-\left(\frac{x-b_2}{c_2}\right)^2} \quad (2)$$

where estimation of fitting parameters with their 95% confidence interval is given by

$$a_1 = 1.6\text{e}+04(1.32\text{e}+04, 1.87\text{e}+04), \quad b_1 = 230.2(228.4, 232)$$
$$c_1 = 15.84(12.58, 19.1), \quad a_2 = 7.64\text{e}+04(7.41\text{e}+04, 7.9\text{e}+04)$$
$$b_2 = 226.2(225.2, 227.2), \quad c_2 = 66.53(64.62, 68.45)$$

The goodness of fitting status of the daily confirmed case is obtained as

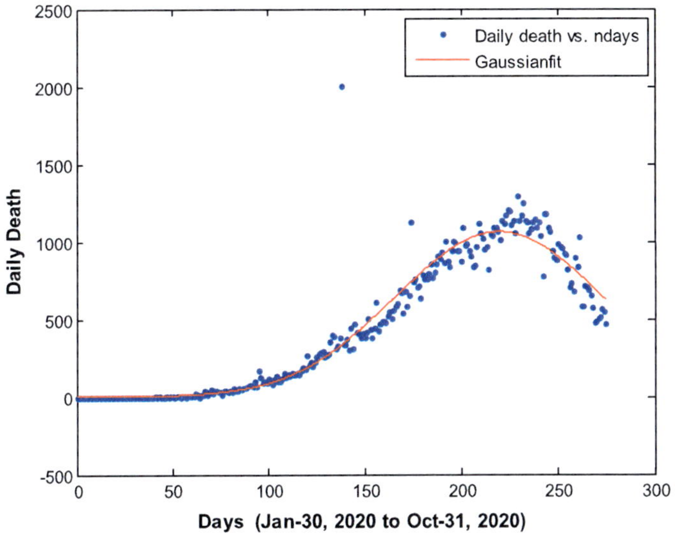

Fig. 7 Single Gaussian model of daily death cases of India

$$\text{sse: } 3.5157\text{e}+09, \quad \text{rsquare: } 0.9877, \quad \text{dfe: } 270$$
$$\text{adjrsquare: } 0.9875, \quad \text{rmse: } 3.6085\text{e}+03$$

$$f(x) = a_1 e^{-\left(\frac{x-b_1}{c_1}\right)^2} \tag{3}$$

where estimation of fitting parameters with 95% confidence interval is given by

$$a_1 = 1062\ (1030, 1095), \quad b_1 = 1.03\ (0.9988, 1.061)$$
$$c_1 = 0.9487\ (0.9004, 0.997)$$

The goodness of fitting status of the single Gaussian model is obtained as

$$\text{sse: } 4.38\text{e}+06, \quad \text{rsquare: } 0.9123, \quad \text{dfe: } 273$$
$$\text{adj rsquare: } 0.9117, \quad \text{Rmse: } 126.63$$

Peak of Covid-19 in India has reached on September,12, 2020 which is indicated in Fig. 6. Case fatality rate (CFR) is computed, and the corresponding result is as shown in Fig. 8.

Epidemic outcome of our networked model is as shown in Fig. 9.

Data range taken into account for predictive analysis as shown in Fig. 8 is up to May 9, 2020. Parameters estimated in this case are as follows: $R =$ reproduction number $= 1.107, R_0 = 1.40$, the rate of infection, $\beta = 0.316$, rate of recovery, and $\gamma =$

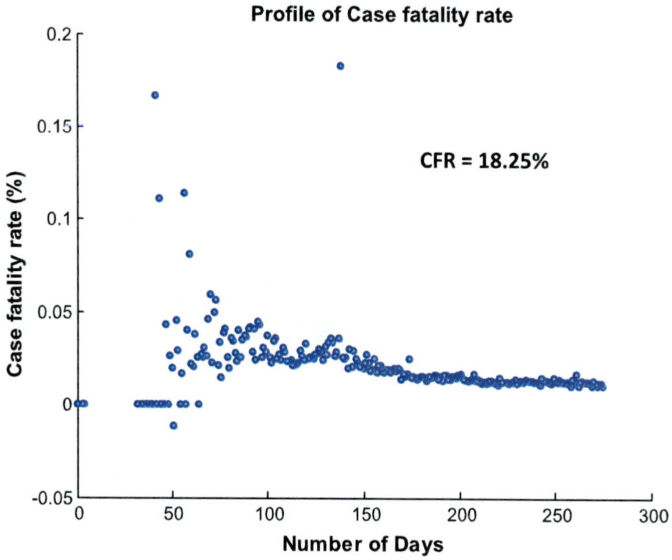

Fig. 8 Case fatality rate

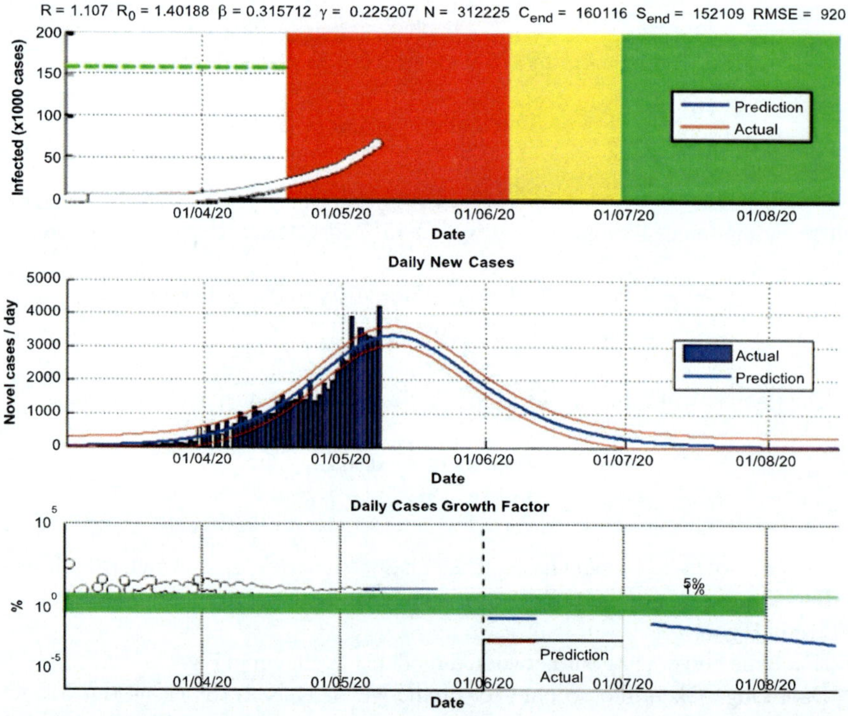

Fig. 9 Outcome of networked model

0.225. On the basis of data available (Jan 22, 2020 to May 9, 2020) during that time, predictive analysis also has been carried out to know the turning day, acceleration phase, deceleration phase, total growth phase duration, and total epidemic duration which are 72, 23 days, 25 days, 48 days, and 238 days, respectively. Based on our predictive analytics, it can be reported that start of acceleration was Apr 19, 2020, turning point was May 12, 2020, start of steady growth was Jun 6, 2020, and start of ending phase was Jun 30, 2020, giving the decision of unlock 1.0. The analysis presented so far is based on deterministic epidemiological models even though the models are networked as per protocol of graph theory. However, in practice, uncertainty present in the data invites stochastic epidemiological models. In this context, it is mandatory to forecast the pandemic situation using entropy. Section 5.1 provides the main attraction of entropy-based decision on pandemic situation.

5.1 Entropy-Oriented Pandemic Forecasting

It is essential to develop reliable predictions of the evolution of an infectious disease such as the present outbreak of COVID-19. Generally, a statistical learning of the time of maximum diffusion of infected carriers is fundamental for preparing healthcare system and creating a robust public health response. Therefore, a thermodynamic approach based in the infection statistics following time arrow has been developed. The computation carried out for this purpose is based on the cumulative probability of infection versus time as a logistic shape given by Eq. (4).

$$P(t) = \frac{\exp(\beta t)}{1 + \exp(\beta t)} \tag{4}$$

where P is the probability of infection, t is the time, and α and β are two constants to be fitted. Now, following usual statistical approach, we can have the Gibbs dimensionless entropy as

$$S(t) = -f(t) \ln f(t) \tag{5}$$

where $f(t)$ is the frequency of the infected people on the respective country. The model of $f(t)$ can be written by its definition as

$$f(t) = \frac{n(t)}{N}, \tag{6}$$

where N represents the population size.

In this part of our Pandemic Analytics, we have following decisions:

- To find the occurrence frequency distribution in time;

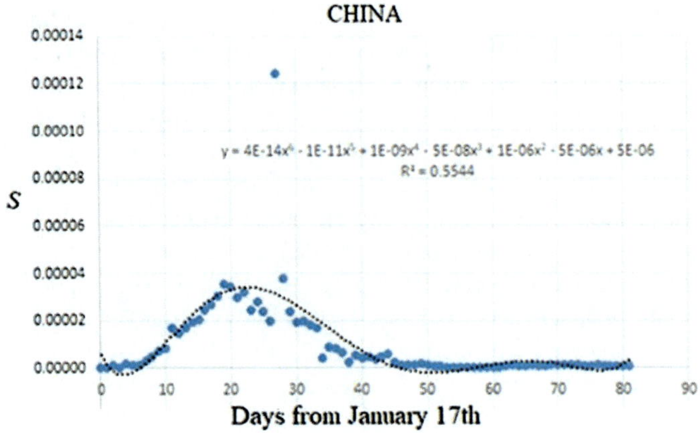

Fig. 10 Entropy model for China

- To find the cumulative value of the occurrence frequency distribution in time;
- To evaluate the entropy using Eq. (5).
- To evaluate the best fit for the entropy obtained at the previous point;
- To determine its maximum and the related time, directly by the shape or by mathematical methods [14, 15].

Since the complete dataset of China and Italy was available, we have presented the shapes of evolution of entropy for these two countries. For China: the time of maximum expansion of the coronavirus infection results 23 days after January 17 (around February 11). The slight discrepancy with the value reported in the table (i.e., February 13) is due to the function used for fitting (the better the fit, the more accurate is the forecasting); Moreover, China declared a correction on April 17, 2020. For Italy, it results 34 days after February 22 (March 27), which corresponds exactly with the observed time point reported in Figs. 10 and 11.

6 Conclusions

Pandemic Analytics is a new branch of data analytics covering the health analytics as well as various predictive models. We have also provided the descriptive analytics model on the basis of data for country India. Deterministic epidemiological models and data-driven models have networked to develop our predictive model. We also have shown various decisions including the estimated value of parameters of epidemiological model. Estimation of parameters of a system of coupled ordinary differential equations describing epidemiological model is an optimization problem, and this part we have carried out using nonlinear least square optimization algorithm. We have applied Levenberg–Marquardt algorithm. Our entropy-based predictive model suggested a novel thermodynamic approach for forecasting outbreaks of large-scale

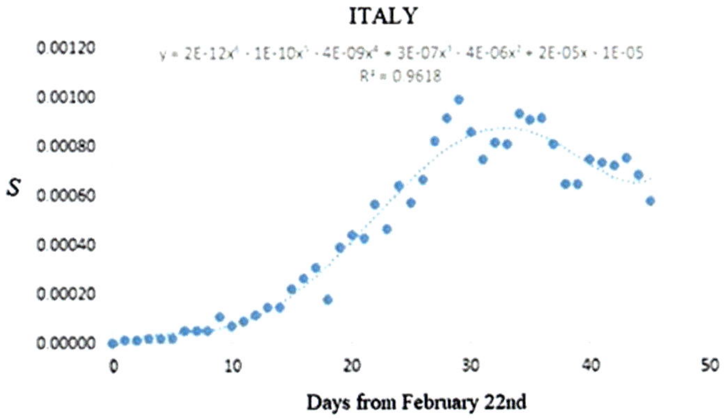

Fig. 11 Entropy model of Italy

infectious disease. Principle of maximum entropy approach can unfold the pandemic situation in the sense of analyzing uncertainty present in the dataset. Entropy-based models are strongly dependent on the statistics used. Our model accurately predicted the date of maximum expansion of COVID-19 infections in China and Italy. The similar approach for India could have been tested provided the data should have been available during that moment. It is also worth to note that peak of COVID-19 situation in India was not available, and now it is available which has been presented in the paper. Entropy-based prediction can also empower the regulatory authority to estimate the prevalence and incidence of risk of infectious disease.

References

1. Huang, C., Wang, Y., Li, X., Ren, L., Zhao, J., Hu, Y., Zhang, L., Fan, G., Xu, J., Gu, X., et al.: Clinical features of patients infected with 2019 novel coronavirus in Wuhan, China. Lancet **395**(10223), 497–506 (2020)
2. Bedford, J., Enria, D., Giesecke, J., Heymann, D.L., Ihekweazu, C., Kobinger, G., Lane, H.C., Memish, Z., Oh, M.-D., Schuchat, A., et al.: Covid-19: towards controlling of a pandemic. Lancet **395**(10229), 1015–1018 (2020)
3. Guo, Y.-R., Cao, Q.-D., Hong, Z.-S., Tan, Y.-Y., Chen, S.-D., Jin, H.-J., Tan, K.-S., Wang, D.-Y., Yan, Y.: The origin, transmission and clinical therapies on coronavirus disease 2019 (COVID-19) outbreak—an update on the status. Mil. Med. Res. **7**(1), 1–10 (2020)
4. Sohrabi, C., Alsafi, Z., O'Neill, N., Khan, M., Kerwan, A., Al-Jabir, A.: World Health Organization declares global emergency: a review of the 2019 novel coronavirus (COVID-19). Int. J. Surg. **76**, 71–76 (2020)
5. WHO director-general's opening remarks at the media briefing on COVID-19—3 March 2020. Available at: https://www.who.int/dg/speeches/detail/who-director-general-s-opening-remarks-at-the-media-briefing-on-covid-19-3-march-2020. Accessed 11 March 2020
6. Watts, J., Kommenda, N.: Coronavirus pandemic leading to huge drop in air-pollution. The Guardian (2020). Retrieved from: https://www.theguardian.com/environment/2020/mar/23/coronavirus-pandemic-leading-to-huge-drop-in-air-pollution. 4 April 2020

7. Chowdhury, S., Dey, S., Di Girolamo, L., Smith, K.R., Pillarisetti, A., Lyapustin, A.: Tracking ambient PM2.5 build-up in Delhi national capital region during the dry season over 15 years using a high-resolution (1 km) satellite aerosol dataset. Atmos. Environ. **204**, 142–150 (2019). https://doi.org/10.1016/j.atmosenv.2019.02.029
8. McMahon, J.: Study: coronavirus lockdown likely saved 77,000 lives in China just by reducing pollution. Forbes (2020). https://www.forbes.com/sites/jefmcmahon/2020/03/16/coronaviruslockdownmay-have-saved-77000-lives-in-china-just-from-pollution-reduction/#2e94bd3f34fe. Accessed 19 March 2020
9. Mitra, A., Chaudhuri, T.R., Mitra, A., Pramanick, P., Zaman, S., Mitra, A., Chaudhuri, T.R., Mitra, A., Pramanick, P., Zaman, S.: Impact of COVID-19 related shutdown on atmospheric carbon dioxide level in the city of Kolkata. Sci. Educ. **6**(3), 84–92 (2020)
10. Sharma, S., Zhang, M., Gao, J., Zhang, H., Kota, S.H.: Effect of restricted emissions during COVID-19 on air quality in India. Sci. Total Environ. **728**, 138878 (2020). https://doi.org/10.1016/j.scitotenv.2020.138878
11. Lane, J., Garrison, M.M., Kelley, J., Sarma, P., Katz, A.: Strengthening policy coding methodologies to improve COVID-19 disease modeling and policy responses: a proposed coding framework and recommendations. BMC Med. Res. Methodol. Article number: 298 (2020)
12. Kriston, L.: Predictive accuracy of a hierarchical logistic model of cumulative SARS-CoV-2 case growth until May 2020. BMC Med. Res. Methodol. **20**, Article number: 278 (2020)
13. Radanliev, P., De Roure, D., Walton, R.: Data mining and analysis of scientific research data records on Covid-19 mortality, immunity, and vaccine development. In the first wave of the Covid-19 pandemic, Computers and Society (cs.CY). Digital Libraries (cs.DL), arXiv:2009.05793 [cs.CY]
14. Pauli, W.: Statistical Mechanics. MIT Press, Cambridge (1973)
15. Schrodinger, E.: Statistical Thermodynamics. Cambridge University Press, Cambridge (1952)

Application of Intrinsic Mode Functions, Linear and Neural Regression in Forecasting of Summer Monsoon Rainfall over Assam and Meghalaya, India

Pijush Basak and Joyanta Kumar Roy

1 Introduction

The rainfall in the months comprising of June, July, August, and September, named South West Monsoon (SWM) rainfall is a substantial component of annual rainfall in North-East India including the meteorological subdivision number 3 covering the region of Assam and Meghalaya (AM). The economy, agriculture, and industrialization are considerably counted on the characteristics of SWM rainfall. The spatial and temporal scales are the major concern to the scientists [1–3]. The meteorological subdivisions of India along with AM are shown in Fig. 1.

The connection between SWM rainfall and different atmospheric phenomenon was a concern to the scientist previously. The linkage between SWM rainfall and Sea Surface Temperature (SST) was studied by Sahai et al. [4]. Efforts were made with simply modeling considering any linkage also. With the help of Fourier decomposition, the latent period in data was examined by different researchers, namely, Quasi-biennial Oscillation (QBO) [5], tidal forcing [6], El Nino-Southern Oscillation (ENSO) [7, 8], Sunspot Cycle [9] and intra-seasonal periodicities [7, 10]. The general method of time series analysis rests on stationarity of data with Gaussian property. The examination of SWM rainfall data that auto-correlation and power spectral densities are too weak for modeling as linear time series whilst the form of the non-linear model is unknown and very difficult to understand.

Decomposition of SWM rainfall into principal component components and understanding the nature and forecasting with components maybe a possible alternative approach [11–13].

P. Basak (✉)
RCC Institute of Information Technology, Kolkata, India

J. K. Roy
Eureka Scientech Research Foundation, Kolkata, India

J. K. Mandal and J. K. Roy (eds.), *Proceedings of International Conference on Computational Intelligence and Computing*, Algorithms for Intelligent Systems,
https://doi.org/10.1007/978-981-16-3368-3_6

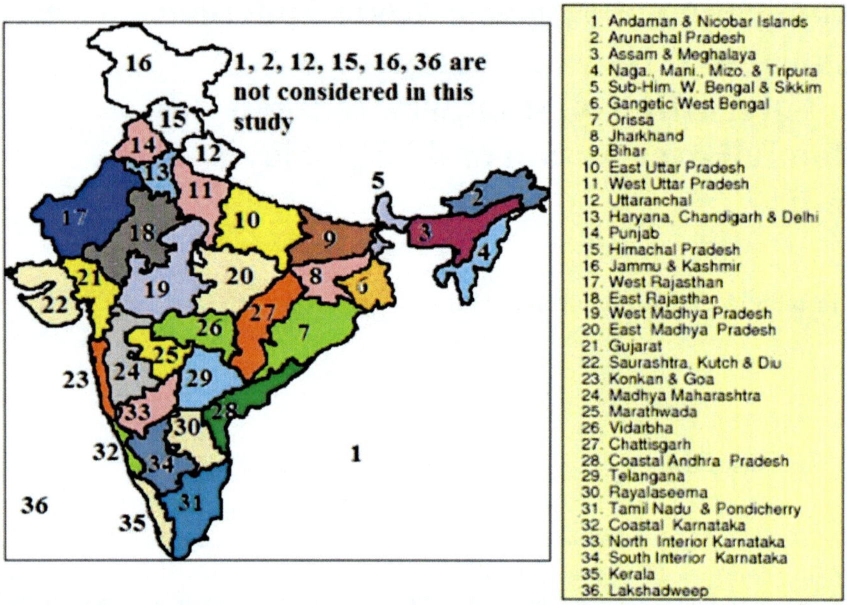

Fig. 1 Meteorological Subdivisions including Assam and Meghalaya (AM)

However, it may be supposed that rainfall series carries its inherent causes as focused in the works of [14, 15]. In the last decade, it is indicated that rainfall data may be decomposed into hierarchical Intrinsic Mode Functions (IMFs) as signals if basic data is not purely white noise [16].

This aspect has been indicated by Iyenger and Raghukant [15] at All India level and Zvarevashe et al. [17] and Sabzehee et al. [18] in other countries like Australia and Caspian catchment area respectively.

However, the variability study and forecasting exercise has received little attention in the region of AM meteorological subdivision No. 3 in the southern plain in the eastern fringe of the Himalayas. It is to be noted that the region under study has profound importance in civilization, industry, and agriculture. In the present paper, a new representation of the SWM rainfall of AM in terms of narrowband IMF series is examined.

The structure of the paper is formulated as follows. Firstly, the Empirical Modes called IMFs of AM would be discussed with a possible forecasting strategy. A combination of multiple linear regression analysis and Generalized Regression Neural Network (GRNN) architecture would be discussed followed by an analysis on the performance of model would be provided.

1.1 Rainfall Data

The monthly rainfall for the period 1871–2001 of subdivision number 3 of Assam and Meghalaya is extracted from the website 'www.tropnet.net' of Indian Institute of Tropical Meteorology, Pune (IITM). The SWM data are sorted out from the collected monthly data.

The SWM data of AM are presented in Fig. 2 for initial preliminary idea. Some basic statistics of the data are presented in Table 1.

2 Intrinsic Mode Functions (IMFs)

The SWM rainfall series of AM is decomposed into hierarchical empirical mode, named Intrinsic Mode Functions (IMFs) as per Huang et al. [19]. The IMFs such extracted are uncorrelated with each other at lag zero but correlated with SWM data and identify dominant period and amplitude [19].

Seven IMF series, such as IMF1, IMF2, ..., IMF7 extracted hierarchically until the data extracted signals almost zero or no oscillatory trace. Each IMF is a narrow-band

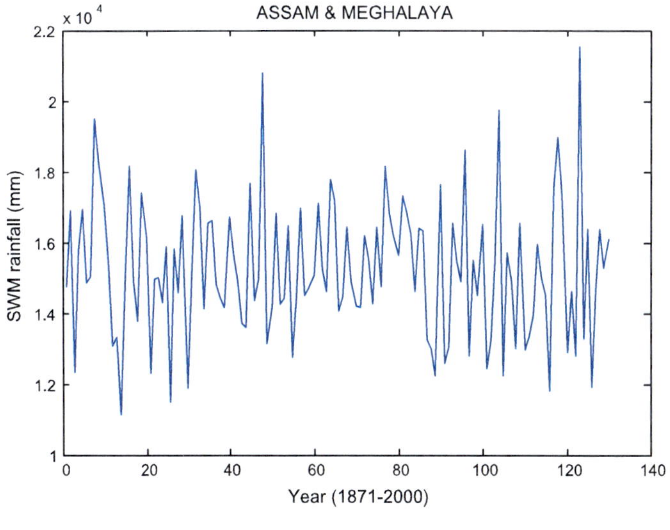

Fig. 2 SWM rainfall of AM for modeling period (1871–2000)

Table 1 SWM rainfall data (1871–2001)

Region	km^2	Mean (mm) (mR)	Std. dev. (6R)	Skewness	Kurtosis
AM	78,438	15,250	102,734	– 0.4425	2.9845

AM Assam and Meghalaya

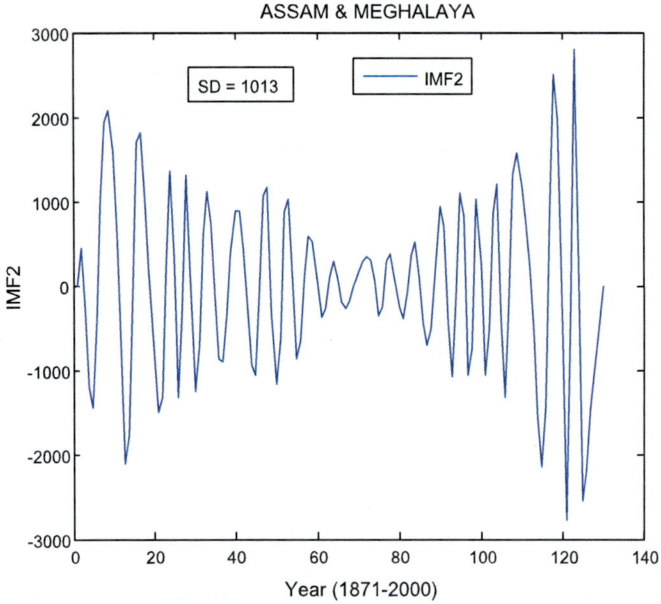

Fig. 3 IMF1 of SWM rainfall of AM

process with definite periods. Climatic trends with specific period and non-stationary characteristics show as the seven IMFs. Figures 3, 4, 5, 6, 7, 8 and 9 presents seven IMFs of SWM rainfall. It is indicated that the last IMF, that is IMF7 is undoubtedly positive with slowly varying mode over the long-term averages [19]. The sum of all the IMF series at a particular time is equal to the original SWM data series upto higher level of accuracy.

It is established from Figs. 3, 4, 5, 6, 7, 8 and 9, IMFs display slowly various amplitudes and frequencies indicated a narrow band process. The technique of counting zeros and extreme in an IMF is adopted for obtaining dominant period of oscillation. Table 2 contains period (years) and the percentage contribution of each IMF to IAV. The first IMF is most prominent with a dominant period of 3.25 years with contribution to IAV of 52.61%; the second most important one IMF2 displaying with corresponding figures as 6.19; years and 21.35% respectively. The two IMF modes are closely connected with Quasi-biennial Oscillation (QBO) and El Niño–Southern Oscillation (ENSO) phenomenon respectively as evidenced in case of All India level [16]. In the same way, IMF3 yields the corresponding figures as 16.25 years and 5.83% linked with the Sunspot cycle of about 12–16 years as per Bhalme and Jadav [9]. In the similar way, IMF4 having period of 21.66 years is in relevance with quasi-cycle of Indian monsoon with Indian monsoon presented in the works of Campbell et al. [6]. IMF5 depicts an extended period of 43.33 years at par with the works of Huang et al. [19]. IMF6 is connected to the extended period of 65 years, contributing approximately 6.75% of IAV is the representative of six quasi-cycle modes of Indian

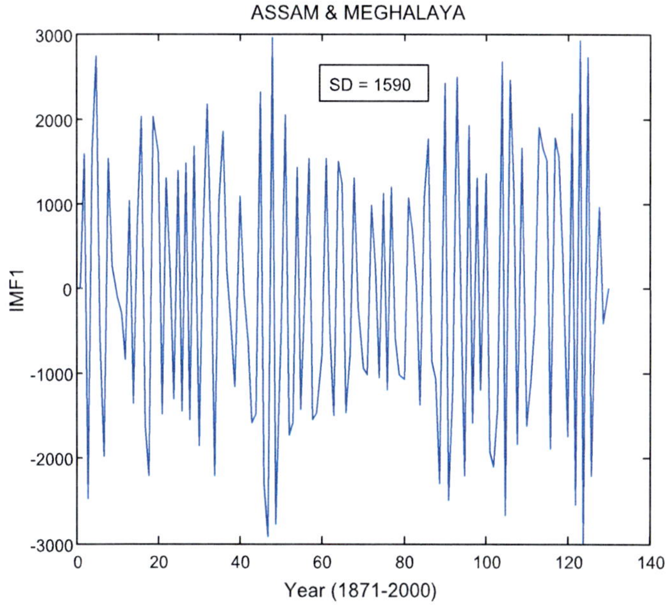

Fig. 4 IMF2 of SWM rainfall of AM

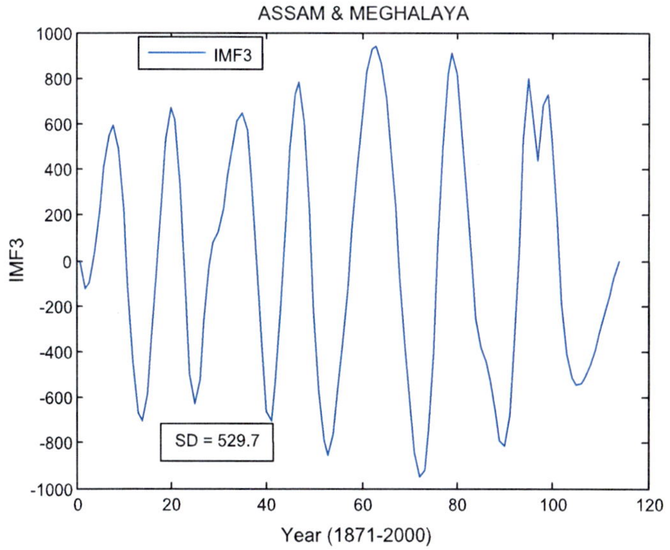

Fig. 5 IMF3 of SWM rainfall of AM

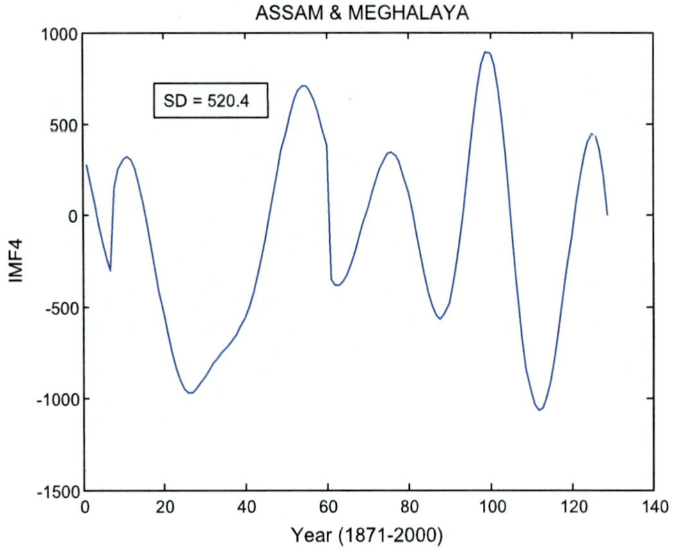

Fig. 6 IMF4 of SWM rainfall of AM

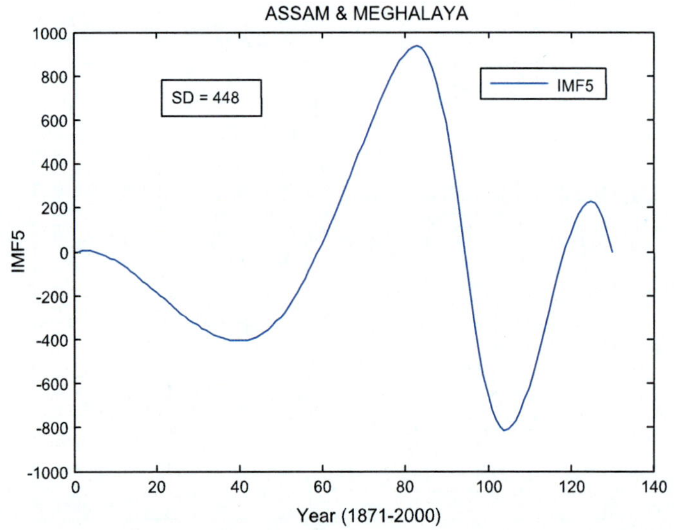

Fig. 7 IMF5 of SWM rainfall of AM

monsoon rainfall as contributed by Narashima and Kailash [7] utilizing wavelet analysis. IMF7 with undetectable period may be understood to be the deterministic shift of the monsoon at par with the Indian monsoon rainfall [15].

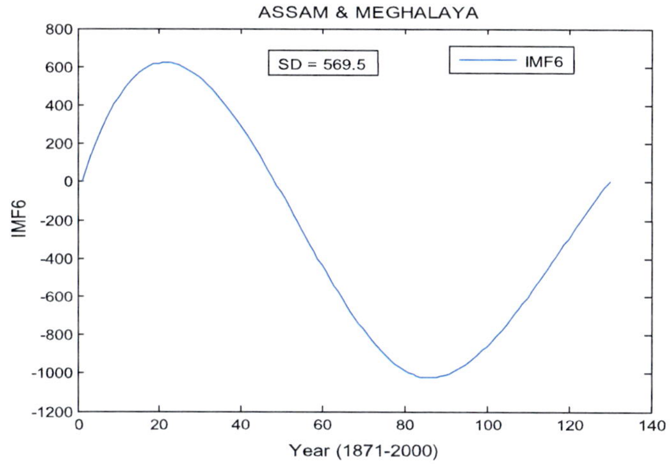

Fig. 8 IMF6 of SWM rainfall of AM

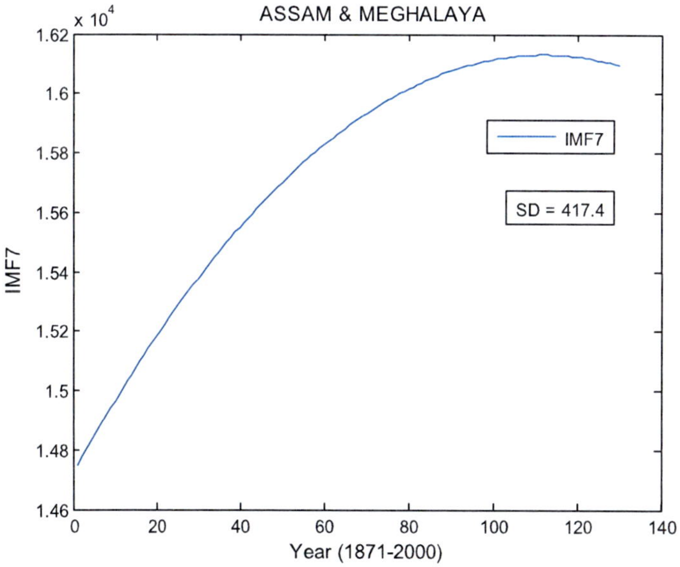

Fig. 9 IMF7 of SWM rainfall of AM

3 IMF Statistics

In the next stage, it is very much necessary to understand how the SWM rainfall series and the IMFs are correlated. Table 3 displays the correlation matrix or order 7 $\times$ 7 of SWM AM data and the seven IMFs. From the table, it is inspected that IMFs are uncorrelated among themselves as desired except few IMFs of higher order. The

Table 2 Periodicity in years (T) of the IMF's and percentage of variance contributed to IAV

Region		AM
IMF1	T	3.25
	IAV%	52.6
IMF2	T	6.19
	IAV%	21.3
IMF3	T	16.25
	IAV%	5.8
IMF4	T	21.66
	IAV%	5.6
IMF5	T	43.33
	IAV%	4.1
IMF6	T	65
	IAV%	6.75
IMF7	T	Not detectable
	IAV%	16.9

AM Assam and Meghalaya

Table 3 Matrix of correlation among SWM rainfall and IMFs

	Data	IMF1	IMF2	IMF3	IMF4	IMF5	IMF6	IMF7
Data	1.0000	0.7873*	0.4433*	0.3054*	0.0686	0.0989	– 0.0103	– 0.0231
IMF1		1.0000	– 0.0472	0.0350	0.0999	– 0.0006	0.0443	– 0.0421
IMF2			1.0000	0.0032	– 0.0176	– 0.0430	– 0.0051	+ 0.0573
IMF3				1.0000	0.0333	– 0.0131	0.0617	– 0.0526
IMF4					1.0000	0.1290	– 0.2960	0.1349
IMF5						1.0000	– 0.4446*	0.1874
IMF6							1.0000	– 0.8063*
IMF7								1.0000

*Significant at 5% level

data is significantly correlated to IMFs which is theoretically expected. It is pointed out, as expected, the sum of the variances of the IMFs is almost equal to the total data variance.

4 Forecasting Strategy

Forecasting is connected with the preparation of a model that closely coincides with the data, of course, with minimum error. Usually, a data series is modeled with simple

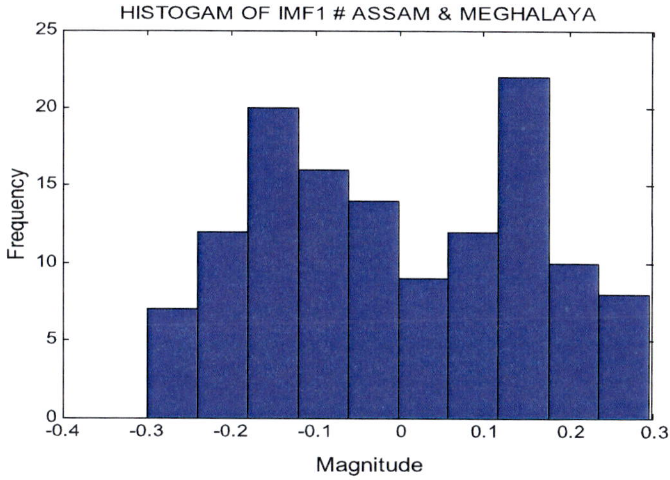

Fig. 10 Histogram of IMF1 showing bi-modality

function, the forecasting exercise may be carried out with Taylor's series expansion. However, as rainfall data is highly erratic, such an effort would not be successful. In this situation, the modeling of decomposed IMF series and thereafter adding the forecasts would put forth the modeling of SWM rainfall. The procedure is certainly a simpler approach than modeling the original data. As a random variable, SWM rainfall is Gaussian except for a few stations [3, 12, 16]. As indicated, forecasting of SWM rainfall is now switched over to 7 hierarchical IMFs. The first IMF is dominant with period 3.25 years explained 50–55% of IAV. It explores the highest frequency end of the information and as expected to be more random than others.

A feature of bi-modality emerges for the first IMF as eminent in the histogram (Fig. 10) and turns down the possibility of Gaussianness supported by Chi-square test. This aspect of bi-modality indicates strong non-linearity in the process and negates the possibility of linear auto-regressive representation.

On the other hand, as tested with standard run test of decadal variance, IMF1 is found to be non-stationary. So, the part of data, that consists of. IMF1 is the non-linear part. Excluding the first IMF, the remaining part (R_j − IMF1_j) is tested for Gaussianness and stationarity as detailed. The remaining part is an indicative hardly Gaussian but an indicative of non-stationary as per run test of decadal variance. Further, IMF2 cannot be treated as Gaussian process. The remaining other part $y_i =$ (R_j – IMF1_i – IMF2_i) is nearly Gaussian and stationary and it is possible to model through multiple linear regressions from its own past values.

The linear part $y_{n+1}$is presented as [15]

$$y_{n+1} = C_1 R_n + C_2 R_n - 1 + C_3 y_n - 2 + C_4 y_n - 3 + C_5 y_n - 4 + C_6 \quad (1)$$

It is observed that Eq. (1) indicates good fit for the SWM data base.

Table 4 Coefficients of regression of Eq. (1)

Region	C_1	C_2	C_3	C_4	C_5	C_6	бy (e)	Correlation coefficient (CC)
AM	0.1098	1.8275	− 2.1554	1.6061	0.5927	4.1076	0.7394	0.8944

AM Assam and Meghalaya

The regression coefficients $C_1, C_2, \ldots, C_6$ are found from the data series of 1871–2000, leaving first 4 years data as regression equation contains up to y_{n-4} terms by least square method and the resulting standard deviation of the error б*y*(e) and correlation coefficient (CC) between data and the model fitted are presented in Table 4. The correlation is highly significant indicating the appropriateness of identifying y_n as the linear part of SWM rainfall of AM.

4.1 Generalized Regression Networks Architecture Connected to Non Linear IMF1 and IMF2

We understand that the first two IMFs are non-Gaussian, non-stationary, and as such non-linear processes. In this case of complex problem, the Generalized Regression Networks architecture (GRNN), an improved version of Neural Network built with non-parametric regression [20] is applied.

5 Architecture of GRNN

The model contains two hidden layers with summation neurons and pattern neurons. The computations are executed among pattern neurons of GRNN with the formulation $\exp(-Dj^2/2б2)$, *Dj* the distance among training sample and б, smoothness parameter. The normal distribution is adapted in training samples. Denominator neuron contains signals of the pattern neuron, are weighted with corresponding values of the training samples y_j and the weights on the signals going into the Numerator are one. Each sample from the training data influences every point as predicted by GRNN.

The author [20] indicates that GRNN works for modeling and extending regression, prediction, classification, and function approximation keeping in view that every training sample will represent mean to a radial basis neuron. After several trials with number of previous values of IMF1, a GRNN with hidden layer is utilized as shown in Fig. 11.

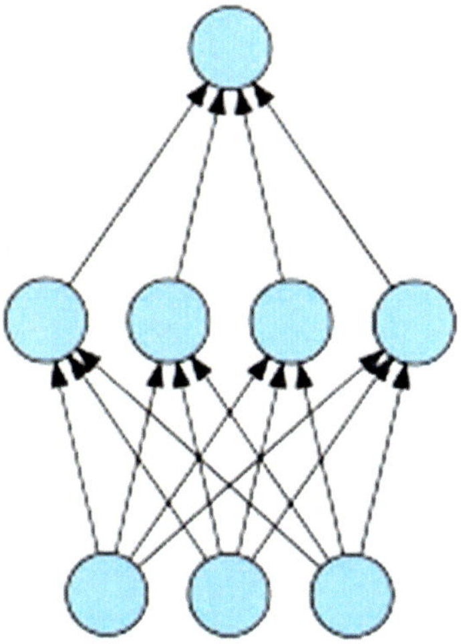

Fig. 11 Generalized regression neural network with radial basis functions

Table 5 GRNN model for IMF1 and IMF2: statistics for training period (1871–2000)

Region	бy(e)	Correlation coefficient (CC)
AM	89.21	0.79615

AM Assam and Meghalaya

5.1 Results of IMF1 and IMF2with GRNN

The computation has been done using MATLAB toolbox on GRNN algorithms, with 1871–2000 as the training period. With the help of previous IMF1 and IMF2 values, the GRNN model is capable of predicting IMF1 and IMF2 for the year $(n + 1)$. In Table 5, the standard deviation б$y(e)$ of the errors is constructed on the training period data is shown along with the correlation coefficient (CC) between the actual IMF1 and IMF2 and the GRNN results. It is observed that GRNN is quite useful in capturing the latent nonlinear structure evidenced by the high correlation (0.7965) between the actual and simulated IMF1 and IMF2 values.

6 Forecasting

The successful modeling of $\mathrm{IMF1}_j$ and $\mathrm{IMF2}_j$ and y_j can be extended by one year, to make a forecast of the next year's rainfall. Firstly, for y_{n+1} and then for $\mathrm{IMF1}_{i,n+1}$ and $\mathrm{IMF2}_{j,n+1}$ are computed from the models mentioned above. The sum of the two values y_n, $\mathrm{IMF1}_n$ and $\mathrm{IMF2}_n$ produces a forecast for R_{n+1}. Here, the performance of the forecast strategy is investigated by considering for the period 2001–2013, the data which was left out of the modeling exercise. The quantification of modeling R_j

Table 6 Modeling and forecasting period: performance

Region	(1871–2000; modeling period)			(2001–2013; forecasting period)		
	б*m* (e)	CCm	PPm	б*f* (e)	CCf	PPf
AM	105.06	0.7063	0.87	41.25	0.8580	0.98

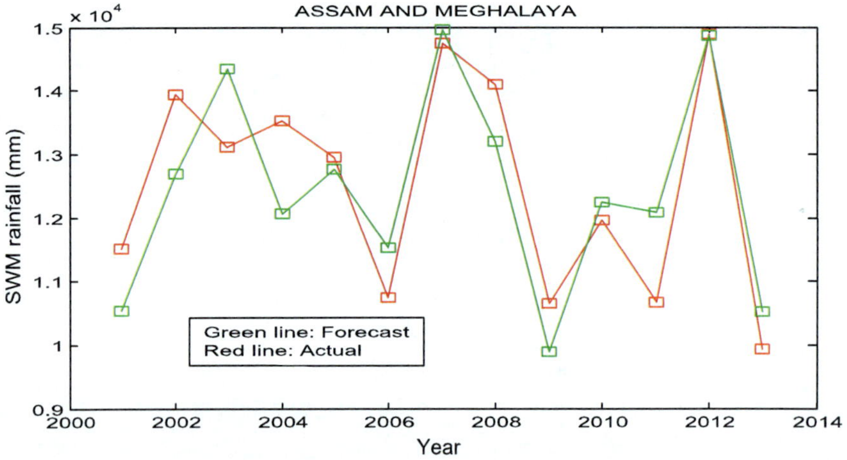

Fig. 12 The actual and predicted SWM rainfall of AM for the testing period (2001–2013)

in the training period (1875–2000) and the efficiency of one-step-ahead forecasting in the testing period (2001–2013) are presented in Table 6.

Figure 12 elaborates the actual rainfall data and predicted rainfall data for testing period (2001–2013). It may be noted that forecast is nothing but expected value and may be slightly different from observation. It is evidenced that the strategy for forecasting SWM rainfall presently works well within certain limits.

7 Performance of the Model

The performance of the model proposed is verified with three statistical parameters. The first two are the Root Mean Square Error (RMSE) and the correlation coefficient (CCm) between the given data and the simulated values out of the model and the third one is a statistic called Performance Parameter [4], namely, $PP_m = 1 - (б m^2)/ (б^2)$, where $б m^2$, and $б^2$ is the mean square error and the actual data variance, respectively, have also been computed. In a perfect model, $б m^2$ will be zero and both CC_m and PP_m would approach unity. Table 7 endorses the efficiency of the present model with correlation coefficient between forecasted and actual data being 0.87, which is sufficiently high. It may be mentioned that under the non-ideal condition of not

Table 7 Independent forecasting for Assam and Meghalaya

Year	Actual (mm)	Forecast (mm)
2001	11,498	10,528
2002	13,932	12,699
2003	13,110	14,347
2004	13,522	12,056
2005	12,942	12,756
2006	10,732	15,529
2007	14,733	14,951
2008	14,094	13,195
2009	10,641	9897
2010	11,961	12,241
2011	10,664	11,674
2012	14,886	14,871
2013	9927	10,520

updating each time by actual data, IMF1 and IMF2 the performance produced by the model is good enough. For a sample size of $N = 13(2001–2013)$, the correlation coefficient (CC_f) in the test period has to be at least 0.6 to be taken as significant whereas it is found from Table 7 that CC_f is well above 0.6.

8 Discussion

IAV of monsoonal rainfall of AM has been studied in this paper with some interesting results. It is identified that the seasonal SWM rainfall time series of AM can be decomposed into seven statistically almost orthogonal modes; the summation of which gives back the original data. The seventh mode is identified easily connected with the overall climatic variation persistent. The remaining six empirical modes (IMFs) are narrow band random processes, with defined periods linked to specified meteorological phenomenon. The first two IMFs which account for the highest variability are strongly non–Gaussian and as such successfully predicted using GRNN techniques. The remaining part of the SWM rainfall after removing the first two IMFs is accessible for a linear multiple regressive representation. With two separate representations; a methodology has been undertaken to forecast rainfall. Nevertheless, the analysis does not account for other variabilities, such as intraannual, interseasonal or intraseasonal variability persistent in the monsoon rainfall. The forecast of SWM rainfall for AM for the year 2012 and 2013 14,871 mm and 10,520 mm respectively corresponding to the actual SWM rainfall of 14,886 mm 9927 mm respectively which are within one standard deviation of mean rainfall. Among the first six IMFs, it has been identified that first three IMFs contributed nearly 90% of the variability. It may be interpreted that if IMF1 and IMF2 those are simultaneously negative, the chances of drought are high. For flood-like situation those are highly positive which are in agreement with [15].

9 Conclusion

It is established that SWM rainfall time series, sampled annually, is decomposable into seven statistically orthogonal modes. Seventh mode is associated with the overall climatic variation whilst the remaining six empirical modes are associated with narrow-band random processes having specified central periods and are connected to important meteorological phenomenon parameters. The approach indicates that first mode IMF1 and IMF2 accounting for highest variability is strongly non-Gaussian and is modeled by GRNN technique; whereas the remaining part of the rainfall is suitable for linear auto-regressive representation is an interesting approach. The combination of two techniques completes the forecasting exercise of the rainfall prediction is developed for AM. The particular approach is general enough and efforts are on to include the analysis in other regions of India.

Acknowledgements The authors sincerely thank National Data Centre, ADGM (R) Pune for providing SWM rainfall data of Assam and Meghalaya.

References

1. Moley, D.A., Parthasarathy, B.: Fluctuations in all-India summer monsoon rainfall during 1871–1978. Clim. Change **6**(3), 287–301 (1978)
2. Rupa Kumar, K., Sahai, A.K., Krishna Kumar, K., Patwardhan, S.K., Mishra, P.K., Revade Kar, J., Kamala, K., Pant, G.B.: High resolution climate change scenarios for India for the 21st century. Curr. Sci. **90**(3), 334–345 (2006)
3. Iyenger, R.N., Basak, P.: Regionalization of Indian monsoon rainfall and long term variability signals. Int. J. Climatol. **14**, 1095–1114 (1994)
4. Sahai, A.K., Grimm, A.M., Satyan, V., Pant, G.B.: Long-lead prediction of Indian summer monsoon rainfall from global SST evolution. Climate Dyn. **20**, 855–863 (2003)
5. Raja Rao, K.S., Lakhole, N.T.: Quasi-biennial oscillation of summer southwest monsoon. Ind. J. Meteorol. Hydrol. Geophys. **29**, 403–411 (1978)
6. Campbel, W.H., Blechman, J.B., Bryson, R.A.: Long-period tidal forcing of Indian monsoon rainfall: a hypothesis. J. Clim. Appl. Meteorol. **22**, 287–296 (1983)
7. Narasimha, R., Kailas, S.V.: A wavelet map of monsoon variability. Proc. Ind. Nat. Sci. Acad. **67**(3), 327–341 (2003)
8. Shukla, J., Paolino, D.A.: The southern oscillation and long-range forecasting of the summer monsoon rainfall over India. Mon. Wea. Rev. **111**, 1830–1837 (1983)
9. Bhalme, H.N., Jadhav, S.K.: The double (Hale) sunspot cycle and floods and droughts in India. Weather 39, 112–116 (1984)
10. Hartmann, D.L., Michelsen, M.L.: Intraseasonal periodicities in Indian rainfall. J. Atmos. Sci. **46**(18), 2838–2862 (1989)

11. Iyenger, R.N.: Application of principal component analysis to understand variability of rainfall. Proc. Ind. Acad. Sc. (Earth Planet Sci.) **100**(2), 105–126 (1991)
12. Basak, P.: Variability of south west monsoon rainfall in West Bengal: an application of principal component analysis. Mausam **65**(4), 559–568 (2014)
13. Basak, P.: Southwest monsoon rainfall in Assam: an application of principal component analysis. Mausam **68**(2), 357–366 (2017)
14. Sahai, A.K., Soman, M.K., Satyan, V.: All India summer monsoon rainfall prediction using an artificial neural network. Climate Dyn. **16**, 291–302 (2000)
15. Iyengar, R.N., Raghu Kanth, S.T.G.: Empirical modeling and forecasting of Indian monsoon rainfall. Curr. Sci. **85**(8), 1189–1201 (2003)
16. Iyengar, R.N., Raghu Kanth, S.T.G.: Intrinsic mode function and a strategy for forecasting Indian monsoon rainfall. Meteorol. Atmos. Phys. 9017–9036 (2005)
17. Zvarevashe, W., Krishnanair, S., Sivkumar, V.: Analysis of rainfall and temperature data using ensemble empirical mode decomposition. Data Sci J. **18**(1), 46 (2019)
18. Sabzehee, F., Nafisi, V., Pour, S.I., Vishwakarma, B.D.: Analysis of the precipitation climate signal using empirical mode decomposition (EMD) over the Caspian catchment area. In: The International Archives of the Photogrammetry, Remote Sensing and Spatial Information Sciences, XLII-4/W18, pp. 923–929 (2019)
19. Huang, N.E., Shen, Z., Long, S.R., Wu, M.C., Shih, H.H., Zheng, Q., Yen, N., Tung, C.C., Liu, H.H.: The empirical mode decomposition and the Hilbert spectrum for nonlinear and non-stationary time series analysis. Proc. Royal Soc. London **A454**, 903–995 (1998)
20. Specht, D.F.: A general regression neural network. IEEE Trans. Neural Netw. **2**(6), 568–576 (2002)

Electrical Load Clustering Using K-Means Algorithm for Identification of Village Clusters to Draw Optimal Power from Distributed Solar Generating Plants

Sanjay Goswami, Dipu Sarkar, and Paushali Majumder

1 Introduction

Nowadays, inclusive development of a society involves development of the villages. Village electrification is one of the important aspects of such development. Cheap electrification can be realized through adoption of renewable energy resources, like solar power. Since solar generating plants are low-capacity power generators, optimal clustering of the villages under each generating plant is necessary. This work focuses on development of such a procedure to optimize village clusters to harness renewable energy sources to the maximum. Works [1–15] focus on the same aspect of power system optimization and restoration to harness energy resources to maximum capacity.

S. Goswami (✉)
Center for Disaster Preparedness and Management, Jadavpur University, Kolkata, India

D. Sarkar
Department of Electrical and Electronics Engineering, NIT Nagaland, Dimapur, India

P. Majumder
Department of Computer Applications, Narula Institute of Technology, Kolkata, India

J. K. Mandal and J. K. Roy (eds.), *Proceedings of International Conference on Computational Intelligence and Computing*, Algorithms for Intelligent Systems,
https://doi.org/10.1007/978-981-16-3368-3_7

2 Objective

The objective of this work is to identify a method of clustering village households into electrical load clusters to harness renewable power generating facilities like solar cells, etc. installed in the vicinity of the villages under village electrification schemes. The next section pictorially describes the workflow that can be adopted for clustering and re-clustering of villages for optimal DG plant utilization.

3 Clustering Workflow

Figure 1 describes how villages located around a handful of DG plants may look like in the beginning. They may be in the vicinity of some DG plants, but they may not be suitable to draw electricity from few of them based on their electrical load needs.

Fig. 1 Unclustered village household loads

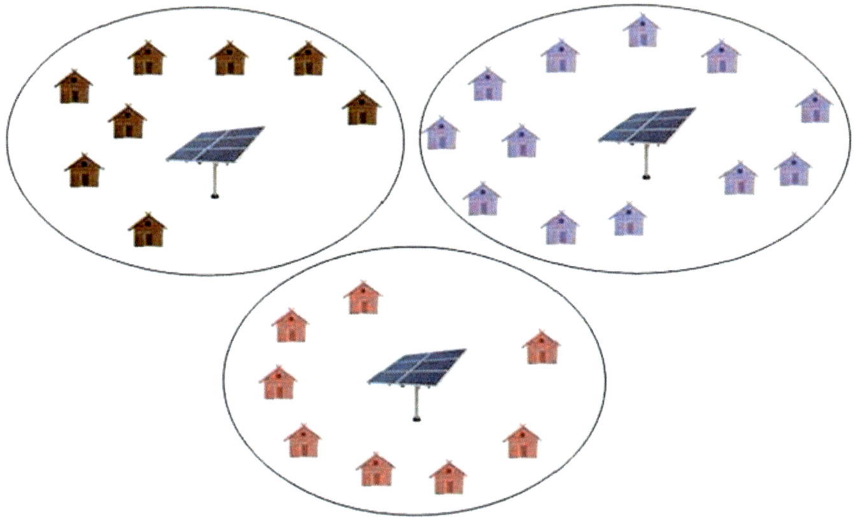

Fig. 2 Clustered village households under various DG plants–based on location

In the first phase of the technique, the villages may be clustered into groups based on their location from the DG plants. They may be grouped into clusters surrounding the DG plants nearest to them (Fig. 2).

In the 2nd phase, the village households are re-clustered into groups based on their electrical load needs. They are rearranged into groups around the DG plants satisfying their load constraints, i.e. their generating capacities (Fig. 3).

4 Load Constraint Definition

The Load Constraint may be defined as follows: *the total load of village households under a given cluster should be less than or equal to the maximum power generating capacity of the DG plant* (Fig. 4).

5 System Logic

5.1 Flowchart

The Fig. 5 shows the flowchart of the system logic. At the start of the program, we input the initial load data of the villages. Then the program clusters the villages into optimal clusters based on their location from the nearest DG plants. Cluster wise total

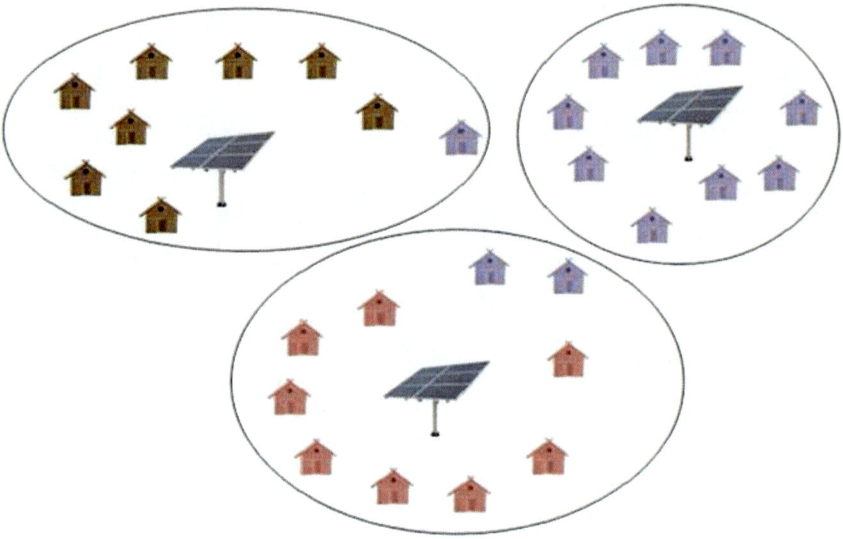

Fig. 3 Re-clustered households among the DG plants based on load constraints

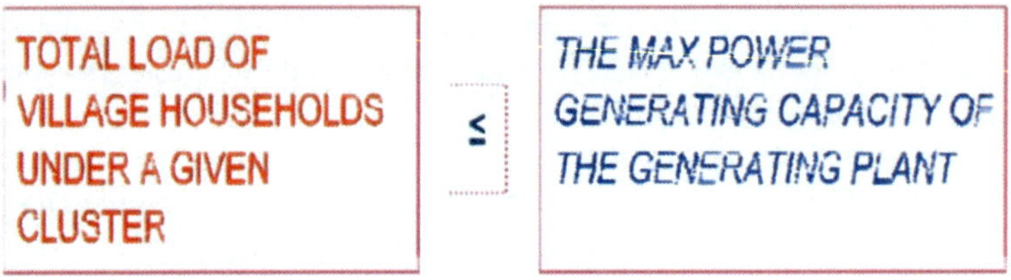

Fig. 4 Load constraint definition

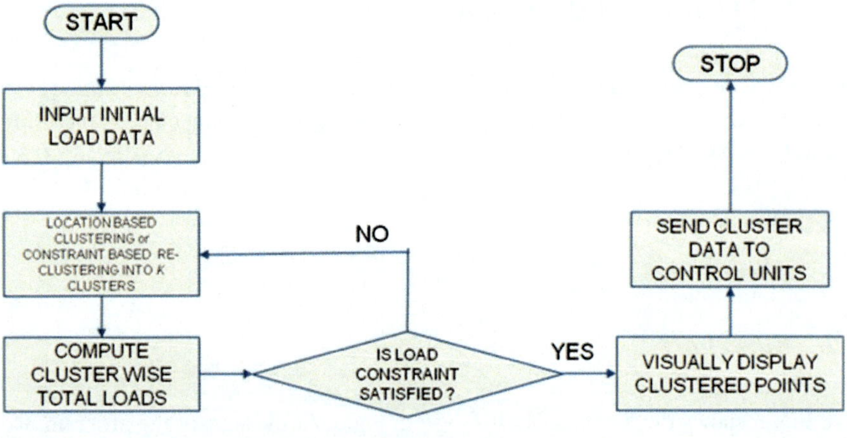

Fig. 5 Flowchart of the system logic

loads of the villages are then calculated. If they do not satisfy the load constraints of the DG plants, the villages are again reclustered based on satisfaction of the load constraints. If the load constraints are satisfied, the output is sent for visual display, and cluster data are sent to the control units of the power system.

5.2 K-Means Algorithm

K-Means clustering is an established clustering algorithm popular for statistical machine learning and data analysis. This is very effective in understanding hidden associations and patterns within data. Initially, the value of K, the number of clusters, is given to the algorithm. Then, mean centroids of K clusters are calculated/initialized. These centroids represent each cluster. Data points are then taken one by one and their distances from the centroids are calculated. Each data object is then assigned to the cluster nearest to them. The centroids are then again calculated, and the whole process is repeated until there is no variation in cluster centroids, i.e. no new additions in the groups through re-clustering [16] (Fig. 6).

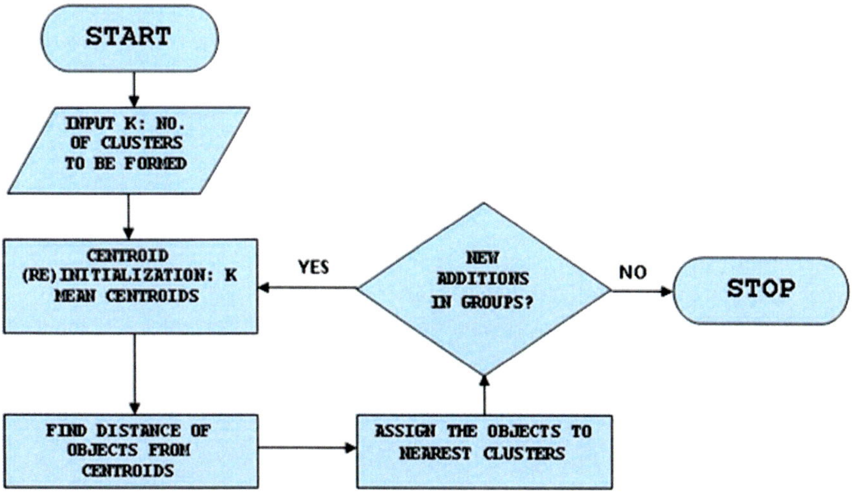

Fig. 6 Visual flowchart of the K-means algorithm

6 Simulated Experiment, Results, and Discussions

6.1 Structure of the Data to Be Clustered

The structure of the data set that is considered to identify each village in the cluster is [*LONGITUDE, LATITUDE, LOAD*]. Each village has a center with certain *Longitude-Latitude* coordinate and a central distribution station keeping track of the total *Load* of the village households. E.g. [10.73608558, 71.55736206, 1359.30] represents a village located at 10.73608558° Longitude and 71.55736206° Latitude and having Load of 1359.30 W or [19.33578019, 86.28478017, 1022.15] represents 19.33578019° Longitude and 86.28478017° Latitude with 1022.15 W load.

6.2 Load Constraint Revisited

$$\sum_{i} \text{load}_{i,j} \leq \text{Cap}_j \tag{1}$$

where $\text{load}_{ij} \rightarrow i$th load of the jth cluster, $\text{Cap}_j \rightarrow$ capacity of the jth Generating Plant. As per Eq. (1) the *Total load of the jth cluster should be less than or equal to the Capacity of the jth Generating Plant.*

6.3 Sample Runs

In the sample runs, number of clusters is considered to be $k = 5$. Figure 7 shows the visual distribution of the villages before clustering.

Figures 8, 9 and 10 show the distribution of the village clusters after 1st, 2nd, and 3rd runs of the algorithm.

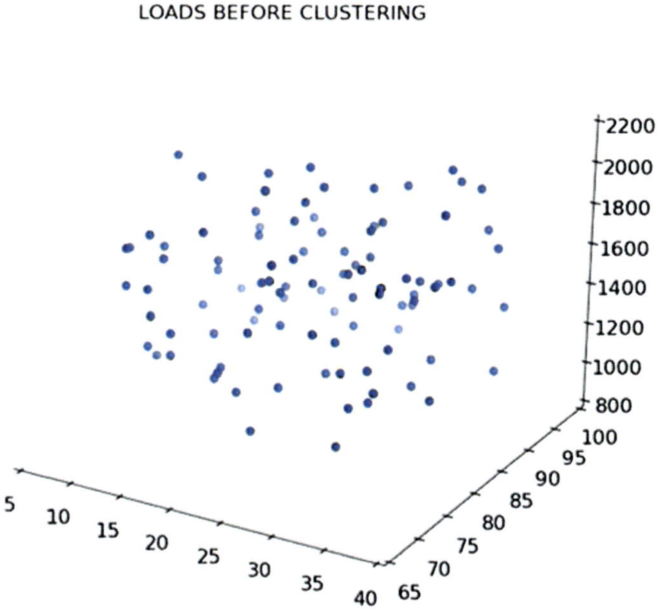

Fig. 7 Villages distribution before clustering

7 Conclusions

The *simulated* runs show that the algorithm is *working successfully* on *dummy data.* Both the phases—*Clustering* and *Re-clustering* are working fine. Load *constraints* are *satisfied* before showing the *final* cluster data and sending the information to the control units for relevant switching purposes. If this is *successful* in real life, *it can lead to a feasible technique* to harness renewable *energy* effectively in villages, *especially with low capacity power generating plants.*

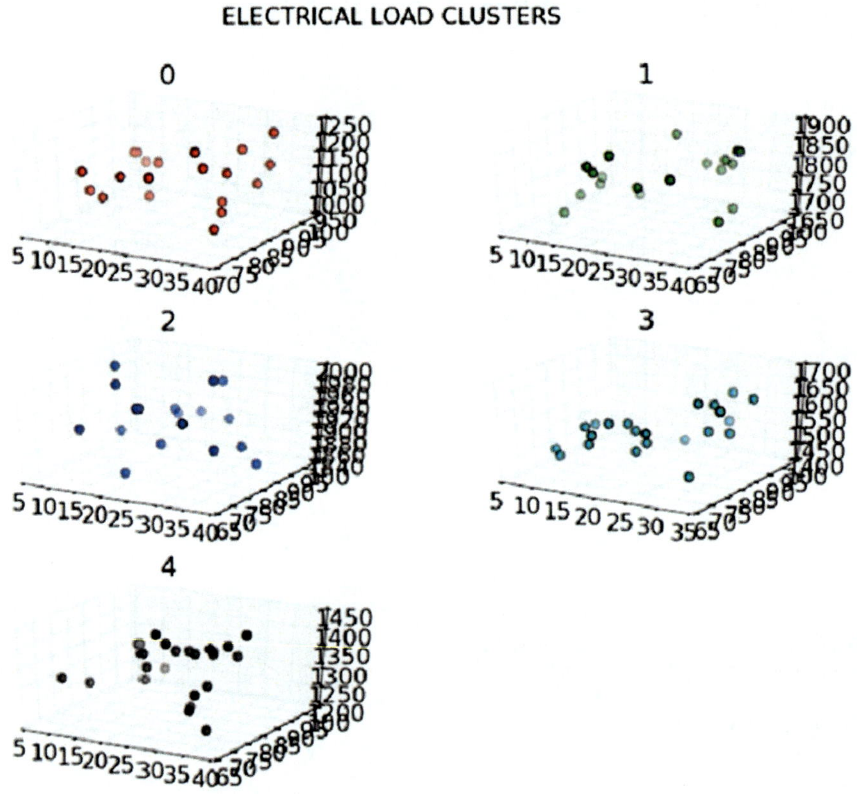

Fig. 8 Villages distribution after clustering—*Sample Run 1*

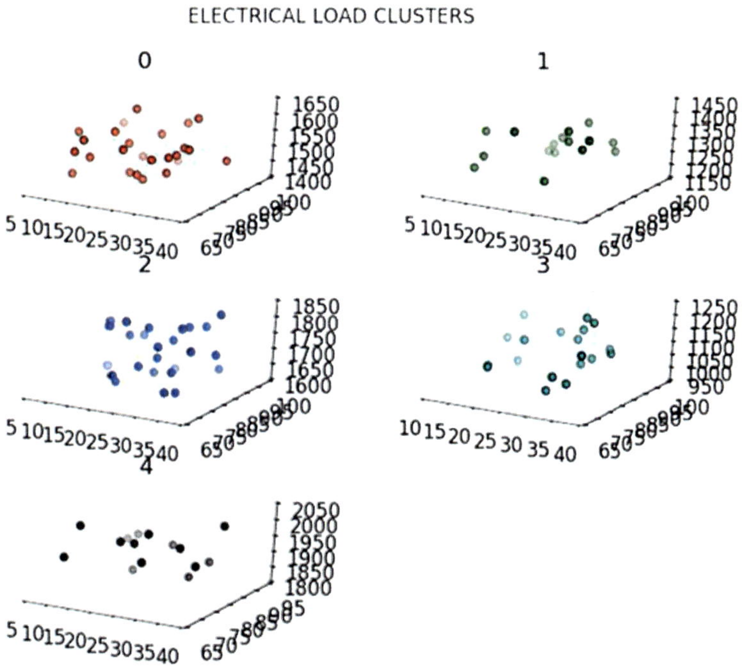

Fig. 9 Villages distribution after clustering—*Sample Run 2*

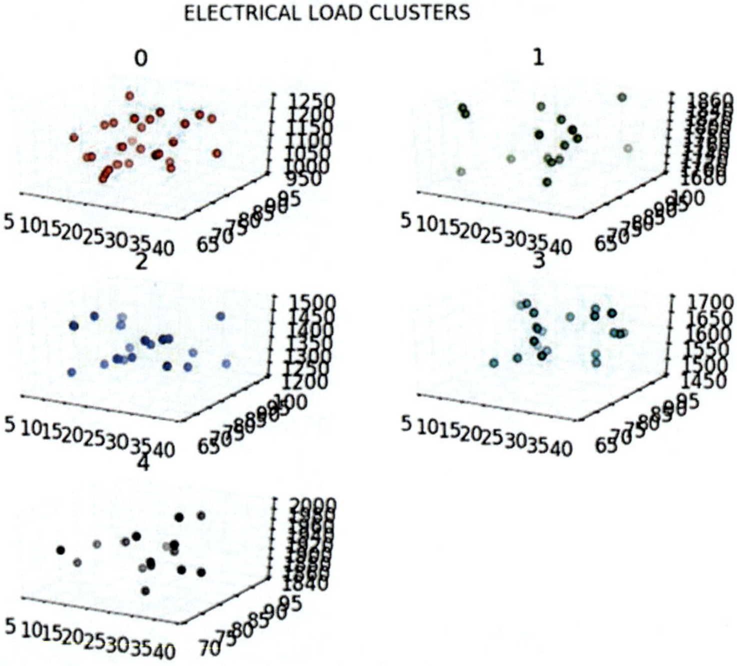

Fig. 10 Villages distribution after clustering—*Sample Run 3*

References

1. Resende, F.O., Gil, N.J., Lopes, J.A.P.: Service restoration on distribution systems using multi-microgrids. Eur. Trans. Elec. Power **21**, 1327–1342 (2011)
2. Justo, J.J., Mwasilu, F., Lee, J., Jung, J.W.: AC-microgrids versus DC-microgrids with distributed energy resources: a review. Renew. Sustain. Energy Rev. **24**, 387–405 (2013)
3. Costa, P.M., Matos, M.A., Lopes, J.A.P.: A regulatory framework for microgeneration and microgrids. PowerTech (2007)
4. Venkataramanan, G., Illindala, M.: Microgrid and sensitive loads. In: Power Engineering Society Winter Meeting, vol. 1, pp 27–31. IEEE (2002)
5. Llaria, A., Curea, O., Jiménez, J., Camblong, H.: Survey on microgrids: unplanned islanding and related inverter control techniques. Renew. Energy **36**(8), 2052–2061 (2011)
6. Moreira, C., Resende, F., Lopes, J.: Using low voltage microgrids for service restoration. IEEE Trans. Power Syst. **22**(1), 395–403 (2007)
7. Wang, J., Mu, L., Zhang, F., Zhang, X.: A parallel restoration for black start of microgrids considering characteristics of distributed generations. Energies **11**(1), 1 (2017)
8. Guerrero, J.M., Vasquez, J.C., Matas, J., de Castilla, M., Vicuna, L.G.: Control strategy for flexible microgrid based on parallel line-interactive UPS systems. IEEE Trans. Industr. Electron. **56**(3), 726–736 (2009)
9. Ronay, K., Bica, D., Munteanu, C.: Micro-grid development using artificial neural network for renewable energy forecast and system control. Procedia Eng **181**, 818–823 (2017)
10. Brearley, B.J., Prabu, R.R.: A review on issues and approaches for microgrid protection. Renew. Sustain. Energy Rev. **67**, 988–997 (2017)
11. Rocabert, J., Luna, A., Blaabjerg, F., Rodriguez, P.: Control of power converters in AC microgrids. IEEE Trans Power Electron **27**(11), 4734–4749 (2012)

12. Cho, C., Jeon, J., Kwon, S., Park, K., Kim, S.: Active synchronizing control of a microgrid. IEEE Trans. Power Electron. **24**, 12 (2011)
13. Balaguer, I.J., Suppatti, U., Lei, Q., Choi, N.S., Peng, F.Z.: Intelligent control for intentional islanding operation of microgrids. In: IEEE International Conference on Sustainable Energy Technologies, Singapore, pp. 898–903 (2008)
14. Zhao, B., Dong, X., Bornemann, J.: Service restoration for a renewable-powered microgrid in unscheduled island mode. IEEE Trans. Smart Grid **6**(3), 1128–1136 (2015)
15. Li, P., Song, B., Wang, W., Wang, T.: Multi-agent approach for service restoration of microgrid. In: 5th IEEE Conference on Industrial Electronics and Applications, pp. 962–966 (2010)
16. Hartigan, J.A., Wong, M.A.: Algorithm AS 136: A k-means clustering algorithm. J. Royal Stat. Soc. Series C **28**(1), 100–108. JSTOR 2346830 (1979)

IoT-Based Laser-Inscribed Sensors for Electrochemical Detection of Phosphate Ions

Anindya Nag, Md Eshrat E. Alahi, Nasrin Afsarimanesh, and Subhas Mukhopadhyay

1 Introduction

The inclusion of sensing systems with daily life activities has improved the quality of human life to a great extent. Different kinds of sensors being conjugated with the regularly used materials have brought about the smoothening of daily courses. After the popularization of commercial sensors three decades ago [1, 2], researchers in the academic scale have started working on their quality to a great extent. Large research funds have been granted to form laboratories operating with efficient equipment for this purpose. The sensors being fabricated and characterized vary in their working mechanism, processed materials, and structural dimensions. Although the reduction in the size of the prototypes has been a growing trend to obtain higher sensitivity, the consideration in the type of the prototypes primarily depends on their application. Initially, when the microelectrochemical systems (MEMS)-based sensors got popularized [3, 4], sensors were mainly developed using silicon substrates [5]. These types of sensors have additional advantages of small size, low power, and the possibility of large-scale fabrication in comparison to the previously used very-large-scale integrated devices [6]. These silicon sensors have been employed for a wide

A. Nag (✉)
Department of Electrical Engineering, CEMSE Division, King Abdullah University of Science of Technology, Thuwal 23955, Saudi Arabia

M. E. E. Alahi
Institute of Biomedical and Health Engineering, Shenzhen Institutes of Advanced Technology, Chinese Academy of Sciences, Shenzhen 518055, China

N. Afsarimanesh
DGUT-CNAM Institute, Dongguan University of Technology, Dongguan 523106, China

S. Mukhopadhyay
Faculty of Science and Engineering, School of Engineering, Macquarie University, Sydney, NSW 2109, Australia

J. K. Mandal and J. K. Roy (eds.), *Proceedings of International Conference on Computational Intelligence and Computing*, Algorithms for Intelligent Systems,
https://doi.org/10.1007/978-981-16-3368-3_8

range of biomedical [7, 8], industrial [9, 10] and environmental [11, 12] applications. Although these silicon sensors served a great purpose for sensing applications, there were still certain drawbacks related to these prototypes, which compelled the researchers to opt for alternative options. Instead of sensors with silicon substrates and nanoparticles-based electrodes, the sensors were then fabricated using flexible materials. The flexible sensors have enhanced properties with lightweight, high electrical conductivity, high mechanical flexibility as compared to that of the MEMS-based sensors. The additional wearable nature of the flexible sensors provides high additional advantages for these sensors.

Various kinds of polymers and conductive materials have been employed so far for the fabrication of these sensors. Among the polymers, some of the common ones are polydimethylsiloxane (PDMS) [13, 14], Polyethylene terephthalate (PET) [15, 16], Polyimide (PI) [17, 18] and Poly (3,4-ethylenedioxythiophene) polystyrene sulfonate (PEDOT: PSS) [19, 20]. Among the conductive parts, other types of nanomaterials like nanotubes [21, 22], nano-spheres [23, 24] and quantum dots [25, 26] have opted. Some of the common ones are Carbon Nanotubes (CNTs) [27, 28], Graphene [29, 30], Aluminum [31, 32] and Copper [33, 34]. In this paper, we describe the fabrication and implementation of sensors formed using metalized PET films. One side of the PET film was coated using aluminum with a thickness of 400 microns. In addition to the advantages imparted by the polymer and conductive material, the advantages of using these materials are quick fabrication time, easy processing, and easy operating principle. The processing of the raw materials has been done in different ways, depending on the electrical, mechanical, and thermal properties of the materials. Some of the common techniques used to fabricate the flexible sensors are 3D printing [35, 36], screen printing [37, 38], inkjet printing [39, 40], laser inscription [41, 42] and gravure printing [43, 44]. This paper showcases the use of the laser inscription technique to process the metalized films. Low processing time, high-quality sensors, and minimized post-processing steps are some of the advantages of the laser inscription process. Similar to the MEMS-based sensors, the use of flexible sensors has been done for a wide range of applications [45–47]. Each of these uses is based on their dual working mechanism, namely strain [48, 49] and electrochemical [50, 51] sensing. The flexible sensors have been successful to function using both these mechanisms for their respective applications. This paper explains the functional capability of the fabricated sensors, the electrochemical detection of phosphate ions in the water bodies. The significance of the detection of the ions in water bodies is related to the harmful effects caused by the increase in its concentration. For example, the excess amount of phosphates in water bodies leads to the growth of algae and weeds, which results in the choking of waterways. It also leads to the use of large amounts of oxygen, which causes the death of the flora and fauna present in the water bodies [52]. The present work conduct experiments in laboratory conditions with solutions of different phosphate concentrations.

2 Fabrication of the Sensor Patches

Figure 1 shows the schematic diagram of the fabrication process. The metalized PET films were initially attached to glass substrates to restrict its movement during the formation of the electrodes. The films were attached to the glass substrates using biocompatible adhesive tapes. The differences between the two sides were marked by higher conductivity and shine of the metallic side in comparison to the PET side.

These samples were taken to the laser inscription system for inscription on the aluminum side. The size and shape of the electrodes were designed in the Corel-Draw software that was associated with the laser inscription system. Three of the laser parameters, namely power, speed, *z*-axis, was optimized to form the electrodes. During the laser-cutting process, the area beside the electrode lines dissolved and regenerated simultaneously due to the heat generated by the laser beam. The electrode lines formed on thin films were also optimized as excess heat burnt the lines, whereas dearth of heat did not form the lines. The fabrication process was very quick, with around six prototypes developed in two minutes. Figure 2 [53] shows the front and rear view of the fabricated sensors. The thickness of the PET substrates was around 800 μm, whereas that of the Aluminum electrodes was 400 μm. Around 16 pairs of interdigitated electrode fingers were formed on each of the prototypes. Figure 3 [54] shows the operating principle of the sensor patches. These sensors functioned in the principle of parallel-plate capacitors. When a low-frequency signal was given as an input, an electrode was generated between the electrodes of opposite polarity. Due to the planar nature of the sensors, the electric field bulged from one electrode to another. When any material is kept in proximity or contact with the electrodes, the electric field penetrates through it, thus changing its properties. This change in its characteristics is studied in order to determine the nature of the material. The materials tested using this phenomenon consist of dielectric substances that vary the relative permittivity of the response of the sensors.

The advantages of this technique include its one-sided, non-invasive measurement process. It has been utilized in different kinds of electrochemical applications [55].

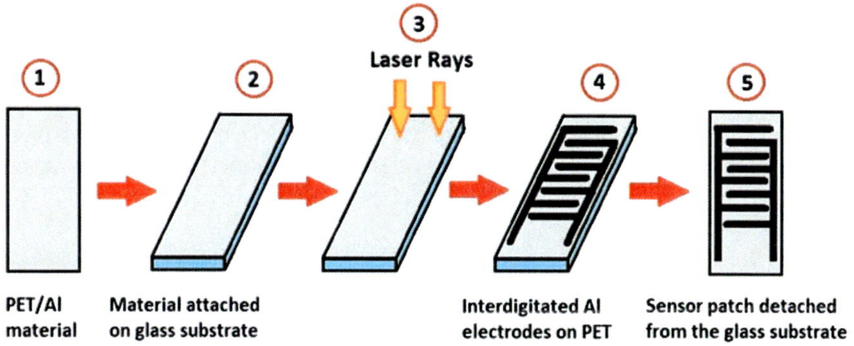

Fig. 1 Schematic diagram of the fabrication process

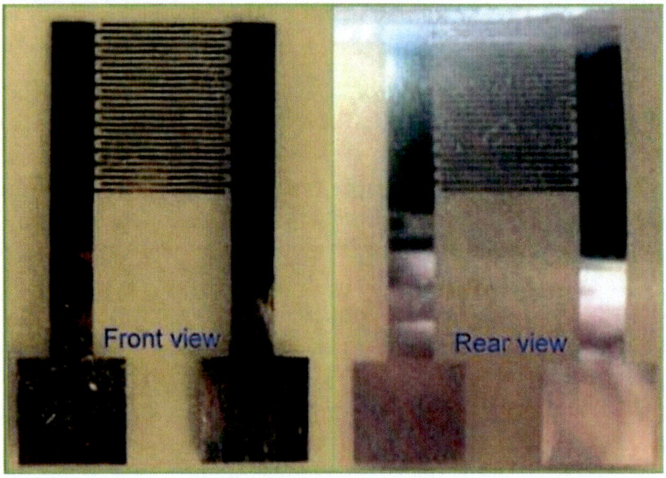

Fig. 2 Front and rear views of the developed Al-PET sensors [53]

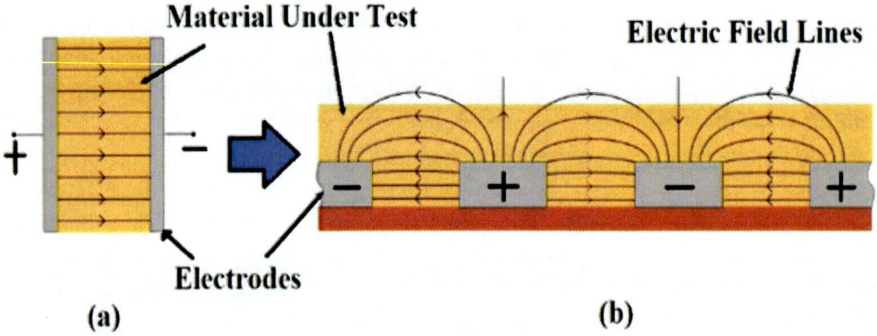

Fig. 3 Working mechanism of the Al-PET sensors [54]

In order to process the signal, the electrochemical impedance spectroscopic (EIS) technique [56, 57] was associated with the sensors. The responses of the sensors were determined in terms of the change in resistive and reactive parts of the impedance. The resistance of the sensors changed due to the corresponding changes in the concentrations of the solutions. The reactance of the sensors changed due to two reasons, namely the interdigital nature of the electrodes and formation of the double-layer capacitance of the electrode–electrolyte interface.

3 Experimental Results

The sensors were tested for the phosphate concentrations ranging between 0.1 and 1000 ppm. This range was chosen to cover the entire range of phosphate concentrations existing in the natural water bodies. These solutions were prepared using the serial dilution process. Although the frequency range was swept between 1 Hz and 1 kHz, the responses were clearly distinguishable between 1 and 200 Hz. These experiments were carried out using an LCR meter HIOKI 3536, which was connected to the sensor via Kelvin probes. The impedance analyzer was, in turn, connected to a computer to collect the sensed data. The sensors were fixed on a wiring board to restrict its movement inside the samples. The data was collected in Microsoft Excel Sheet via an automated data acquisition algorithm. Figures 4 and 5 show the responses of the Al-PET sensors in terms of resistive and reactive values, respectively. It is seen that the sensors were clearly capable of differentiating the five tested concentrations. As mentioned in the previous section, these differences in the responses were possible due to the change in the phosphate concentrations of the solutions and nature of the electrodes. The increase in the concentrations of the solutions led to a simultaneous increase in the ionic current, thus reducing the solution resistance values. The differences in the resistance values gradually increased with the increase in the concentration samples. This was due to the flow of higher current densities as a result of increased number of ions. The reactance of the sensors increased with the increase in the concentrations of the solutions as a result of the decrease in faradic current. The decrease in the current led to a corresponding decrease in the capacitance values, which subsequently increases the reactive values.

Followed by the experiments done using impedance analysis, one of the frequencies (50 Hz) was chosen to obtain the sensitivity of the fabricated Sensors. Figure 6

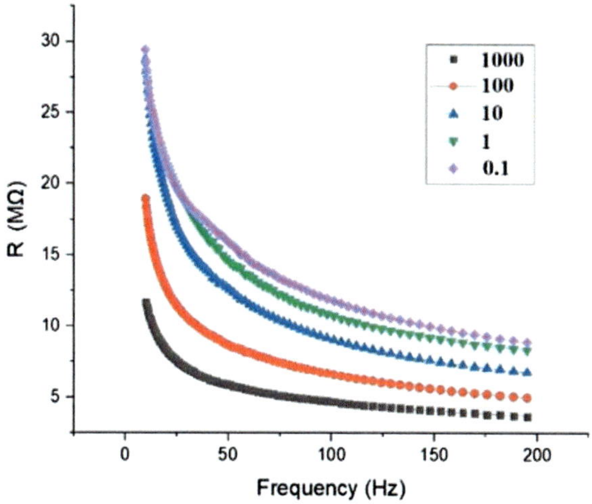

Fig. 4 Response of the sensors towards the phosphate concentrations in terms of resistance values

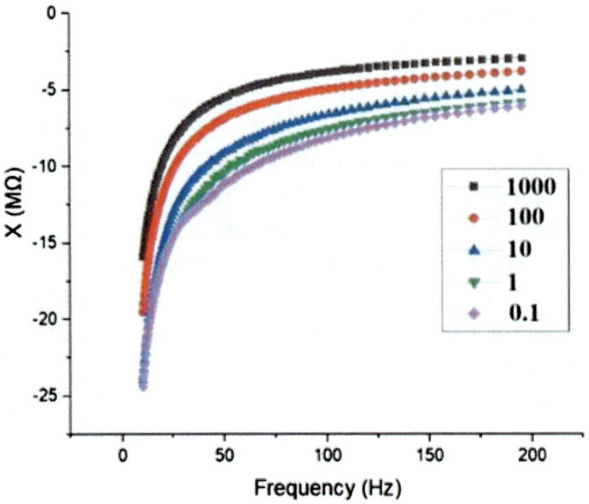

Fig. 5 Response of the sensors towards the phosphate concentrations in terms of reactive values

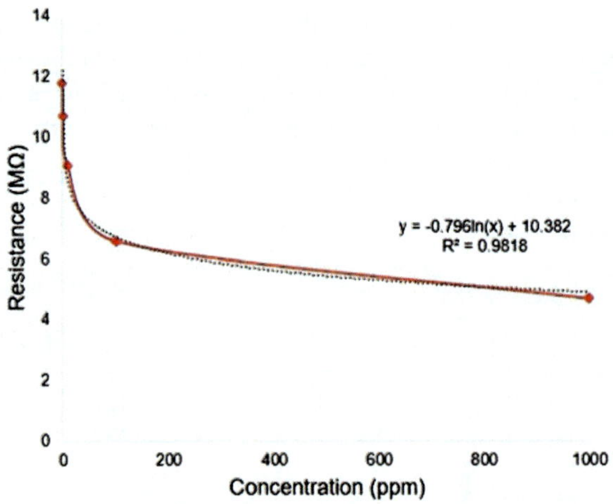

Fig. 6 Sensitivity of the Al-PET sensors for the tested phosphate concentrations (0.1–1000 ppm)

shows the sensitivity graph obtained for the resistance values with respect to the tested concentrations. It is seen that the sensitivity of the sensors is 0.796 Ω/ppm. The response of the sensors was around one second. This frequency was used in the microcontroller-based system for wireless detection of the response of the sensors. The microcontroller-based system consisted of an impedance analyzer that was used to determine the resultant resistance with respect to the tested concentration.

Figure 7 shows the output of the microcontroller-based system, where the resistance values were obtained for phosphate-spiked solutions. Thus, it is seen that a portable sensing system consisting of sensors developed by the laser-induction of the

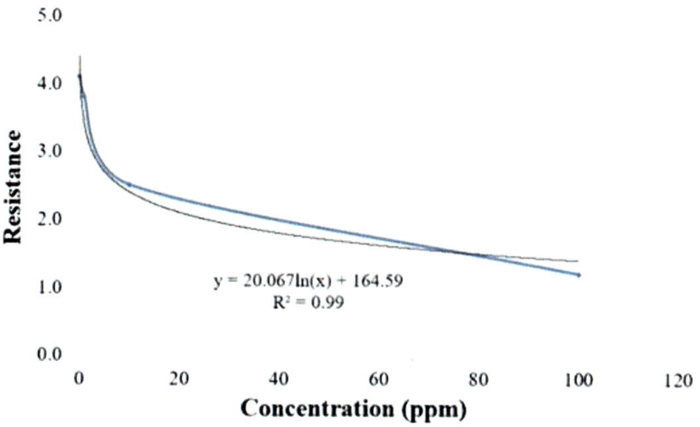

Fig. 7 Response of the microcontroller-based sensing system

metalized polymer films was able to detect the presence of phosphate concentrations in real-time applications.

4 Conclusion

The paper showcases the fabrication and implementation of laser-inscribed metalized polymer films for electrochemical applications. A laser inscription was done on PET films to obtain Al-PET sensors. These prototypes were used for the detection of phosphate ions in the water bodies. Laboratory-made samples were prepared and tested using the EIS technique to validate their functionality. This was followed by using a microcontroller-based system to determine the resistance values of the phosphate-spiked solutions to ensure its functionality. Further work would be done on these sensors to determine their capability of detecting multiple ions. This would be done by the inclusion of selectivity on the sensing area to differentiate between the impedance values for the respective ions.

References

1. Jaaniso, R., Tan, O.K.: Semiconductor Gas Sensors. Elsevier (2013)
2. Sze, S.M.: Semiconductor Sensors. Wiley (1994)
3. Grayson, A.C.R., et al.: A BioMEMS review: MEMS technology for physiologically integrated devices. Proc. IEEE **92**(1), 6–21 (2004)
4. Bryzek, J.: Impact of MEMS technology on society. Sens. Actuators A: Phys. **56**(1–2), 1–9 (1996)
5. Tilli, M., Paulasto-Krockel, M., Petzold, M., Theuss, H., Motooka, T., Lindroos, V.: Handbook of Silicon Based MEMS Materials and Technologies. Elsevier (2020)

6. Hampel, D.: Application of very large scale integrated (VLSI) circuits to "Smart Sensors". In: Smart Sensors II, vol. 252, pp. 2–9 International Society for Optics and Photonics (1980)
7. Xu, Y., et al.: Silicon-based sensors for biomedical applications: a review. Sensors **19**(13), 2908 (2019)
8. Gabriel, G., et al.: Manufacturing and full characterization of silicon carbide-based multi-sensor micro-probes for biomedical applications. Microelectron. J. **38**(3), 406–415 (2007)
9. Nag, A., Zia, A.I., Li, X., Mukhopadhyay, S.C., Kosel, J.: Novel sensing approach for LPG leakage detection: part I—operating mechanism and preliminary results. IEEE Sens. J. **16**(4), 996–1003 (2015)
10. Kim, K., et al.: A silicon-based flexible tactile sensor for ubiquitous robot companion applications. J. Phys. Conf. Series **34**(1), 399 (2006)
11. Afsarimanesh, N., Nag, A., Alahi, M.E.E., Han, T., Mukhopadhyay, S.C.: Interdigital sensors: biomedical, environmental and industrial applications. Sens. Actuators A: Phys. 111923 (2020)
12. Alahi, M.E.E., Xie, L., Mukhopadhyay, S., Burkitt, L.: A temperature compensated smart nitrate-sensor for agricultural industry. IEEE Trans. Industr. Electron. **64**(9), 7333–7341 (2017)
13. Nag, A., et al.: A transparent strain sensor based on PDMS-embedded conductive fabric for wearable sensing applications. IEEE Access **6**, 71020–71027 (2018)
14. Nag, A., Feng, S., Mukhopadhyay, S., Kosel, J., Inglis, D.: 3D printed mould-based graphite/PDMS sensor for low-force applications. Sens. Actuators, A **280**, 525–534 (2018)
15. Yaqoob, U., Phan, D.-T., Uddin, A.I., Chung, G.-S.: Highly flexible room temperature NO2 sensor based on MWCNTs-WO3 nanoparticles hybrid on a PET substrate. Sens. Actuators, B Chem. **221**, 760–768 (2015)
16. Altenberend, U., et al.: Towards fully printed capacitive gas sensors on flexible PET substrates based on Ag interdigitated transducers with increased stability. Sens. Actuators, B Chem. **187**, 280–287 (2013)
17. Han, T., et al.: Gold/polyimide-based resistive strain sensors. Electronics **8**(5), 565 (2019)
18. Nag, A., Mukhopadhyay, S.C., Kosel, J.: Sensing system for salinity testing using laser-induced graphene sensors. Sens. Actuators, A **264**, 107–116 (2017)
19. Thana, K., Petchsang, N., Jaisutti, R.: Electrical and mechanical properties of PEDOT: PSS strain sensor based microwave plasma modified pre-vulcanized rubber surface. MS&E **773**(1), 012049 (2020)
20. Ramírez, J., Rodriquez, D., Urbina, A.D., Cardenas, A.M., Lipomi, D.J.: Combining high sensitivity and dynamic range: wearable thin-film composite strain sensors of graphene, ultrathin palladium, and PEDOT: PSS. ACS Appl. Nano Mater. **2**(4), 2222–2229 (2019)
21. Daňová, R., Olejnik, R., Slobodian, P., Matyas, J.: The piezoresistive highly elastic sensor based on carbon nanotubes for the detection of breath. Polymers **12**(3), 713 (2020)
22. Yuan, Z., et al.: Wrinkle structured network of silver-coated carbon nanotubes for wearable sensors. Nanoscale Res. Lett. **14**(1), 1–8 (2019)
23. Chen, C., et al.: An efficient flexible electrochemical glucose sensor based on carbon nanotubes/carbonized silk fabrics decorated with Pt microspheres. Sens. Actuators, B Chem. **256**, 63–70 (2018)
24. Gu, H., et al.: Carbon nanospheres induced high negative permittivity in nanosilver-polydopamine metacomposites. Carbon **147**, 550–558 (2019)
25. Huang, Q., Lin, X., Tong, L., Tong, Q.-X.: Graphene quantum dots/multiwalled carbon nanotubes composite-based electrochemical sensor for detecting dopamine release from living cells. ACS Sustain. Chem. Eng. **8**(3), 1644–1650 (2020)
26. Zhou, X., et al.: Dopamine fluorescent sensors based on polypyrrole/graphene quantum dots core/shell hybrids. Biosens. Bioelectron. **64**, 404–410 (2015)
27. Han, T., Nag, A., Mukhopadhyay, S.C., Xu, Y.: Carbon nanotubes and its gas-sensing applications: a review. Sens. Actuators, A **291**, 107–143 (2019)
28. Nag, A., Mukhopadhyay, S.C., Kosel, J.: Flexible carbon nanotube nanocomposite sensor for multiple physiological parameter monitoring. Sens. Actuators, A **251**, 148–155 (2016)
29. Han, T., et al.: Multifunctional flexible sensor based on laser-induced graphene. Sensors **19**(16), 3477 (2019)

30. Nag, A., Mitra, A., Mukhopadhyay, S.C.: Graphene and its sensor-based applications: a review. Sens. Actuators, A **270**, 177–194 (2018)
31. Meng, X., Yang, J., Liu, Z., Lu, W., Sun, Y., Dai, Y.: Non-contact, fibrous cellulose acetate/aluminum flexible electronic-sensor for humidity detecting. Compos. Commun. (2020)
32. Lamanna, L., et al.: Flexible and transparent aluminum-nitride-based surface-acoustic-wave device on polymeric polyethylene naphthalate. Adv. Electron. Mater. **5**(6), 1900095 (2019)
33. Zhang, Y., et al.: A flexible non-enzymatic glucose sensor based on copper nanoparticles anchored on laser-induced graphene. Carbon **156**, 506–513 (2020)
34. Zhang, S., et al.: A flexible bifunctional sensor based on porous copper nanowire@ IonGel composite films for high-resolution stress/deformation detection. J. Mater. Chem. C **8**(12), 4081–4092 (2020)
35. Liu, C., et al.: 3D printing technologies for flexible tactile sensors toward wearable electronics and electronic skin. Polymers **10**(6), 629 (2018)
36. Han, T., Kundu, S., Nag, A., Xu, Y.: 3D printed sensors for biomedical applications: a review. Sensors **19**(7), 1706 (2019)
37. Knoll, M., Offenzeller, C., Mayrhofer, B., Jakoby, B., Hilber, W.: A screen printed thermocouple-array on a flexible substrate for condition monitoring. Multidisc. Dig. Pub. Inst. Proc. **2**(13), 803 (2018)
38. Abellán-Llobregat, A., et al.: A stretchable and screen-printed electrochemical sensor for glucose determination in human perspiration. Biosens. Bioelectron. **91**, 885–891 (2017)
39. Nayak, L., Mohanty, S., Nayak, S.K., Ramadoss, A.: A review on inkjet printing of nanoparticle inks for flexible electronics. J. Mater. Chem. C **7**(29), 8771–8795 (2019)
40. Andò, B., Baglio, S., Bulsara, A.R., Emery, T., Marletta, V., Pistorio, A.: Low-cost inkjet printing technology for the rapid prototyping of transducers. Sensors **17**(4), 748 (2017)
41. Han, T., Nag, A., Afsarimanesh, N., Mukhopadhyay, S.C., Kundu, S., Xu, Y.: Laser-assisted printed flexible sensors: a review. Sensors **19**(6), 1462 (2019)
42. You, R., Liu, Y.Q., Hao, Y.L., Han, D.D., Zhang, Y.L., You, Z.: Laser fabrication of graphene-based flexible electronics. Adv. Mater. **32**(15), 1901981 (2020)
43. Wang, H., et al.: Stamp-assisted gravure printing of micro-supercapacitors with general flexible substrates. In: 2019 IEEE 32nd International Conference on Micro Electro Mechanical Systems (MEMS), pp. 950–953. IEEE (2019)
44. Bariya, M., et al.: Roll-to-roll gravure printed electrochemical sensors for wearable and medical devices. ACS Nano **12**(7), 6978–6987 (2018)
45. Alahi, M.E.E., Nag, A., Mukhopadhyay, S.C., Burkitt, L.: A temperature-compensated graphene sensor for nitrate monitoring in real-time application. Sens. Actuators, A **269**, 79–90 (2018)
46. Nag, A., Menzies, B., Mukhopadhyay, S.C.: Performance analysis of flexible printed sensors for robotic arm applications. Sens. Actuators, A **276**, 226–236 (2018)
47. Chung, S., Lee, T.: Towards flexible CMOS circuits. Nat. Nanotechnol. **15**(1), 11–12 (2020)
48. Nag, A., Afasrimanesh, N., Feng, S., Mukhopadhyay, S.C.: Strain induced graphite/PDMS sensors for biomedical applications. Sens. Actuators, A **271**, 257–269 (2018)
49. Pan,S., et al.: A highly stretchable strain sensor based on CNT/graphene/fullerene-SEBS. RSC Adv. **10**(19), 11225–11232 (2020)
50. Aparicio-Martínez, E., Ibarra, A., Estrada-Moreno, I.A., Osuna, V., Dominguez, R.B.: Flexible electrochemical sensor based on laser scribed Graphene/Ag nanoparticles for non- enzymatic hydrogen peroxide detection. Sens. Actuators B: Chem. **301**, 127101 (2019)
51. Manjakkal, L., Sakthivel, B., Gopalakrishnan, N., Dahiya, R.: Printed flexible electrochemical pH sensors based on CuO nanorods. Sens. Actuators, B Chem. **263**, 50–58 (2018)
52. Water Research Center. Available: https://www.water-research.net/index.php/phosphates#:~:text=If%20too%20much%20phosphate%20is,are%20consumed%20by%20aerobic%20bacteria. Last accessed on 27th Oct 2020
53. Nag, A., Mukhopadhyay, S.C., Kosel, J.: Tactile sensing from laser-ablated metallized PET films. IEEE Sens. J. **17**(1), 7–13 (2016)

54. Nag, A., Mukhopadhyay, S.C.: Fabrication and implementation of printed sensors for taste sensing applications. Sens. Actuators, A **269**, 53–61 (2018)
55. Rahman, M.S.A., Mukhopadhyay, S.C., Yu, P,-L.: Novel planar interdigital sensors. In: Novel Sensors for Food Inspection: Modelling, Fabrication and Experimentation, pp. 11–35. Springer, Berlin (2014)
56. Orazem, M.E., Tribollet, B.: Electrochemical Impedance Spectroscopy. New Jersey (2008)
57. Lasia, A.: Electrochemical impedance spectroscopy and its applications. In: Modern Aspects of Electrochemistry, pp. 143–248. Springer, Berlin (2002)

Internet of Things (IoT)-Enabled Pedestrian Counting in a Smart City

Md Eshrat E. Alahi, Fowzia Akhter, Anindya Nag, Nasrin Afsarimanesh, and Subhas Mukhopadhyay

1 Introduction

With the technical advancements and rapid urbanization, smart cities' development has been emerged over a decade [1–3] to improve the quality of life of human beings. Smart cities' concept evolves to properly utilize the available resources to ensure efficiency and sustainability in life. As the world population rises, the shortage of non-renewable resources increases [4]. This can have its effects on the ambiance, like climatic changes along with the human capital. Some of the major urban cities in developed countries are already facing tremendous pressure due to the unprecedented growth of people from urban areas for a better quality of living [5]. Thus, it has become necessary to develop innovative solutions where the available infrastructure would be integrated cohesively for developing intelligent systems. This has caused some governmental and private sectors to invest in smart city projects to gather information in a structured manner. Internet of Things (IoT) is used for remotely monitoring data from multiple sensors for various applications. The IoT's growth [6–8], has also helped to store and process data in the cloud. The centralization of the information helps to determine the denser parts of the cities. Figure 1 [5, 9] shows the intrinsic relationship between the smart cities' different services with the

M. E. E. Alahi (✉)
Institute of Biomedical and Health Engineering, Shenzhen Institutes of Advanced Technology, Chinese Academy of Sciences, Shenzhen 518055, China

F. Akhter · S. Mukhopadhyay
Faculty of Science and Engineering, The School of Engineering, Macquarie University, Sydney, NSW 2109, Australia

A. Nag
School of Information Science and Engineering, Shandong University, Jinan 251600, China

N. Afsarimanesh
DGUT-CNAM Institute, Dongguan University of Technology, Dongguan 523106, China

J. K. Mandal and J. K. Roy (eds.), *Proceedings of International Conference on Computational Intelligence and Computing*, Algorithms for Intelligent Systems,
https://doi.org/10.1007/978-981-16-3368-3_9

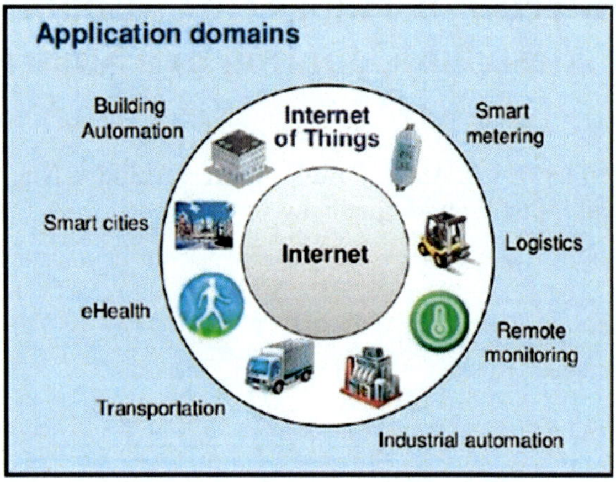

Fig. 1 Relationship between the Internet of things and different applications of smart cities [5]

individuals via IoT. The sensing services employed for different sensors help resolve the current problems as a common goal.

Some of Europe's cities have developed smart systems in their cities [10–14] to encompass modern urban production factors. For all the theoretical ideologies [15–17] that have been outlaid about smart cities, very few realistic approaches have been taken to implement them. It is state-of-the-art to use smart sensing systems in a defined area to understand the walking behavior of the pedestrian people [18–21]. The counting of pedestrians is an essential factor in urban places where population density is too high. It is essential to determine the people's density at pick hours in public places like bus stops, train stations, etc. This can help the city council to go for alternative approaches to minimize and possibly avoid traffic jams and hindrances to the ordinary people. Some of the research works [22–25] have been done earlier related to determining the relationship between the city's economic prosperity and the pedestrian's safety. However, all these works are related either to a theoretical model or based on the survey done specifically for a particular city. Researchers need to come up with generic ideas that can be implemented globally. This will help better planning of the cities and create a better place to live and work. The significance of pedestrian counting is also related to affect the emotional situation of a person. For example, if a person needs to be at work on an urgent basis, he would be hoping to take a comparatively empty road. If that person has limited options, he would be stuck on a busy road, which would create panic and tension in him. If he would have to use public transport, there are chances of accidents due to ignorance or rash decisions [26, 27]. Thus, it is essential to determine the pedestrians are crossing the busier parts of the city to ensure the avoidance of anomaly. This paper presents a real-time monitoring system deployed inside a university campus to count the pedestrians walking through different locations. The sensor nodes have also been functionalized for temperature and humidity sensing for determining pedestrians' congestion by the ambient conditions. The data collected in the sensor nodes have been further analyzed to validate the developed system's functionality for pedestrian counting in a defined location.

The chapter is organized in the following manner: Sect. 1 explains the introduction, Sect. 2 discusses the applied methods. The proposed method and system are explained in these sections. Section 3 demonstrates the collected results and discussions, and Sect. 4 explained the conclusion and commented on some points for future works.

2 Materials and Methods

This section presents the prototype (sensors and microprocessors) used for building the proposed pedestrian counting system and discusses the data collection and processing method.

2.1 Low-Cost Pedestrian Counting System Description

The system consists of low-cost sensors, a microcontroller, a LoRa shield for the data transmission, a power converter, and a solar panel. All the sensor node's electronic components were placed inside a box that is printed using Poly Lactic Acid (PLA) using a 3D printer. A 6000 mAh battery and 6 V solar panel were included in the system that allows five power autonomy days. The temperature and humidity sensor was fixed with screws on the outside of the box. The unique holder was designed for holding the PIR sensors related to screws on the box. The connected electronic systems are placed inside the box.

The block diagram of the pedestrian counting system based on LoRa for IoT-enabled network is shown in Fig. 2. LoRaWAN (Long-range, low-power Wireless Area Network) [28] covers the long-range, consumes the least power, and has a low bit rate excellent solution IoT enabled WSN network. It consists of two layers,

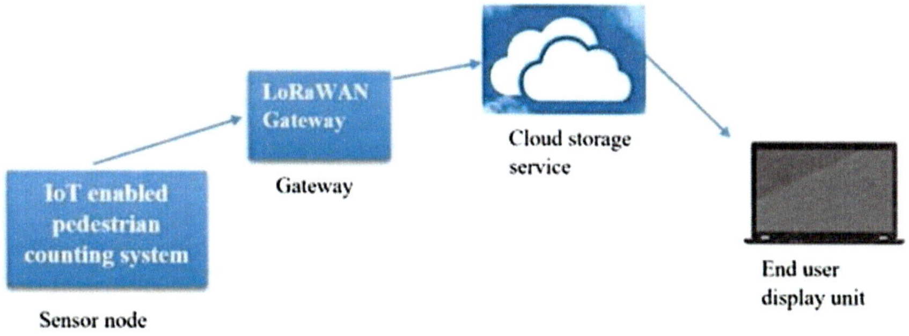

Fig. 2 Block diagram of the LoRa-based smart pedestrian counting system

where a physical layer is responsible for radio modulation called CSS (Chirp Spread Spectrum) [29], and a MAC-layer that is used to get access to LoRa network [30]. LoRa communication follows the FSK (Frequency Shift Keying) modulation technique [31] while communicating between the sensor node and the gateway. It also saves battery life, covers wider distance, enriches the network's capacity and the QoS (Quality of Service) by the defined LoRa protocol.

2.2 *Method of Pedestrian Detection*

The system is mounted on the pole with 3.25 m height. The project limitations required system installation on existing infrastructure and did not directly affect or influence its configuration. The width of the footpath was considered 5 m to cover the maximum range of footpaths. Figure 3 illustrates the coverage region of 3 PIR sensors. It has also shown the angles of the PIR sensors and their corresponding coverage regions. The angles are created based on the principle of FOV of the Fresnel lens. From Fig. 3, it is seen that the PIR sensor 1 was installed with an angle of 15° from the *y*-plane. The height of the mounting system is 3.25 m. From the right triangle, the tangent of an angle is the opposite side's length divided by the adjacent side's length. Therefore,

$$\tan\theta = \frac{\text{Opposite}}{\text{Adjacent}}$$

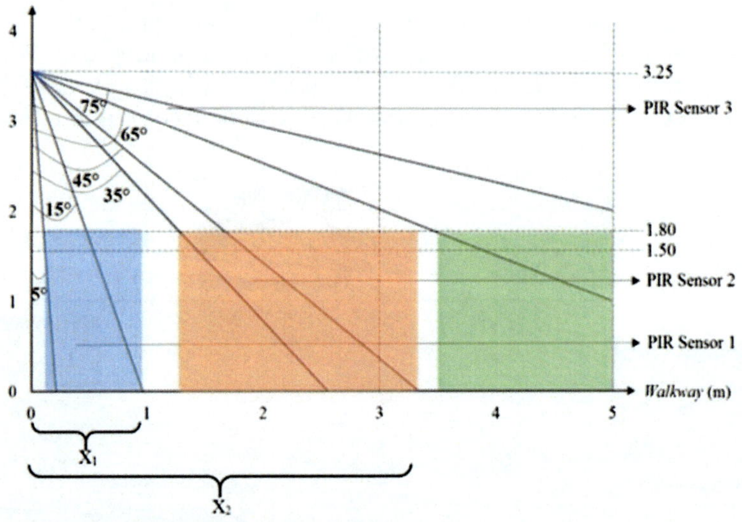

Fig. 3 Height of the system and counting mechanism

$$\text{or, } \tan 15^\circ = \frac{X_1}{3.25}$$

$$\text{or, } X_1 = 0.87 \text{ m} \tag{1}$$

$X1$ is the distance of the maximum coverage region of PIR sensor 1 from the Pole. The blue region covers it in Fig. 3. If a person passes that region, he/she will be detected by PIR sensor 1 only. PIR sensors 2 and 3 would not detect any movement. Similarly, the orange zone and green zone have been created for sensors 2 and 3, respectively. There is some undetected region between blue and orange zones and orange and green zones, which are kept intentionally. Using this empty area is to avoid detecting the same person by two sensors at the same time.

From Fig. 4, it is seen the principle of Fresnel Lenses, which is part of the PIR sensor module. The Fresnel lens provides a short focal length and large aperture without significantly changing its mass and size. The focal point can be divided into the ranges from *A*–*F*, which can cover the length of 1–5 m depending on the region. The sensor module used region *A* as an active region, covering the 5 m distance from the installation location (Fig. 5).

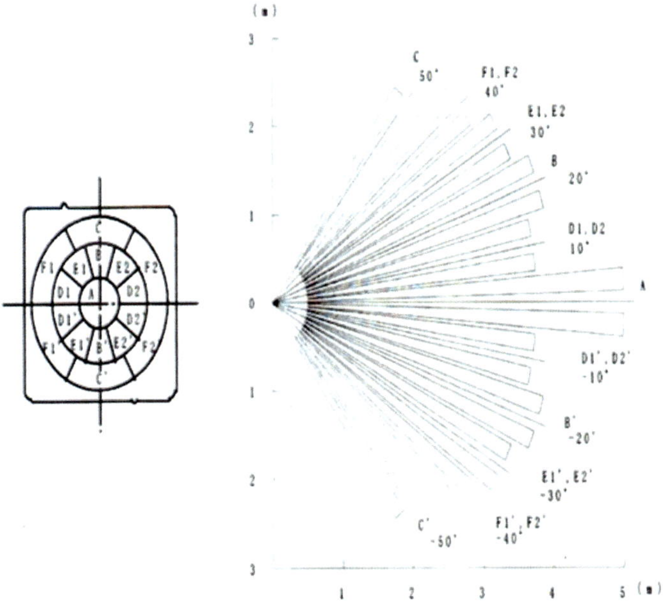

Fig. 4 Fresnel lens' Field-of-view (FOV) [32]

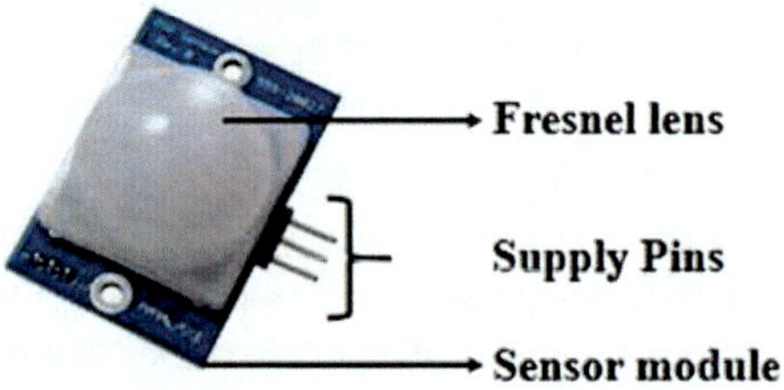

Fig. 5 PIR sensor module. PIR Sensor, "PIR Sensor with LED Signal - Parallax"; [Accesses: 21/06/2021]

2.3 Designing of the System

The proposed prototype was designed using Autodesk Fusion 360 software. Figures 6 and 7 show the design of the pedestrian counting system. The specific proposed

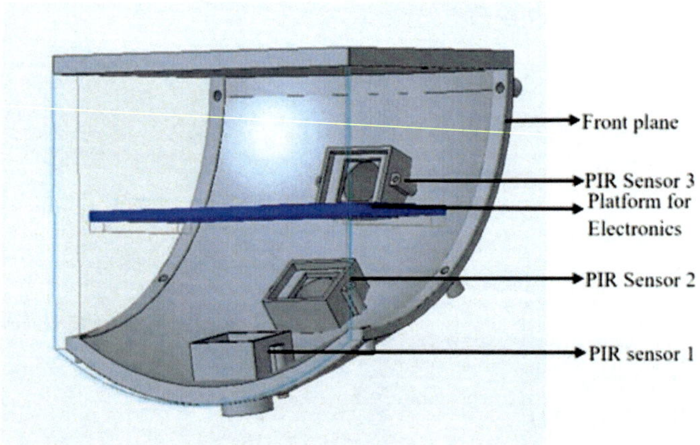

Fig. 6 See-through design of the system

Fig. 7 Frontal oval shape for PIR sensors

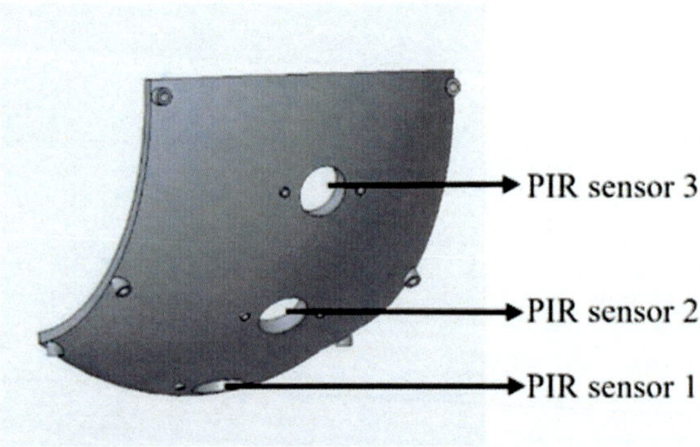

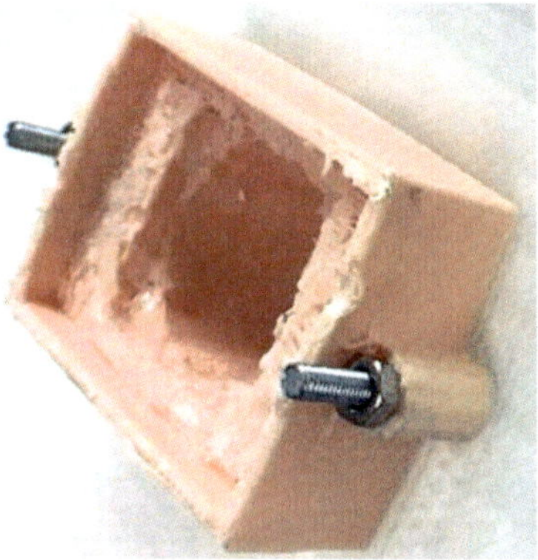

Fig. 8 Back view of the PIR sensor holder

angles on the system were accommodating on a curvature plane created as Fig. 7. The radius of the circle is 150 mm, close to the Fresnel lens's radius. The holder of the PIR sensor was also designed, which was like Fig. 8. A small platform was also designed for holding electronic components inside of the system. The holders were screwed on the surface, which can be seen from Fig. 6. The PIR holder allows the Fresnel lens to be fitted appropriately and protect from the direct sunlight. As a result, the performance of the sensor is not affected.

Additional mechanical considerations were also added for assembling the system. A flat surface is provided for the screws fitting correctly by adding an extrusion along with the curvature's faces. This was done to avoid any incident occurring around the screw and damaging the prototype. The position and ability to remove the platform from the system were also considered for human and maintenance factors.

2.4 Circuit Diagram of the System

Figure 9 displays the internal connection of the sensor node. Analog inputs' of Arduino UNO are connected with the DHT 11 sensor and PIR sensors. Arduino Uno is a low-powered microcontroller which has been used as the main microcontroller. LoRa shield and power converter are also connected with the Arduino UNO. A rechargeable battery is connected with the power management block, connected with the solar panel. The solar panel is harvesting the constant energy to the developed system and can sustain 5 days. All the sensors are connected to the 5 V supply to get the required power from the microcontroller.

A long-range transceiver, Dragino LoRa shield, is used for communicating with the gateway. It promotes sending low-rate data to a very long range. It applies RFM95W/RFM98W standard, and a frequency of 915 MHz transmission/reception frequency is used in Australia. LG01S [33] is used as a gateway for communicating between the cloud server and the proposed systems. The gateway is also responsible for data transferring to the cloud server. Thingspeak [34] is used as an IoT-based

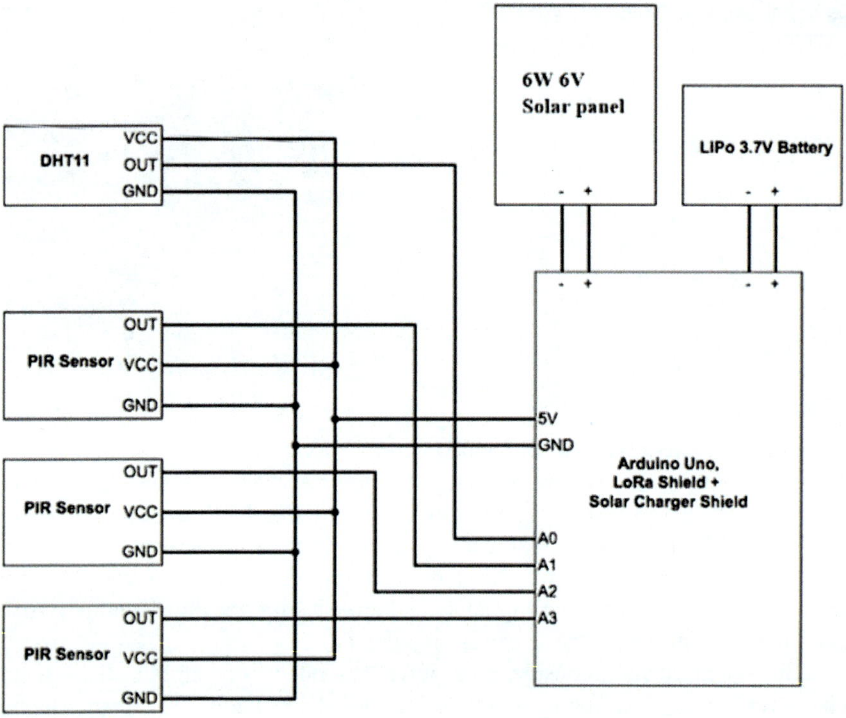

Fig. 9 Circuit diagram of the system

cloud server for the data storage unit. Figure 10 shows the connected sensors and the internal of the system.

2.5 *Installation of the System*

One of the proposed system's advantages was that it did not alter the city's existing infrastructure. The bandit bracket was used due to its flared leg shaped around the pole, while the stainless-steel hose gear clamp is corrosion-resistant, durable, and intended for outdoor applications. The solar bracket was also used for mounting the solar panel with the existing pole. Figure 11 shows the installation system at the electric pole.

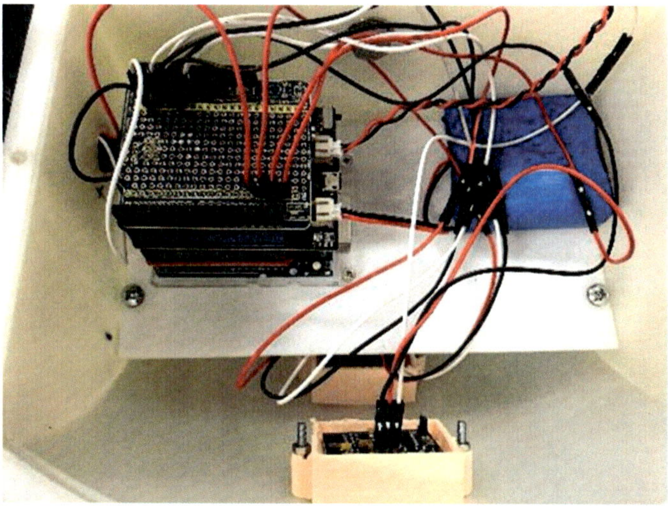

Fig. 10 Internals of the system

Fig. 11 Final system mounted on the pole

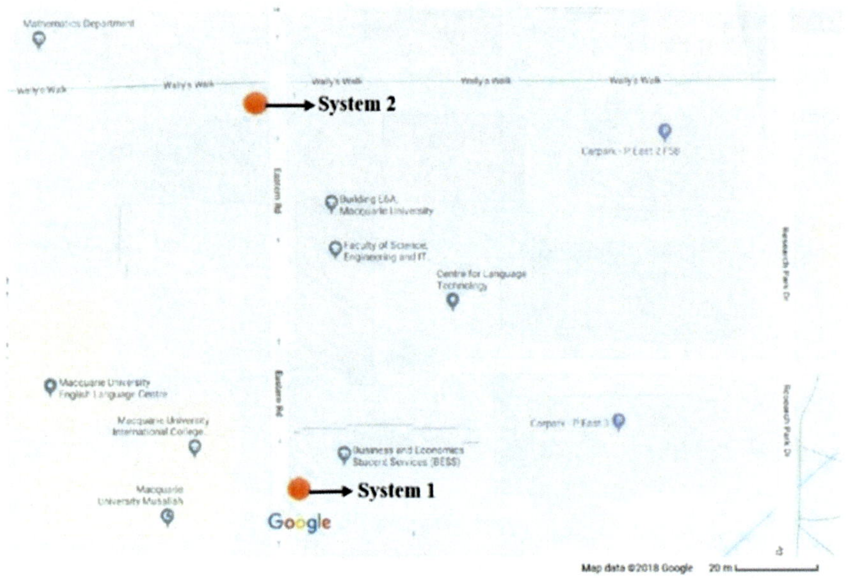

Fig. 12 Study location of the two systems for pedestrian counting

2.6 *Study Location*

The study location (− 33.775314, 151.115583) has chosen inside the campus of Macquarie University, Australia. Two systems were installed to count the pedestrian on both sides of a road. Figure 12 illustrates the two systems' location where the orange markers indicate the proposed systems' location.

2.7 *Collecting the Data*

It is required to collect the data from the developed system and validate the results. System 1 was installed as per Fig. 11. To validate the results, it was allowed to collect pedestrian counting and manual counting simultaneously. The counting was started at 8:00 am and finished the counting at 10:00 pm. The flow of the pedestrian was different, depending on the time of the day. The data were collected for three consecutive days and performed data analysis using the average data.

3 Results and Discussions

3.1 Collecting the Data

Table 1 shows the actual counting and the counting done automatically from the system. The table represents the data collection time from 8:00 am to 10:00 pm. Figure 13 represents the relationship between the system counting and the actual counting. The linear regression analysis shows the excellent coefficient of determination ($r^2 = 0.97$). The following equation is used for calculating the original counting from the system's counting.

Table 1 Comparative study of the actual and system counting at various time of the day

Time	Manual counting	Sensor node counting
08:00–09:00	35	42
09:00–10:00	70	62
10:00–11:00	84	102
11:00–12:00	108	122
12:00–13:00	101	105
13:00–14:00	164	185
14:00–15:00	219	210
15:00–16:00	245	225
16:00–17:00	142	159
17:00–18:00	80	87
18:00–19:00	62	61
19:00–20:00	40	46
20:00–21:00	50	57
21:00–22:00	74	81

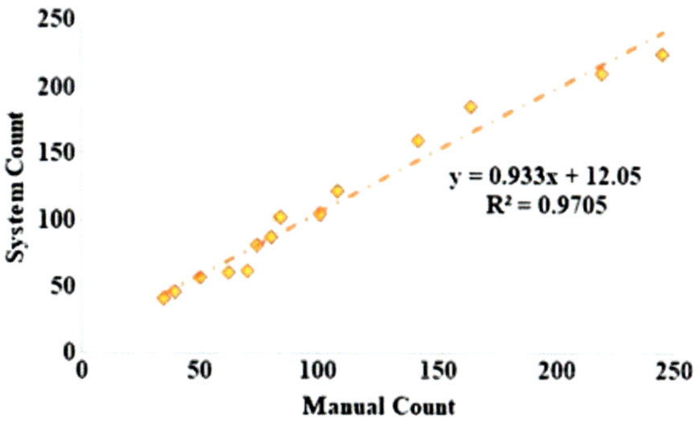

Fig. 13 Relationship between the actual counting and system counting

$$x = \frac{y - 12.05}{0.93} \tag{2}$$

where x is the actual count, y shoes the counting done by the sensor node. This equation was used as a fitted line for both system 1 and system 2.

3.2 Data Collection

A free IoT cloud server, Thingspeak, is used as data storage. Both the systems are trialed in their designated place (as per Fig. 11) to evaluate the performance. The sampling time of the data collection was 15 min—the Arduino programming control sampling time.

Figure 14 represents the Thingspeak data for system 1 and system 2, which were live data. The different field was defined for the pedestrian count, temperature, and humidity for system 1 and 2. The system data are transferred to the Thingspeak channel via the gateway.

3.3 Power Consumption and Energy Harvesting

The sensor node is programmed to collect the data and transmit the data every fifteen minutes. Lora Shield and DHT11 sensors are kept in idle conditions to reduce power consumption and increase battery life. Each current device consumption is estimated by voltage determination across a 1 Ω resistance, which is connected with the Arduino UNO. Table 2 shows the current consumed by the devices in a cycle (15 min).

The overall power consumed by the sensing system for a cycle (15 min) is calculated as,

$$P_s = V_s \times I_n + V_s \times I_{td} \times t_{td} \times f_{td} \tag{3}$$

Here,
Vs = Supply voltage = 5 V.
In = Current consumed by the system in one hour = 30.21 mA.
Itd = Current consumed while transmitting data = 120 mA.
ttd = data transmission duration = 60 ms.
ftd = data transmission interval.

$$\frac{1}{15\ \text{min}} = \frac{1}{15 \times 60} = \frac{1}{900}\text{Hz}$$

The total current drawn by the system in one cycle is found as,

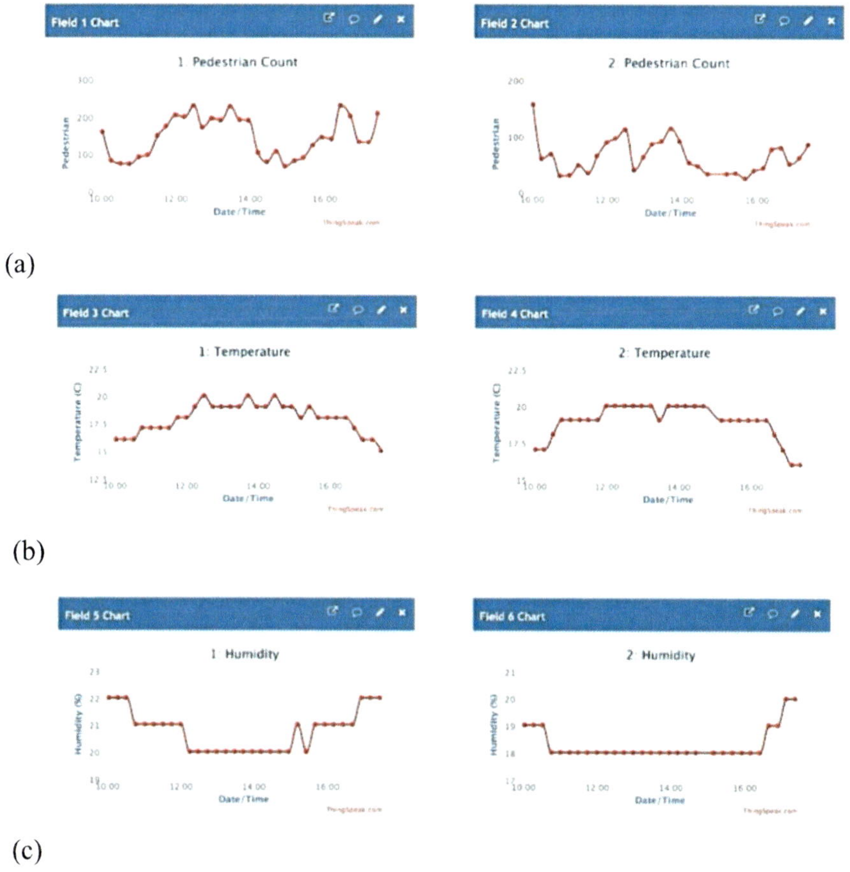

Fig. 14 Transferred **a** Pedestrian data, **b** temperature (°C) data, and **c** humidity (%) data from system 1 and system 2 to *Thingspeak* cloud server

Table 2 Consumption of current in 1 cycle

Component	Time	Current consumption (mA)	Current drawn in 1 cycle (mA s)
Arduino Uno	15 min	20	$20 \times 15 \times 60 = 18{,}000$
3 PIR sensors	15 min	9	$9 \times 15 \times 60 = 8100$
DHT11 (idle)	14 min 58 s	0.15	$0.15 \times (14 \times 60 + 58) = 134.7$
DHT11 (active)	2 s	1	$1 \times 2 = 2$
LoRa shield (idle)	15 min 59 s 940 ms	1	$1 \times (15 \times 60 + 59 + 940/1\ 000) = 959.94$

$$I_n = \frac{18{,}000 + 8100 + 134.7 + 2 + 959.94}{15 \times 60} = 30.21\ \text{mA}$$

Power consumption, *P*s is calculated as

$$\Rightarrow P_s = \left(5 \times 30.21 + 5 \times 120 \times \frac{600}{10^3}\right)$$

$$\Rightarrow P_s = 151.11\ \text{mVA}$$

If the power converter efficiency is 0.85%, the minimum mVA required from the battery,

$$P_{bs} = \frac{151.11}{0.85} = 177.77\ \text{mVA}$$

As the voltage of the battery is 5 V, the required discharge from the battery is,

$$I_{dis} = \frac{177.77}{3.7} = 48.04\ \text{mA}$$

Hence, the life of 3.7 V 6000 mAh battery is found as:

$$B_{Life} = \frac{6000}{48.04} = 124.88\ \text{h}$$

If the battery is 100% discharged, the 3.7 V 6000 mAh battery can sustain up to 124 h. The 6 V 6 W solar panel provides continuous power to the system, which can sustain 5 days without interruption. The LoRa module helps to consume less power compared to other WSN protocols such as WIFI, ZigBee, or 3G/4G during the data transmission. The generating energy is clean and environmentally friendly.

4 Conclusion

IoT-enabled pedestrian counting system was developed in a smart city scenario. Movement detection PIR sensors were used to detect the human movement and convert that to pedestrian counting. Two systems are tested simultaneously and showed 90% accuracy. The detection method is novel and tested at a different time of the day to evaluate the performance. 3D printer was used to develop the prototype of the system. The temperature and humidity sensor was also added to provide extra feature during pedestrian counting. Two systems were tested in two different locations to know pedestrian trends. The result shows that the proposed system can help develop a WSN for monitoring the pedestrian numbers in real-time.

Acknowledgements The authors would like to thank the School of Engineering, Macquarie University, to provide the laboratory facilities and support the field trials while conducting the research. This work is supported by the National Natural Science Fund (61950410613) from the National Science Foundation of China (NSFC) and CAS President International Fellowship Initiative (2019PT0008) from the Chinese Academy of Sciences.

References

1. Shapiro, J.M.: Smart cities: quality of life, productivity, and the growth effects of human capital. Rev. Econ. Stat. **88**(2), 324–335 (2006)
2. Boulos, M.N.K., Al-Shorbaji, N.M.: On the internet of things, smart cities and the WHO healthy cities. BioMed Central (2014)
3. Pirbhulal S., Zhang, H., E Alahi, M.E., Ghayvat, H., Mukhopadhyay, S.C., Zhang, Y.-T., Wu, W.: A novel secure IoT-based smart home automation system using a wireless sensor network. Sensors 17(1), 69 (2017)
4. Hancke, G.P., Hancke, G.P.: The role of advanced sensing in smart cities. Sensors **13**(1), 393–425 (2012)
5. Perera, C., Zaslavsky, A., Christen, P., Georgakopoulos, D.: Sensing as a service model for smart cities supported by internet of things. Trans. Emerg. Telecommun. Technol. **25**(1), 81–93 (2014)
6. Nag, A., Alahi, M.E.E., Afsarimanesh, N., Prabhu, S., Mukhopadhyay, S.C.: IoT for smart homes. In: Sensors in the Age of the Internet of Things: Technologies and Applications, pp. 171 (2019)
7. Alahi, M.E.E., Xie, L., Mukhopadhyay, S., Burkitt, L.: A temperature compensated smart nitrate-sensor for agricultural industry. IEEE Trans. Industr. Electron. **64**(9), 7333–7341 (2017)
8. Akhter, F., Khadivizand, S., Siddiquei, H.R., Alahi, M.E.E., Mukhopadhyay, S.: IoT enabled intelligent sensor node for smart city: pedestrian counting and ambient monitoring. Sensors **19**(15), 3374 (2019)
9. Akhter, F., Khadivizand, S., Lodyga, J., Siddiquei, H.R., Alahi, M.E.E., Mukhopadhyay, S.: In design and development of an IoT enabled pedestrian counting and environmental monitoring system for a smart city. In: 2019 13th International Conference on Sensing Technology (ICST), pp. 1–6. IEEE (2019)
10. Caragliu, A., Del Bo, C., Nijkamp, P.: Smart cities in Europe. J. Urban Technol. **18**(2), 65–82 (2011)
11. Afsarimanesh, N., Alahi, M.E.E., Mukhopadhyay, S.C., Kruger, M.: Development of IoT-based impedometric biosensor for point-of-care monitoring of bone loss. IEEE J. Emerg. Select. Top. Circ. Syst. **8**(2), 211–220 (2018)
12. Poursafar, N., Alahi, M.E.E., Mukhopadhyay, S.: In Long-range wireless technologies for IoT applications: a review (2018)
13. Nag, A., Alahi, M.E.E., Feng, S., Mukhopadhyay, S.C.: IoT-based sensing system for phosphate detection using graphite/PDMS sensors. Sens. Actuators, A **286**, 43–50 (2019)
14. Alahi, M.E.E., Pereira-Ishak, N., Mukhopadhyay, S.C., Burkitt, L.: An internet-of-things enabled smart sensing system for nitrate monitoring. IEEE Internet Things J. **5**(6), 4409–4417 (2018)
15. Schaffers, H., Komninos, N., Pallot, M., Trousse, B., Nilsson, M., Oliveira, A.: In smart cities and the future internet: towards cooperation frameworks for open innovation. In: The Future Internet Assembly, pp. 431–446. Springer, Berlin (2011)
16. Fernandez, M.: Smart cities of the future. In: Smart Citizens. Manchester: FutureEverything, pp. 43–46 (2013)

17. Vlacheas, P., Giaffreda, R., Stavroulaki, V., Kelaidonis, D., Foteinos, V., Poulios, G., Demestichas, P., Somov, A., Biswas, A.R., Moessner, K.: Enabling smart cities through a cognitive management framework for the internet of things. IEEE Commun. Mag. **51**(6), 102–111 (2013)
18. Zook, J.B., Lu, Y., Glanz, K., Zimring, C.: Design and pedestrianism in a smart growth development. Environ. Behavior 44(2), 216–234 (2012)
19. Mone, G.: The new smart cities. Commun. ACM **58**(7), 20–21 (2015)
20. Talen, E.: Pedestrian access as a measure of urban quality. Plan. Pract. Res. **17**(3), 257–278 (2002)
21. Oakes, J.M., Forsyth, A., Schmitz, K.H.: The effects of neighborhood density and street connectivity on walking behavior: the Twin Cities walking study. Epidemiol. Perspect. Innov. **4**(1), 16 (2007)
22. Jin, D., Hannon, C., Li, Z., Cortes, P., Ramaraju, S., Burgess, P., Buch, N., Shahidehpour, M.: Smart street lighting system: a platform for innovative smart city applications and a new frontier for cyber-security. Electr. J. **29**(10), 28–35 (2016)
23. Cervero, R., Duncan, M.: Walking, bicycling, and urban landscapes: evidence from the San Francisco Bay Area. Am. J. Public Health **93**(9), 1478–1483 (2003)
24. Ameli, S.H., Hamidi, S., Garfinkel-Castro, A., Ewing, R.: Do better urban design qualities lead to more walking in Salt Lake City, Utah? J. Urban Des. **20**(3), 393–410 (2015)
25. Ewing, R., Hajrasouliha, A., Neckerman, K.M., Purciel-Hill, M., Greene, W.: Streetscape features related to pedestrian activity. J. Plan. Educ. Res. **36**(1), 5–15 (2016)
26. Turner, S., Francis, T., Roozenburg, A., Transport, N.L.: Predicting Accident Rates for Cyclists and Pedestrians. Citeseer (2006)
27. Bu, F., Greene-Roesel, R., Diogenes, M.C., Ragland, D.R.: Estimating Pedestrian Accident Exposure: Automated Pedestrian Counting Devices Report (2007)
28. Adelantado, F., Vilajosana, X., Tuset-Peiro, P., Martinez, B., Melia-Segui, J., Watteyne, T.: Understanding the limits of LoRaWAN. IEEE Commun. Mag. **55**(9), 34–40 (2017)
29. Wang, X., Fei, M., Li, X.: In performance of chirp spread spectrum in wireless communication systems. In: 11th IEEE Singapore International Conference on Communication Systems, ICCS 2008, pp 466–469. IEEE (2008)
30. Alahi, MEE., Pereira-Ishak, N., Mukhopadhyay, S.C., Burkitt, L.: An internet-of-things enabled smart sensing system for nitrate monitoring. IEEE Internet of Things J (2018)
31. Usselman, G.L.: Frequency shift keying. Google Patents (1949)
32. Miyamoto, K.: The phase Fresnel lens. JOSA **51**(1), 17–20 (1961)
33. Dragino LG01-S IoT Gateway featuring LoRa® technology. http://www.dragino.com/products/lora/item/119-lg01-s.html (accessed 15/11/2018)
34. Thingspeak Thingspeak. https://thingspeak.com/ (accessed 15/11/2018)

JK Flip-Flop Design Using Layered T Logic: A Quantum-Dot Cellular Automata-Based Approach

Chiradeep Mukherjee, Saradindu Panda, Asish Kumar Mukhopadhyay, and Bansibadan Maji

1 Introduction

The CMOS-based semiconductor technology has been facing two simultaneous discontinuities in the last two years. The industry gives its specific countenance to the physical limits of existing semiconductor materials and Moore's prediction's impending end [1, 2]. These challenges have forced scientists to rethink novel computational devices and architectures. The quantum-dot cellular automata (QCA), a potential alternative to CMOS technology, has a particular appeal for its high operating speed, extreme-low energy dissipation, and ultra-high packing density [3]. The QCA-based computing architecture operates on quantum–mechanical tunneling between the electrons rather than the conventional voltage and current flow in the circuit. Although the researchers have introduced several logic reduction techniques for realizing sequential logic blocks in QCA, there is a need for an efficient flip-flop, which would be extremely helpful in successive type central processing unit (CPU) designs.

C. Mukherjee (✉)
University of Engineering and Management, Jaipur, India

S. Panda · A. K. Mukhopadhyay
Narula Institute of Technology, Kolkata, India

C. Mukherjee · B. Maji
National Institute of Technology, Durgapur, India

J. K. Mandal and J. K. Roy (eds.), *Proceedings of International Conference on Computational Intelligence and Computing*, Algorithms for Intelligent Systems,
https://doi.org/10.1007/978-981-16-3368-3_10

The existing logic reduction techniques of QCA include Majority Voter (MV), Universal Quantum Cellular Automate Logic (UQCAL), Universal Logic Gate (ULG), Coupled Majority Voter Minority (CMVMIN), And-Or-Invert (AOI), FNZ, and Layered T (LT) gate [4]. This work utilizes the LT logic reduction-based cost-efficient J-K flip-flop to establish the LT Gate's scalability, reproducibility, and feasibility in the realization of high-speed synchronous systems.

The rest of the entire work is framed as follows: Sect. 2 provides the background of QCA and reviews the existing JK flip flop designs. In Sect. 3, LT logic-based JK flip flop is proposed, and a comparative analysis with current designs is investigated. Section 4 highlights the improvements and concludes the entire work.

2 Background of QCA and Existing Works on JK Flip-Flop

2.1 Background of QCA

The Quantum cell is an elementary unit of QCA. The Quantum Cell with four quantum dots accommodates two excess electrons through the electrostatic force of attraction and repulsion. These electrostatic forces bound these two extra electrons on two diagonal corners of a quantum cell. These stable orientations of two electrons represent polarizations $P = +1$ or $P = -1$ to convey logic 1 and logic 0, as demonstrated in Fig. 1a, b [5].

If a group of cells is placed close to each other, then one cell's polarization may influence the others. This mechanism is exploited for the realization of the logic elements like majority voter, binary wire, and seven-cell inverter, as shown in Fig. 2a–c, respectively. The majority voter takes three inputs, *A*, *B*, and *C*, to assesses the output as $Z = AB + BC + CA$. The binary wire is utilized to bring the fan-in

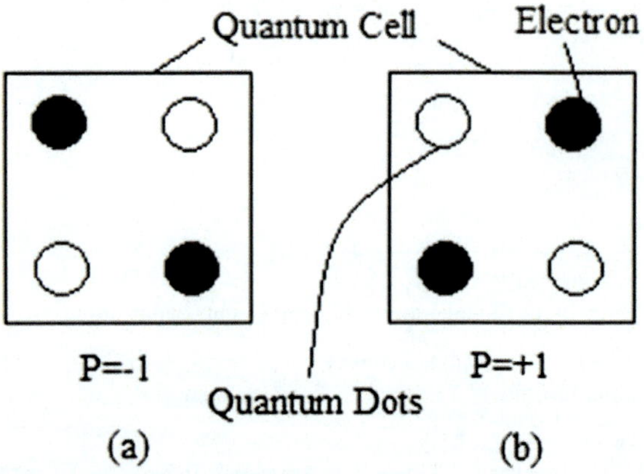

Fig. 1 **a** Quantum cell with $P = -1$, **b** quantum cell with $P = +1$

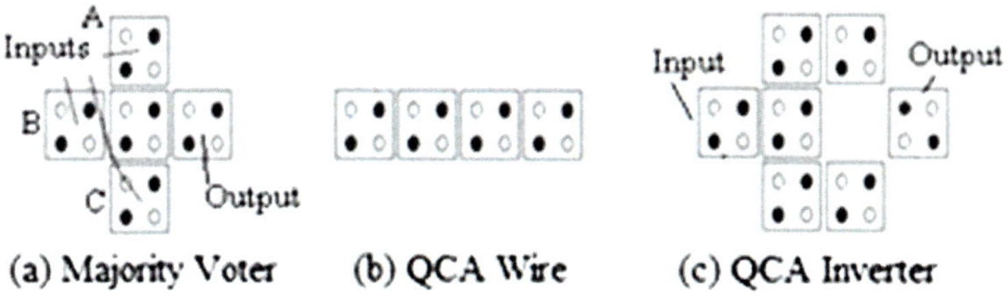

Fig. 2 **a** QCA majority voter, **b** QCA wire, and **c** QCA inverter

and fan-outs to a specific point of a circuit. The inverter, which is composed of seven QCA cells, is used to invert the input A.

The driver force of QCA circuits is a clock with four phases: switch, hold, release, and relax. In the switch phase, a QCA cell becomes polarized due to a raised barrier between the quantum dots. Moreover, in the hold state, the barrier remains at a high state so that the particular cell retains its polarization. The release phase starts decreasing the barrier, and finally, the relax phase pushes the cell into an un-polarized state [6].

2.2 Existing Works on JK Flip-Flop

The efficient and effective design of the QCA logic circuits requires an extensive systematic review of the existing logic reduction techniques to develop future nano processors. Table 1 attempts to present a systematic survey on the QCA layouts of existing JK flip flops.

Such a survey claims the need for QCA designs of JK flip flop that consider the cost functions and efficient space utilization.

3 JK Flip-Flop Design Using LT Logic Reduction Methodology

This section emphasizes the JK flip flop based on the LT logic reduction technique [13–15], which can be further utilized to implement multilevel sequential circuits. Generally, a clock signal (Clk) controls the flow of information, and consequently, claims a pivotal role in designing flip-flops. The clock is implicitly implemented in QCA circuits using the four-phase adiabatic switching mechanism, as discussed in the previous section. These four phases of in-built clock signal impose strict timing restrictions on the QCA layouts of sequential circuits.

The JK flip flop can avoid the disadvantage of SR flip flop [16]. If both the inputs, *S*, and *R* become logic 1, then the output of SR flip flop remains unpredicted.

Table 1 Review of the existing JK flip-flops

S. No	JK flip-flop designs	
1	Types of crossover used	Coplanar: [7, 8]
		Multilayer: Not utilized in JK flip-flop designs
		Logical: [9, 10]
		Not used: [11, 12]
2	Types of logic reduction used	3-input MV: [7–10]
		5-input MV: [11]
		Coupled 3-input, and 5-input MVs: [12]
3	Redesignability, and extendability	[7]: MV based designs of JK, SR, *D*, and *T* flip-flops are proposed
		[8]: Dual edge-triggered JK flip flop is proposed
		[9]: Proposed JK flip flop is extended in n-bit counter designs (3-bit and 5-bit counters are instantiated)
		[10]: Proposed JK flip flop is extended in n-bit counter designs (4-bit counters are instantiated)
		[11]: Proposed JK flip flop is extended in 4-bit counter designs
		[12]: MV based hybrid designs of JK, SR, *D*, and *T* flip-flops are proposed
4	The design rules related issues	Exist: [9, 11]
		Not exist: [7, 8, 10, 12]

However, JK flip flop toggles the output in such situations. A JK flip flop has two inputs, J and K, and one output, Q. If both the inputs become logic 0, the result, Q, holds its previous value. The input logic at $J = 1$ sets the output Q to 1. The input logic at $K = 1$ resets the output Q to 0, on the other side. However, the output Q gets toggled or complemented when both J and K's inputs have the same values as logic 1. Therefore, we can generate the truth table of JK flip flop, as given in Fig. 3a. The JK flip flop is constructed by placing three-input NAND gates followed by two cross-coupled two-input NAND gates, as demonstrated in Fig. 3b. The three-input NAND gates fetch the inputs J and K, clock signal Clk, and the present value of output signal Q to feed two-input NAND gates of the next stage.

The characteristic equation and LT equation of JK flip flop are given in Eq. (1) and (2). The LT equation of JK flip flop computes temporary output, $\overline{Q\_\text{old}}$ as $L_T^+, \left(L_T^+(Q\_\text{old}), L_{TA}^+(K, \text{Clk})\right)$.

$$Q = J\overline{(Q\_\text{old})} + \overline{K}(Q\_\text{old}) \tag{1}$$

J	K	Q
0	0	Q_old
0	1	0
1	0	1
1	1	Toggle

(a)

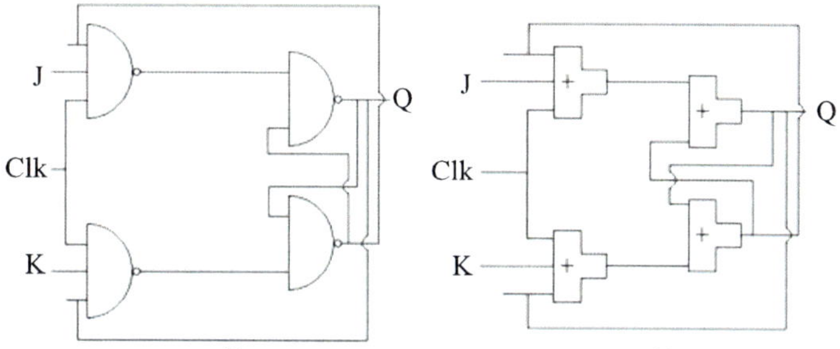

Fig. 3 **a** Truth table, and **b** logic diagram, and **c** LT Schematic of JK flip flop

$$\begin{cases} Q = Q\_\text{old} + \left(\overline{J.\overline{Q\_\text{old}}.\text{Clk}}\right) \\ Q = L_T^+\left(\overline{Q\_\text{old}}, L_T^+\left(Q\_\text{old}, L_{TA}^+(J, \text{Clk})\right)\right) \end{cases} \quad (2)$$

We generate a corresponding logic diagram and LT schematic of JK flip flop as shown in Fig. 3b, c.

The QCA layout of the JK flip flop is generated according to Eq. (2), as shown in Fig. 4. It employs four LT NAND gates to obtain its desired output. Three-input LT NAND gates are constructed by placing a two-input LT AND gate followed by a two-input LT NAND gate. The layout needs a 0.087204 μm^2 effective area, O-Cost of 97, and one clocking zone to generate the output Q.

The QCA layout of the JK flip flop is simulated in QCADesigner [17] using a coherence vector and bistable approximation-based simulation, as demonstrated in Fig. 5a, b, respectively. It can be observed that dotted lines (as denoted by I′, I″, I‴, I⁗ and II′, II″, II‴, II⁗) points to the change of output logic at the negative falling edge of clock 3.

The comparative analysis of the proposed layout in terms of design parameters like effective area, O-Cost, delay, gate counts, IPD, complexity, Cost_α [18] reported in Table 2. The comparison includes the existing designs as cited in Refs. [7–12].

Fig. 4 The QCA layout of JK flip flop

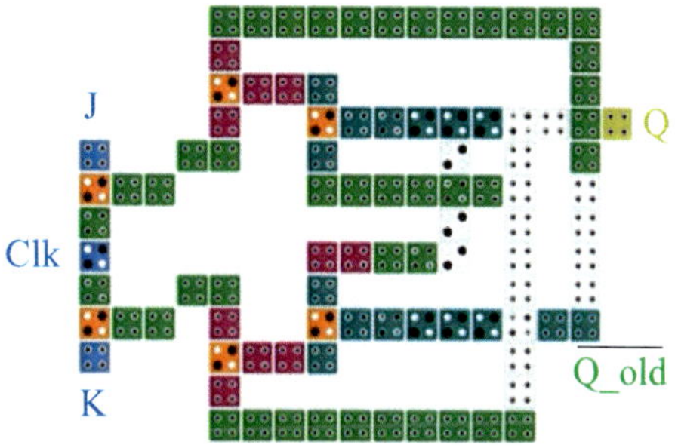

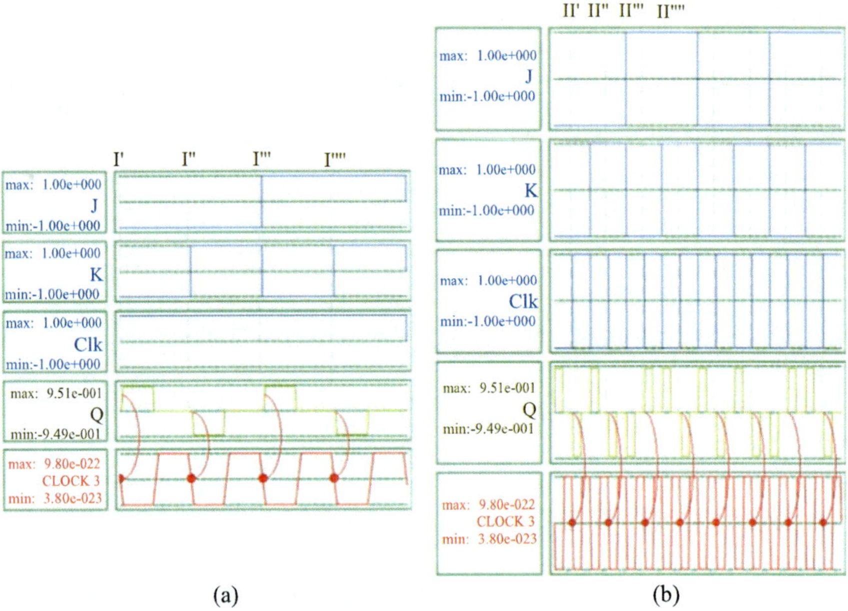

Fig. 5 The outputs of the proposed layout of JKJK flip flop simulated in **a** coherence vector, and **b** bistable approximation-based engines of QCADesigner

4 Result Discussion and Conclusion

As shown in Table 2, the proposed JK flip-flop needs only one clock phase, making LT counterpart faster than Vetteth et al. JK flip-flop by 33.33%. Moreover, the LT implementation of the JK flip-flop requires a 5.6% less effective area and 17.78% more cost-effective than the design of Vetteth et al. JK flip-flop [7]. The rest of the designs have issues in design rules and are not scalable. The LT Gate alleviates the absence of Universal NAND or NOR based logic gate in the Quantum Cellular Automata. The memory design would become highly efficient in essential parameters like effective area, delay, and Costα with JK flip-flop's LT implementation. In the future, higher-order sequential circuit designs can be possible with the help of the proposed JK flip-flop.

Table 2 Comparative analysis of proposed JK flip-flop with the existing designs

S. No	Design		Effective area	O-Cost	Delay	#gate	IPD	Complexity	$Cost_{\alpha}$	Crossover type
1	JK Flip Flop	[7]	0.092364	90	1.5	MV3 = 5 INV = 4	5	10	45	Coplanar
2		[8]	0.268272	208	4.25	MV3 = 7 INV = 6	7	14	238	Coplanar
3		[9] @	0.06	52	0.75	MV5 = 1 MV3 = 1 INV = 4	2.5	6.5	8.4375	NR
4		[10]	0.06	50	1	MV3 = 3 INV = 2	3	6	12	Coplanar
5		[11] @	0.114048	119	2	MV3 = 7 INV = 5	7	10	108	NR
6		[12] @	0.04	31	1	MV3 = 4 INV = 3	4	7	19	NR
7		This work	0.087204	97	1	LT NAND = 6	6	7	37	Logical and multilayer

@ issues of design rules exist

References

1. Radamson, H.H., et al.: Miniaturization of CMOS. Micromachines **10**, 293 (2019)
2. Peercy, P.S.: The drive to miniaturization. Nature **406**, 1023–1026 (2000)
3. Lent, C.S., Douglas, P., Tougaw, W.P.: Quantum cellular automata. IEEE Trans. Nanotechnol. **2**, 758–763 (2015)
4. Mukherjee, C., Panda, S., Mukhopadhyay, A.K., Maji, B.: Synthesis of standard functions and generic Ex-OR module using layered T gate. Int. J. High Perform. Syst. Archit. **7**, 70–86 (2017)
5. Snider, G.S.L., et al.: Quantum-dot cellular automata. Microelectron. Eng. **47**, 261–263 (1999)
6. Campos, C.A.T., Marciano, A.L., Vilela Neto, O.P., Torres, F.S.: USE: a universal, scalable, and efficient clocking scheme for QCA. IEEE Trans. Comput. Des. Integr. Circ. Syst. **35** (2016)
7. Vetteth, A., Walus, K., Dimitrov, V.S., Jullien, G.A.: Quantum-Dot Cellular Automata of Flip-Flops (2003)
8. Xiao, L.R., Chen, X.X., Ying, S.Y.: Design of dual-edge triggered flip-flops based on quantum-dot cellular automata. J. Zhejiang Univ. Sci. C **13**, 385–392 (2012)
9. Kong, K., Shang, Y., Lu, R.: Counter designs in quantum-dot cellular automata. In: 2010 10th IEEE Conference on Nanotechnology, NANO 2010, pp. 1130–1134 (2010). https://doi.org/10.1109/NANO.2010.5698033
10. Sarmadi, S., Azimi, S., Sheikhfaal, S., Angizi, S.: Designing counter using inherent capability of quantum-dot cellular automata loops. Int. J. Mod. Educ. Comput. Sci. **7**, 22–28 (2015)
11. Abdullah-Al-Shafi, M., et al.: Designing single layer counter in quantum-dot cellular automata with energy dissipation analysis. Ain Shams Eng. J. **9**, 2641–2648 (2018)
12. Bahar, A.N., Laajimi, R., Abdullah-Al-Shafi, M., Ahmed, K.: Toward efficient design of flip-flops in quantum-dot cellular automata with power dissipation analysis. Int. J. Theor. Phys. **57**, 3419–3428 (2018)
13. Mukherjee, C., Panda, S., Mukhopadhyay, A.K., Maji, B.: QCA gray code converter circuits using LTEx methodology. Int. J. Theor. Phys. **57**, 2068–2092 (2018)
14. Mukherjee, C., Panda, S., Mukhopadhyay, A.K., Maji, B.: Towards modular binary to gray converter design using LTEx module of quantum-dot cellular automata. Microsyst. Technol. **25**, 2011–2018 (2019)
15. Mukherjee, C., et al.: Implementation of toffoli gate using LTEx module of quantum-dot cellular automata. Adv. Intel. Syst. Comput. **812** (2019)
16. Mano, M.M.: Digital Logic and Computer Design. Pearson (2007)
17. Walus, K., Jullien, G.A.: Design tools for an emerging SoC technology: quantum-dot cellular automata. Proc. IEEE **94**, 1225–1244 (2006)
18. Liu, W., Lu, L., Neill, O., Member, S.: A first step toward cost functions for quantum-dot cellular automata designs. IEEE Trans. Nanotechnol. **13**, 476–487 (2014)

An Intelligent Pattern Recognition Algorithm

Abhijit Bag and Damodar Prasad Goswami

1 Introduction

In today's world, data from different credible sources become more available as the technology emerges, which was speculated and erroneous previously. With the availability of huge datasets, it is evident to identify patterns from historical information to achieve data-driven decision for today and predict tomorrow. It is applicable to all the areas of science, research, academia and most importantly the all business houses towards Industry 4.0 transition. However, this involves lots of manual, complex activities and mostly dependent on human skillset and expertise today. Our work is focussed to automate few of manual efforts.

It is unlikely that every set of data will follow a mathematical rule, but sometimes, we do not know why, some very practical and familiar figures follow a mathematical rule. The greatest historical example of this kind is Kepler's third law. It took years to discover the pattern in planetary motion even by the genius like Johannes Kepler. There are numerous examples in science, economics, finance (and where not?) and in our everyday life. A large volume of literatures are devoted towards curve fitting [1], nonlinear regression [3] and their modern counterpart 'machine learning' [2].

But there is no known procedure to find any pattern in a given data set. Scientists often use their expertise and experience for that. Sometimes, they follow one very modest way. They just plot the data and look for any visual pattern. Then, they try to express that visual pattern in mathematical forms. But unfortunately, our visual ability cannot go beyond three dimensions and plotting cannot help anymore. In this frustrating situation, only one way is left behind to experiment with different functions randomly and see which serves the need. But how long can we go with

A. Bag (✉)
Research Member, Eureka Scietech Research Foundation, Kolkata, India

D. P. Goswami
Asutosh Mookerjee Memorial Institute, Sivatosh Mookerjee Science Centre, Kolkata, India

J. K. Mandal and J. K. Roy (eds.), *Proceedings of International Conference on Computational Intelligence and Computing*, Algorithms for Intelligent Systems,
https://doi.org/10.1007/978-981-16-3368-3_11

this method? Not so far; because there are so many mathematical expressions and this number grows exponentially high when we start to make their combinations. After all, we have very limited human life, capacity and time. At this critical point, scientists try to apply their domain knowledge and facilitate the task as much as they can. But this also can be done in some very limited cases, where the situation is relatively simple and domain knowledge can help making intelligent guesses. But in a completely unknown and complex situation, this cannot be done. We are left behind with the only alternative of making random guesses. Won't it be nice if we can leave out the entire task on the machine? Machines can perform at a larger speed with a greater accuracy than humans and take charge of boring, laborious, time-consuming tasks in an automated fashion. This algorithm is designed with this aim in mind; to simplify the task, save the valuable time of the scientists and assist them with a powerful tool for discovering mathematical patterns in numerical dataset. We have cited a simple yet interesting example to demonstrate the power of this product.

1.1 Material and Methods

Sample Dataset

We have taken a synthetic randomly generated dataset of length equal to 100 following a pure quadratic rule $2 + 3x + 7x^2$ lying in between -5 and 5. We have added randomly generated noise to the datasets to prepare more practical datasets for experiment.

Tools & Technology

We have utilised open source Python programming using jupyter notebook to find rules from the dataset.

Method Applied

For the time being, we shall pretend as if we do not know the rule and use our algorithm to get back the same. There are 100 data points in the list in which we have divided in an 80–20 manner for training and test purposes, respectively. First 80 data points have been taken to build the model and when the model is reached, it has been tested on the last 20. That is, the model that has been used to forecast the last 20% of data to check how efficient the model is. We start with a high degree polynomial, calculate the coefficients of the equation and then decrease the degree of the polynomial by one. Then, it repeats the same thing to calculate the coefficients again and continues in this fashion until we reach the conclusion. In this example, we have started with a fifth-degree polynomial and observed that the coefficients of x^3, x^4 and x^5 are almost zero. From this observation, we make a guess that the underlying rule is a quadratic one. This guess is made from the following observations: in the fourth-degree polynomial, the coefficients of x^3, x^4 and in the third-degree polynomial, the coefficient of x^3 are very small and tend to zero as before. Finally, for the quadratic polynomial, we have all nonzero coefficients. From these experiments, we can conclude that the underlying rule is really a quadratic one. Now, we have calculated the *R*-square value with this quadratic rule to check the goodness of fit on both the training and test data.

In the next phase, we have added a random noise with our data and checked whether the rule can be extracted even in the presence of this noise. We notice that even if the data has been masked with noise, the rule can still be taken out with the same algorithm.

1.2 Process Overview

Below chart talks about the steps performed by us.

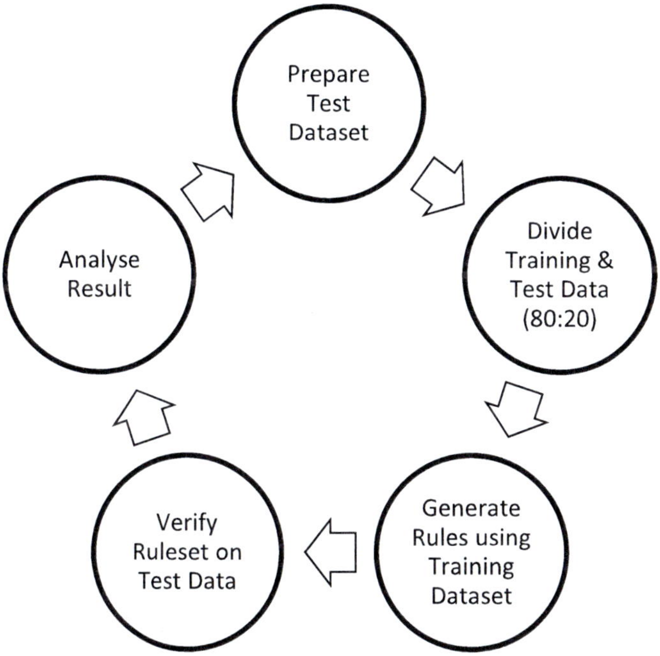

1.3 Results and Discussion

Table 1 shows the equations and coefficients of the fitted curves over the training data. Table 2 represents the corresponding plots calculated on the noisy dataset. If we look at the coefficients in Table 1, we observe that for the fifth-degree polynomial, and the value of $p4$, $p5$ and $p6$ is nearly zero. Similarly, for the fourth-degree polynomial, the value of $p4$, $p5$ is nearly zero, and for the third-degree polynomial, the value of $p4$ is

Table 1 Result without noise

Degree of the polynomial	Equation of the polynomial	Coefficients obtained on training data
5	$f(x) = p1 + p2 * x + p3 * x^2 + p4 * x^3 + p5 * x^4 + p6 * x^5$	$p1 = 2$ $p2 = 3$ $p3 = 7$ $p4 = 2.627e{-}15$ $p5 = -1.099e{-}15$ $p6 = -2.691e{-}16$
4	$f(x) = p1 + p2 * x + p3 * x^2 + p4 * x^3 + p5 * x^4$	$p1 = 2$ $p2 = 3$ $p3 = 7$ $p4 = 7.45e{-}16$ $p5 = 4.608e{-}16$
3	$f(x) = p1 + p2 * x + p3 * x^2 + p4 * x^3$	$p1 = 2$ $p2 = 3$ $p3 = 7$ $p4 = -1.215e{-}15$
2	$f(x) = p1 + p2 * x + p3 * x^2$	$p1 = 2$ $p2 = 3$ $p3 = 7$ *R*-Square training data = 1.0 *R*-Square test data = 1.0

Table 2 Results with noise

Degree of the polynomial	Equation of the polynomial	Coefficients obtained on training data
5	$f(x) = p1 + p2 * x + p3 * x^2 + p4 * x^3 + p5 * x^4 + p6 * x^5$	$p1 = -0.2588$ $p2 = 2.07$ $p3 = 7.126$ $p4 = 0.165$ $p5 = -0.008986$ $p6 = -0.006553$
4	$f(x) = p1 + p2 * x + p3 * x^2 + p4 * x^3 + p5 * x^4$	$p1 = 0.06156$ $p2 = 2.314$ $p3 = 6.844$ $p4 = 0.08366$ $p5 = 0.01882$
3	$f(x) = p1 + p2 * x + p3 * x^2 + p4 * x^3$	$p1 = -0.1805$ $p2 = 2.78$ $p3 = 7.034$ $p4 = 0.01673$
2	$f(x) = p1 + p2 * x + p3 * x^2$	$p1 = -0.04003$ $p2 = 2.925$ $p3 = 6.994$ *R*-Square training data = 0.9982557285123025 *R*-Square test data = 0.9926006392733089

nearly zero. For the second-degree polynomial, we get nonzero $p1, p2$ and $p3$ which matches with the function that generated this data. Finally, the *R*-square value has been calculated on both the training and test data and found to be nearly 1 indicating a very good fit.

For the noisy dataset, we found almost the same results and so the same conclusion followed here also. Of course, in the presence of noise, the results are a little bit distorted as expected. The detailed results can be perceived from Table 2. This shows that our method is also capable of finding rules from a noisy dataset.

2 Conclusion

We have presented an intelligent algorithm which is capable of finding a mathematically closed-form expression in a set of an experimental data. We have confined our discussion within polynomial type functions and with one independent variable. This algorithm can further be extended with other types of functions and with multivariable as well. We hope this algorithm will help scientists to perform some routine work effortlessly and conveniently.

References

1. Scarborough, J.B.: Numerical mathematical analysis. Oxford University Press, Oxford (1930)
2. Zielesny, A: From curve fitting to machine learning. ISBN: 978-3-319-32545-3, Springer International Publishing (2016). https://doi.org/10.1007/978-3-319-32545-3
3. David Garson, G.: Curve Fitting & Nonlinear Regression, ASIN: B00942WWA6, 3rd edn. Statistical Publishing Associates, Asheboro (2012)

A New Approach to Solve Fractional Logistic Growth Model and Its Numerical Simulation

Arnab Gupta

1 Introduction

In the last few decades, fractional-based calculus plays an important role to nurture various aspects of science and engineering field. The concept of fractional derivatives and integration has been studied and receives much attention and interests in the past twenty years; the reader may refer to [1–7] for the applications of fractional calculus. The fractional derivatives viz., Riemann–Liouville, Caputo, Weyl, and Grunworld–Letnikov are worth mentioning. However, Caputo (1967) reformulated the more classic definition of Riemann–Liouville fractional order derivatives in order to solve fractional differential equation (FDE) with integer order condition. Some applications have been studied in [8, 9]. The definitions of Riemann–Liouville and Caputo are as follows:

Definition 1.1 Riemann–Liouville Integration:
For $\alpha > 0$.

(i) *Forward Integration*:

$$ {}_aI_x^\alpha f(x) = \frac{1}{\Gamma(\alpha)} \int_a^x (x-u)^{\alpha-1} f(u) \mathrm{d}u $$

(ii) *Backward Integration*:

$$ {}_xI_b^\alpha f(x) = \frac{1}{\Gamma(\alpha)} \int_x^b (u-x)^{\alpha-1} f(u) \mathrm{d}u $$

A. Gupta (✉)
Department of Mathematics, Prabhu Jagatbandhu College, Andul-Mouri, Howrah-711302, India
e-mail: arnabgupta@pjc.ac.in

J. K. Mandal and J. K. Roy (eds.), *Proceedings of International Conference on Computational Intelligence and Computing*, Algorithms for Intelligent Systems,
https://doi.org/10.1007/978-981-16-3368-3_12

Definition 1.2 Fractional Derivatives:

Riemann–Liouville left hand definition (RL LHD):

Let us select an integer $m > \alpha$, α: fractional number such that

(i) Integrtate the function $(m - \alpha)$-folds in the sense of forward RL
(ii) Differentiate the above result by m

Then, RL LHD is defined as

$$ {}_0D_t^\alpha f(t) = \frac{d^m}{dt^m}\left[\frac{1}{\Gamma(m-\alpha)}\int_0^t \frac{f(u)}{(x-u)^{\alpha+1-m}}du\right] \quad (m-1 \le \alpha < m) $$

Definition 1.3 Caputo right hand definition (Caputo RHD):

Let us select an integer $m > \alpha$, α: fractional number such that

(i) Differentiate the functions m times
(ii) Integrate the above result $(m - \alpha)$-fold by RL LHD integration method.

Thus, Caputo's RHD is defined as

$$ {}_0^CD_t^\alpha = \frac{1}{\Gamma(m-\alpha)}\int_0^t \frac{\frac{d^m}{dt^m}f(u)}{(t-u)^{\alpha+1-m}}du \quad (m-1 \le \alpha < m) $$

But it is to be noted that Caputo derivative of constant function is zero, whereas the left RL derivative of a constant K is non-zero until the Jumarie derivative is developed [10]. Moreover, all the fractional derivative of order α neither satisfies the formula of derivative of the product or quotient of two functions nor the chain rule. Keeping that in mind, a new definition of fractional derivative called conformable fractional derivative of order α $(0 < \alpha \le 1)$ has been established [11–19]. This new definition is the generalization of the usual derivative viz., product/ quotient of two functions, chain rule, etc. Moreover, it is to be noted that using this definition, a modified law of mean (mean value theorem), Taylor's series expansion, etc., has been developed [17, 18].

The present paper highlights the solution of fractional order logistic growth model as an application of conformable fractional derivative. The logistic growth model was named in 1844–1845 by Pierre Francois Verhulst who interpreted it in relation to population growth [20]. The initial stage of growth is approximately exponential then as saturation begins, the growth slows and at maturity, growth stops. The logistic growth model can be applied in various science field viz., biology (especially ecology), biomathematics, geoscience, mathematical psycology, sociology, political science, etc. [21–23] The fractional order logistic equation has been studied by El-Sayed et. al. [24, 25]. Thus, the present paper discusses the solution of a modified fractional order logistic growth model using conformable fractional derivative. In this connection, the model is compared with a fractional order differential equation (where the derivative is conformable one), and a numerical simulation is done by choosing values of parameters present in the model for different values of α ($0 < \alpha < 1$).

The whole matter of the paper is organized in the following manner. Section 1 is the introductory one. Section 2 discusses a new definition of conformable fractional derivative. Some results related to it is also discussed in this section. In Sect. 3, fractional order logistic growth model is solved by using conformable fractional derivative. A numerical simulation is performed by taking different values of parameters involved in the model with different α's ($0 < \alpha < 1$).

2 Conformable Fractional Derivative [17]

We present the following new definitions and theorems of conformable fractional derivative of f of order α, where $0 < \alpha \leq 1$.

Definition 2.1 Let f: $[0, \infty) \longrightarrow R$ and $t > 0$. Then the conformable fractional derivative of f of order α is defined as.

$$T_\alpha(f)(t) = \lim_{\varepsilon \to 0} \frac{f(t + \varepsilon t^{1-\alpha})}{\varepsilon} \tag{2.1}$$

for all $t > 0$ and $0 < \alpha \leq 1$. If f is α-differentiable in some $(0, a)$, $a > 0$ and $\lim_{t\to 0+} f^{(\alpha)}(t)$ exists then we define $f^{(\alpha)}(0) = \lim_{t\to 0+} f^{(\alpha)}(t)$.

The conformable derivative of order α is often denoted by $f^{(\alpha)}(t)$ or $T_\alpha(f)(t)$, and we simply say f is α-differentiable if T_α(f)(t) exists. It is to be noted that this definition coincides with the classical definition of RL and Caputo type on polynomials (upto a constant multiple).

As a consequence of above definition, we obtain the following useful theorem.

Theorem 2.2 *If a function* f:$[0, \infty) \to R$ *is* α*-differentiable at* $t_0 > 0$, $\alpha \in (0, 1]$, *then* f *is continuous at* t_0.

Theorem 2.3 *Let* $\alpha \in (0, 1]$ *and* f, g be α*-differentiable at a point* $t > 0$. Then

(i) $T_\alpha(af + bg) = aT_\alpha(f) + bT_\alpha(g), \forall a, b \in R$
(ii) $T_\alpha(fg) = fT_\alpha(g) + gT_\alpha(f)$
(iii) $T_\alpha\left(\frac{f}{g}\right) = \frac{gT_\alpha(f) - fT_\alpha(g)}{g^2}$, *provided* $g \neq 0$
(iv) If f is differentiable, then $T_\alpha(f)(t) = t^{1-\alpha}\frac{\mathrm{d}f(t)}{\mathrm{d}t}$.

As a consequence of the above theorem, the following are the fractional derivatives of certain functions.

(i) $T_\alpha(t^p) = pt^{p-\alpha}, \forall p \in R$
(ii) $T_\alpha(\lambda) = 0$ for all constant functions $f(t) = \lambda$

(iii) $T_\alpha(e^{ax}) = ax^{1-\alpha}e^{ax}, \forall a \in R$
(iv) $T_\alpha(\sin bx) = bx^{1-\alpha}\cos bx, \forall b \in R$
(v) $T_\alpha(\cos bx) = -bx^{1-\alpha}\sin bx, \forall b \in R$
(vi) $T_\alpha\left(\frac{1}{\alpha}\right) = 1$
(vii) $T_\alpha\left(\sin\frac{1}{\alpha}t^\alpha\right) = \cos\frac{1}{\alpha}t^\alpha$
(viii) $T_\alpha\left(\cos\frac{1}{\alpha}t^\alpha\right) = -\sin\frac{1}{\alpha}t^\alpha$
(ix) $T_\alpha\left(e^{\frac{1}{\alpha}t^\alpha}\right) = e^{\frac{1}{\alpha}t^\alpha}$

Remark 2.4 It is to be noted that an α-differentiable function at a point may not be differentiable therein. For example, let us consider the function $f(t) = \sin\sqrt{t}$. Then, clearly, $T_{\frac{1}{2}}f(0) = \lim_{t\to 0} T_{\frac{1}{2}}f(t) = 1$, where $T_{\frac{1}{2}}f(t) = 1$ for $t > 0$. but $T_1 f(0)$ does not exists. This is not the case for the known classical derivatives.

Definition 2.5 Let $\alpha \in (n, n+1]$ and f be an n-differentiable function at t, where $t > 0$. Then the conformable fractional derivatives of f of order α is defined as.

$$T_\alpha f(t) = \lim_{t\to 0}\frac{f^{\lceil\alpha\rceil-1}(t+\varepsilon t^{\lceil\alpha\rceil-1})}{\varepsilon},$$

where $\lceil\alpha\rceil$: smallest integer greater than or equal to α.

Remark 2.6 From Definition 2.5, it can be shown that.

$$T_\alpha f(t) = t^{(\lceil\alpha\rceil-1)}f^{(\alpha)}(t)$$

where $\alpha \in (n, n+1]$ and f is $(n+1)$-differentiable at $t > 0$.

A modified version of mean value theorem and Rolle's like theorem can also be established in conformable fractional derivatives [17].

Theorem 2.7 (Mean Value like Theorem for Conformable Fractional Derivative) *Then there exists atleast one* $c \in (a, b)$ *such that* $f^{(\alpha)}(c) = \frac{f(b)-f(a)}{\frac{1}{\alpha}b^\alpha - \frac{1}{\alpha}a^\alpha}$.

As a special case of mean value theorem, we have Rolle's like theorem of conformable fractional derivatives.

Theorem 2.8 (Rolle's like Theorem for Conformable Fractional Derivative) *Let* $a > 0$ *and* f:$[a, b] \to R$ be a given function such that.

(i) f *is continuous on* $[a, b]$
(ii) f is *α-differentiable on* (a, b) for some $\alpha \in (0,1)$
(iii) $f(a) = f(b)$

Then there exists atleast one $c \in (a,b)$ such that $f^{(\alpha)}(c) = 0$.

The following proposition is true for conformable fractional derivatives of order α.

Proposition 2.9 Let f:[a,b] $\rightarrow R$ be α-differentiable for some $\alpha \in (0,1)$.

(i) If $f^{(\alpha)}$ is bounded on [a,b] where $a > 0$. Then f is uniformly continuous on [a,b] and hence f is bounded.
(ii) If $f^{(\alpha)}$ is bounded on [a,b] and continuous at a. Then f is uniformly continuous on [a,b] and f is bounded.

Note 2.10: The converse of the result need not be true. Let us consider the function $f(t) = 2\sqrt{t}$ on $I = [0, 1]$. Then, f is uniformly continuous on [0, 1], but $f'(t)$ is not bounded on I. However, boundedness of $f^{(\alpha)}(t)$ for $0 < \alpha < 1$ and the continuity of f on I (continuity of f at 0 in the subspace topology is equivalent to right continuity of f at 0), which implies by the above proposition, the uniform continuity of f on I.

2.1 *Fractional Integral [17]*

Definition 2.11 An α-fractional integral of a function f starting from $a \geq 0$ is defined by

$$I_\alpha^a f(t) = I_1^\alpha (t^{\alpha-1} f) = \int_a^t \frac{f(x)}{x^{1-\alpha}} \mathrm{d}x,$$

where the integral is the usual Riemann improper integral and $\alpha \in (0, 1)$.

Using the above definition, it follows that $I_{1/2}^0\left(\sqrt{t}\cos t\right) = \int_0^t \cos x \mathrm{d}x = \sin t$ and $I_{1/2}^0\left(\cos 2\sqrt{t}\right) = \sin 2\sqrt{t}$.

The following theorem is valid for fractional integral.

Theorem 2.12 *If f is continuous function in the domain of I_α, then $T_\alpha I_\alpha^a(f)(t) = f(t)$ for $t \geq a$.*

We will now solve a fractional logistic growth model using a new approach viz., conformable fractional derivative.

3 Solution of Fractional Logistic Growth Model Using Conformable Fractional Derivative

The fractional order logistic growth model can be represented as

$$D^\alpha x(t) = rx(2-\alpha)\left(1 - \frac{x}{K}\right) \tag{3.1}$$

where $t > 0$, $x(0) = x_0$, $r > 0$, $x_0 > 0$ and $0 < \alpha \leq 1$.

x: population size with respect to time,

r: rate of maximum population growth (intrinsic growth rate).

K: Carrying capacity.

The stability analysis of (3.1) has been thoroughly studied by El-Sayed et al. [24]. Moreover, the existence and uniqueness of the model have been established.

Solution of (3.1) using conformable fractional derivative:

Equation (3.1) can be rewritten as

$$x^{(\alpha)} = (2-\alpha)\left(rx - \frac{rx^2}{K}\right) \tag{3.2}$$

Let us find a solution of homogeneous equation $x^{(\alpha)} = 0, 0 < \alpha \leq 1$.

Let $x_h = e^{rt^{\frac{1}{\alpha}}}$ be a solution of $x^{(\alpha)} = 0$. Then, using the consequence of Theorem 2.3 (ix), we have $T_\alpha(e^{rt^{\frac{1}{\alpha}}}) = \frac{1}{\alpha} r e^{rt^{\frac{1}{\alpha}}}$.

Thus, $x_h^{(\alpha)} = 0$ implies that $r = 0$. Therefore $x_{\mathrm{h}} = 1$.

Now, $T_\alpha = L_\alpha = D^\alpha$.

$$x(t) = x_0 + L_\alpha^{-1}\left[rx(2-\alpha)\left(1 - \frac{x}{K}\right)\right]$$

Again, we have $x^{(\alpha)}(t) = t^{1-\alpha}x'(t) \Rightarrow rx(2-\alpha)\left(1-\frac{x}{K}\right) = t^{1-\alpha}x'(t) \Rightarrow \frac{K\,\mathrm{d}x}{rx(2-\alpha)(K-x)} = t^{\alpha-1}\mathrm{d}t$.

Integrating both sides the above equation leads to

$$x(t) = \frac{K}{1 + A\mathrm{e}^{-\frac{r(2-\alpha)}{\alpha}t^\alpha}} \quad (0 < \alpha \leq 1).$$

At $t = 0$, $x(t) = x_0$, which gives $A = \frac{K}{x_0} - 1$.

Thus, the solution of (3.2) using conformable fractional derivative is given by

$$x(t) = \frac{Kx_0}{x_0 + (K - x_0)e^{-\frac{r(2-\alpha)}{\alpha}t^\alpha}} \tag{3.3}$$

For $\alpha = 1$, the solution of logistic growth model is deduced, and when the time is very large, the density $x(t)$ tends to its carrying capacity K.

4 Numerical Simulation and Discussion

A numerical simulation results of Eq. (3.1) are done for the chosen parameter values $r = 0.5$, $K = 10$ and $\alpha = 0.8$. For numerical simulation, we have chosen $\Delta t = 0.2$ and checked the limiting behaviour of the trajectories for different choices of initial conditions. It is illustrated in Fig. 1. This clearly shows that the rate of convergence solution trajectories to different steady states is not the same for fractional order differential equation (where the derivative is conformable fractional derivative of order α), rather it depends upon the initial point.

Next, we fix the initial condition $x_0 = 0.5$, intrinsic growth rate $r = 0.8$, perform the numerical simulation for different values of α, K, say $\alpha = 0.25, 0.5, 0.75$, and examine the behavior of the trajectories starting from the initial point $x_0 = 0.5$, which is illustrated in Figs. 2, 3, and 4, respectively. This shows that the time required for the convergence of the trajectories increases to the stable steady state K ($K = 1, 5, 8$) as the values of α decreases ($\alpha = 0.75, 0.5, 0.25$). Finally, the population density is calculated for different values of α's. The value of intrinsic growth rate, carrying capacity and initial density at $t = 0$ are taken to be $r = 0.5$, $K = 1$, $x_0 = 0.85$, respectively. A time versus population density curve for the said parameter values

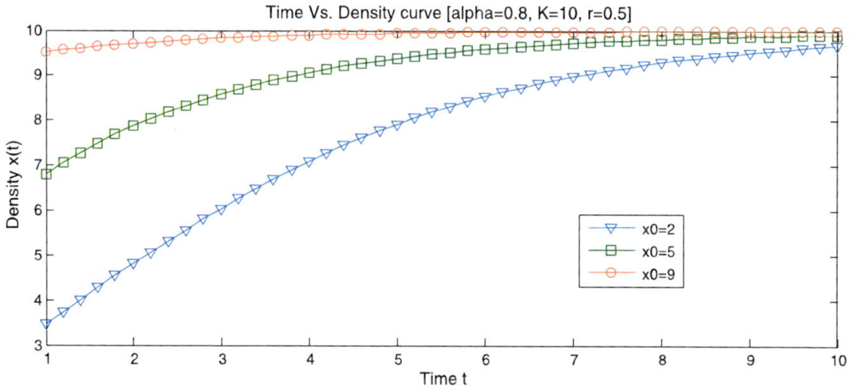

Fig. 1 Time t versus population density $x(t)$ starting from different initial conditions for $\alpha = 0.8$, $r = 0.5$, $K = 10$

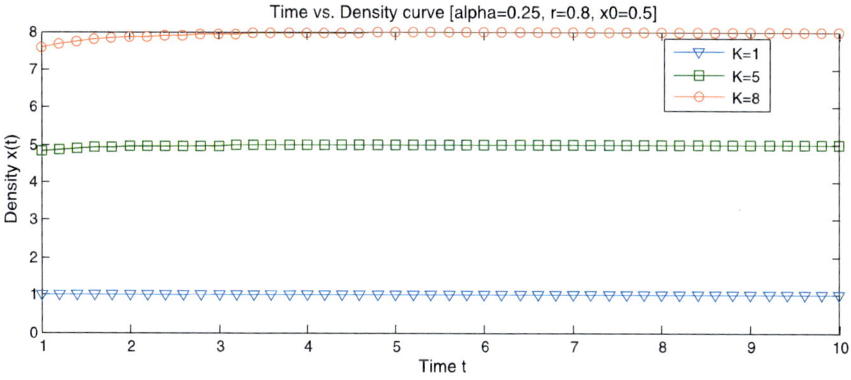

Fig. 2 Time t versus population density $x(t)$ for $\alpha = 0.25$, $r = 0.8$, $x_0 = 0.5$

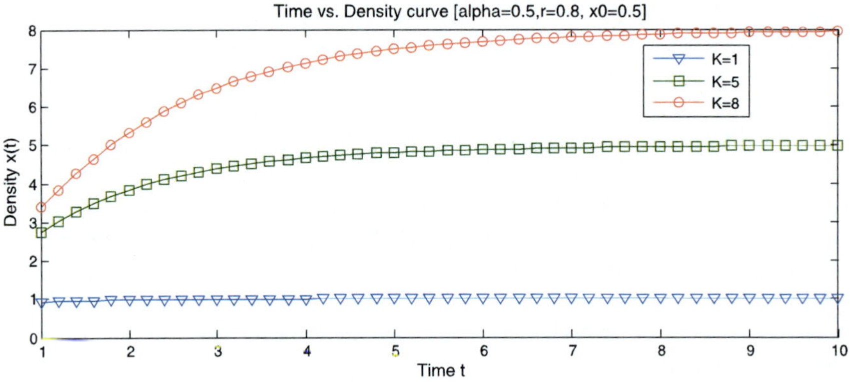

Fig. 3 Time t versus population density $x(t)$ for $\alpha = 0.5$, $r = 0.8$, $x_0 = 0.5$

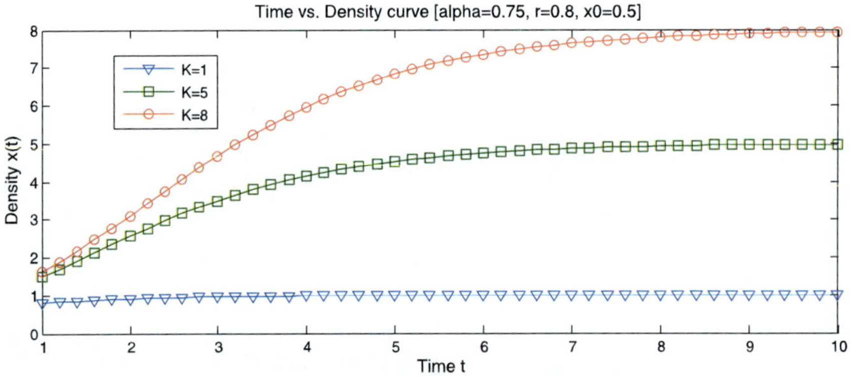

Fig. 4 Time t versus population density $x(t)$ for $\alpha = 0.75$, $r = 0.8$, $x_0 = 0.5$

is shown in Fig. 5. It is clear from Fig. 5, a sharp bent occurs for fractional logistic growth model at α which takes the value starting from 0.155,0.499,0.805,0.955, respectively.

A qualitative analysis viz., stability, asymptotic stability of fractional logistic growth model has been established in [25]. But in the present paper, an alternative solution of fractional logistic growth model is discussed using conformable fractional derivative. The delay effect of derivative is not taken to be considered here. However, it is demonstrated that using the capabilities of MATHEMATICA software, the population density calculation of the fractional logistic growth model (3.1) by taking several values of parameters is given in Table 1. Thus, using conformable fractional derivative, unlike biological model, other fractional differential equation viz., fractional mechanical oscillator, RC circuit, LC circuit, etc. could be solved

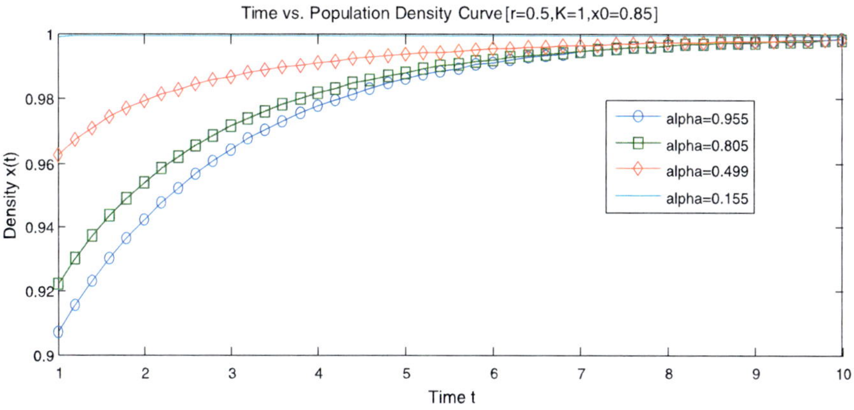

Fig. 5 Solution trajectories converging to $K = 1$ from $x_0 = 0.85$ and different values of α, say $\alpha = 0.955, 0.805, 0.499, 0.155$

using conformable fractional derivative. This indicates the strong influence of population density in the past history as α is away from 1. The closure to the values of 1 indicates that the influence of history is minimal.

Table 1 Density calculations for the model (3.1) with different values of α's with parameter values $r = 0.5$, $x_0 = 0.85$, $K = 1$ [Time step $\Delta t = 0.5$]

α	Time t	Density $x(t)$
0.25	1.0	0.994699
	1.5	0.996345
	2.0	0.997259
	2.5	0.99784
	3.0	0.99824
	3.5	0.998531
	4.0	0.998751
	4.5	0.998923
	5.0	0.99906
	5.5	0.999171
	6.0	0.999263
	6.5	0.99934
	7.0	0.999406
	7.5	0.999462
	8.0	0.99951
	8.5	0.999552
	9.0	0.999589
	9.5	0.999622
	10.0	0.99965

(continued)

Table 1 (continued)

α	Time t	Density $x(t)$
0.5	1.0	0.962116
	1.5	0.972661
	2.0	0.979284
	2.5	0.983798
	3.0	0.987038
	3.5	0.989448
	4.0	0.991291
	4.5	0.992729
	5.0	0.993872
	5.5	0.994792
	6.0	0.995543
	6.5	0.996162
	7.0	0.996676
	7.5	0.997107
	8.0	0.997471
	8.5	0.99778
	9.0	0.998043
	9.5	0.99827
	10.0	0.998466
0.75	1.0	0.928769
	1.5	0.946043
	2.0	0.958357
	2.5	0.967435
	3.0	0.974274
	3.5	0.979508
	4.0	0.983563
	4.5	0.986735
	5.0	0.989239
	5.5	0.991229
	6.0	0.99282
	6.5	0.994101
	7.0	0.995135
	7.5	0.995975
	8.0	0.996661
	8.5	0.997221
	9.0	0.997682
	9.5	0.998062
	10.0	0.998375

References

1. Kilbas, A.A., Srivastava, H.M., Trujillo, J.J.: Theory and applications of fractional differential equations. North Holland Mathematical Studies, 204, Elsevier North Holland Science Publishers, Amsterdam (2006)
2. Cattani, C., Srivastava, H.M., Yang, X.J.: Fractional Dynamics. Emerging Science Publishers (De Gruyter Open), Berlin (2015)
3. Podlubny, I.: Fractional differential equations. Math. Sci. Eng. **198** (1999)

4. Miller, K., Ross, B.: An Introduction to the Fractional Calculus and Fractional Differential Equations. Wiley, New York (1993)
5. Oldham, K., Spanier, J., Ross, B.: The Fractional Calculus: Theory and Applications of Differentiation and Integration to Arbitary Order. Academic, London (1974)
6. Agrawal, O.P., Tenreiro-Machado, J.A., Sabetier, I.: Fractional derivatives and their applications. Nonlinear Dynam. **38** (2004)
7. Das, S.: Functional Fractional Calculus. Springer, Berlin (2011)
8. Gupta, A., Biswas, S.: A new path to reach the solution of fractional mechanical oscillator using adomain decomposition method. Jnanabha **II**, 267–276 (2017)
9. Gupta, A., Roy, J.K., Datta, D.: Fractional calculus in solving engineering problems. In: Proceedings of International Conference on Computing, Communication and Manufacturing, pp 40–47 (2014)
10. Jumarie, G.: Modified Riemann-Liouville derivative and fractional Taylor series of nondifferentiable functions, further results. Comput. Math. Appl. **51**, 1367–1376 (2006)
11. Gokdogan, A., Unal, E., Celik, E.: Existence and uniqueness theorems for sequential linear conformable fractional differential equations. Miskolc Math. Notes **17**, 267–279 (2016)
12. Atangana, A., Baleanu, D., Alsaedi, A.: New properties of conformable derivative. Open Math. **13**, 889–898 (2015)
13. Anderson, D.R., Ulness, D.J.: Newly defined conformable derivatives. Adv. Dyn. Syst. Appl. **10**, 109–137 (2015)
14. Unal, E., Gokdogan, A.: Solution of conformable fractional ordinary differential equations via differential transform method. Opt. Int. J. Light Electron Opt. **128**, pp. 264–273
15. Karayer, H., Demirhan, D., Buyukkilic, F.: Conformable fractional Nikiforov-Uvarov method. Commun. Theor. Phys. **66**, 12–18 (2016)
16. Acan, O., Firat, O., Keskin, Y., Oturanc, G.: Solution of conformable fractional partial differential equations by reduced differential transform method. Selcuk J. Appl. Math. (2016)
17. Khalil, R., Al Horani, M., Yousef, A., Sababheh, M.: A new definition of fractional derivative. J. Comput. Appl. Math. **264**, 65–70 (2014)
18. Abdeljawad, T.: On conformable fractional calculus. J. Comput. Appl. Math. **279**, 57–66 (2015)
19. Cenesiz, Y., Kurt, A.: The solutions of time and space conformable fractional heat equations with conformable Fourier transform. Acta Univ. Sapientiae, Math. **7**, 130–140 (2015)
20. Feller, W.: On the logistic law of growth and its empirical verification. Acta. Biotheor. **5**, 51–56 (1940)
21. Cushing, J.M.: An Introduction to Structured Population Dynamics, vol. 71. SIAM, Philadelphia (1998)
22. Sauer, T.D., Yorke, J.A.: An Introduction to Dynamical Systems. Springer, New York (1996)
23. Ausloos, M.: The Logistic Map and the Route to Chaos: From the Beginnings to Modern Applications. Springer, Berlin (2006)
24. El-Sayed, A.M.A., El-, A.E.M., El-Saka, H.A.A.: On the fractional-order logistic equation. Appl. Math. Lett. **20**(7), 817–823 (2007)
25. Suansook, Y., Paithoonwattanakij, K.: Dynamic of logistic model at fractional order. In: Proceedings of the IEEE International Symposium on Industrial Electronics (IEEE ISIE '09), pp. 718–723 (2009)

A Smart Flow Transmitter Using Ultrasonic Sensors

Praveen Maurya, S. F. Ali, and N. Mandal

1 Introduction

The application of ultrasonic sensors has been growing very fast in recent years, which is widely used in chemical, petroleum, fertilizer, food, dyes, light industry and scientific research in various sectors. Due to the high accuracy, the transit time flow meter is used as a measuring instrument. In conventional glass rotameter [1–3] without transmission system, flow rate can be measured in the industrial field by matching the float tip and scale present on the glass tube. A number of new and modified techniques on the ultrasonic sensor for the measurement of the flow rate of liquid are reported recently. Willatzen [4] has analysed different mathematical procedure to determine the flow rate of the liquid and compared with the transit time method which is used in the ultrasonic-based flow measurement system. Tanyildizi et al. [5] have discussed different manufacturing techniques, design and working principle of two different flow measuring instrument rotameter and venturi tubes. Zhu et al. [6] have developed a mathematical model that has been done in three steps and used to analyse the relationship between the gas flow rates, excitation signals, propagation time, echo amplitudes, etc. Fang et al. [7] have developed an algorithm to recover the ultrasonic pulse from the noise for accurate calculation of time of flight during flow measurement of liquid.

Goh et al. [8] have done the review on ultrasonic tomography (UT)-based flow measurement system. They discussed this for the application of flow measurement in the multiphase system.

In this paper, we have presented a modified rotameter which is able to convey the signal corresponding to the flow rate in electrical form. The modification in the rotameter has been done with the help of a rod attached between a circular disc

P. Maurya (✉) · S. F. Ali · N. Mandal
Department of Electronic Engineering, Indian Institute of Technology (ISM) Dhanbad, Dhanbad, India

J. K. Mandal and J. K. Roy (eds.), *Proceedings of International Conference on Computational Intelligence and Computing*, Algorithms for Intelligent Systems,
https://doi.org/10.1007/978-981-16-3368-3_13

and the float of the rotameter. The movement of the circular disc is sensed and measured with the help of an ultrasonic sensor and microcontroller Arduino board. Arduino board has been used for triggering the ultrasonic sensor and receiving the output of the sensor which is further processed by microcontroller present in Arduino board. The whole system is fabricated, and we have developed a prototype model. After fabrication, the unit has been experimentally tested. We have also given details experimental results in this paper.

2 Theory of Operation

In this technique, we used a guided variable area type flow meter, i.e. rotameter. A carbon fibre rod is attached with rotameter's float, and the other end is attached with a lightweight circular disc made of polyvinyl chloride as shown in Fig. 1a. The rod is properly sealed so that it can move freely inside the sensor holding chamber. When the flow is zero, the distance between circular disc and sensor is maximum. The variation of different parameters with flow rate is shown in Fig. 1b.

The displacement of the float with respect to the flow rate can be written as

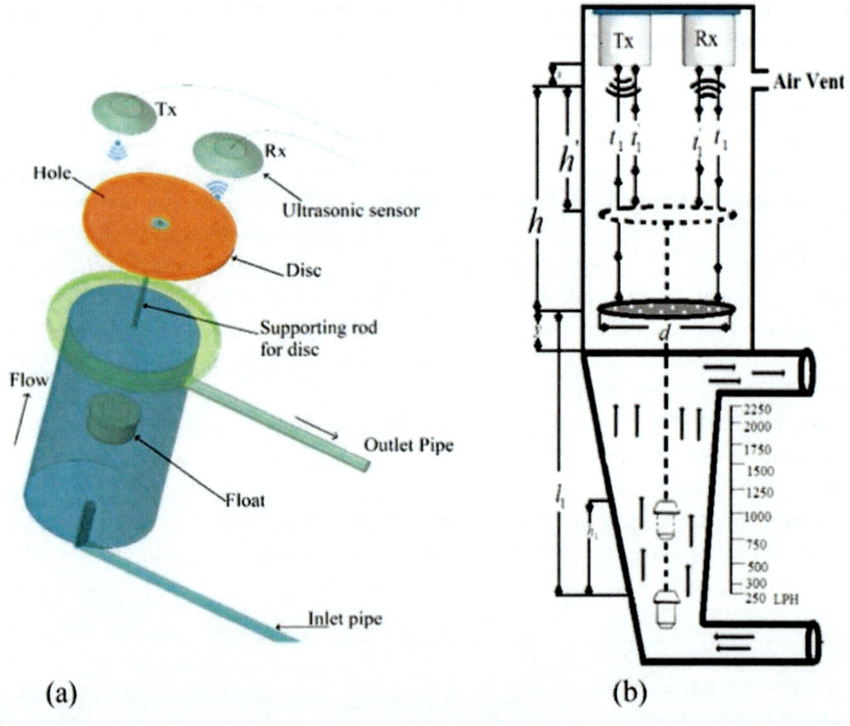

Fig. 1 **a** Schematic diagram of modified rotameter. **b** Variation in the parameter with the flow

$$h_1 = K_1 Q \tag{1}$$

Let at flow rate Q, the position of the bob from its rest location is h_1 and voltage generated by the circuit increases with the change in the height due to change in flow rate. As the flow increases, the position of the float increases from the reference level so the attached rod with rotameter and the circular disc also lifts up. The change in the distance is observed by the ultrasonic sensor. The distance between the ultrasonic sensor and circular disc decreases as the flow increases as shown in Fig. 1b. The time taken by the ultrasonic sound pulse to reach the disc is t_1 when the gap between the ultrasonic sensor and disc is h which corresponds to the zero flow rate of the liquid. As flow rate increase to Q, the position of the float is h_1 and separation between the sensor and disc decreases up to h' time is t_1'. Therefore, from the principle of ultrasonic sensor, t_1 and t_1' can be expressed as

$$t_1 = h/v, \quad t_1' = h'/v, \quad t_1 > t_1' \tag{2}$$

where v is the velocity of the sound in air 340 m/s, h and h' are the positions of the circular disc in cm at flow rate zero and Q, respectively. So round trip time is

$$T_1 = t_{1Tx} + t_{1Rx}, \quad t_{1Tx} = t_{1Rx} = t_1 \tag{3}$$

So, the total time is taken to receive the sound pulse

$$T_1 = t_1 + t_1 \Rightarrow T_1 = 2t_1 \tag{4}$$

Now, putting the value of t_1 from Eq. (2) in Eq. (4)

$$T_1 = 2h/v \Rightarrow h = T_1 0.0034/2 \text{ cm} \tag{5}$$

In similar way, when the flow rate is Q, the total time taken by the ultrasonic pulse is

$$\text{Or,} \quad T_2 = 2h'/v \Rightarrow h' = T_2 0.0034/2 \tag{6}$$

$$h' = K_3 Q \tag{7}$$

This displacement is converted into a voltage signal with the help of pulse width modulation (PWM). The output voltage of PWM varies with the distance between disc and ultrasonic trans-receiver and is measured with the help of microcontroller.

The distance also linearly varies with the flow rate which can be written as,

$$V = K_2 K_3 Q \tag{8}$$

$$V = K_4 Q \tag{9}$$

where K_2, K_3 and $K_4 = K_2K_3$ are the constant terms.

Equation (9) shows the linear relationship between output voltage V and flow rate of fluid Q. The output voltage of Arduino UNO board is intensified using IA and amplified signal is transformed into current signal in the range of 4–20 mA current signal which is suitable for industrial transmission.

3 Experimental Results

The experiment has been performed in the experimental set up as shown in Fig. 2. Three steps have been followed to perform this experiment. The first step is responsible for the conversion of the flow rate in terms of the electrical signal with the help of an ultrasonic sensor, and the second part is with the help of Arduino board which consists of an Atmega16 microcontroller. The maximum distance is detected by the ultrasonic sensor is 21 cm when the flow rate is 0 LPH, and the minimum distance is detected by the sensor is 1 cm when the flow rate is highest which is 2250 LPH. The variation in the separation of the disc towards the ultrasonic sensor is shown in Fig. 3a. The programming in the microcontroller is done in such a way that the distance is converted into its equivalent voltage. The output voltage 57.4 mV is maximum when the flow rate is minimum 0 LPH, and voltage is continuously decreasing with the increased flow rate as shown in Fig. 3b.

In the second step, the signal conditioning circuit is used for the standardization of the output voltage of the transducer. It has been performed with the help of an instrumentation amplifier and Op-amp circuits. The output of the signal conditioning

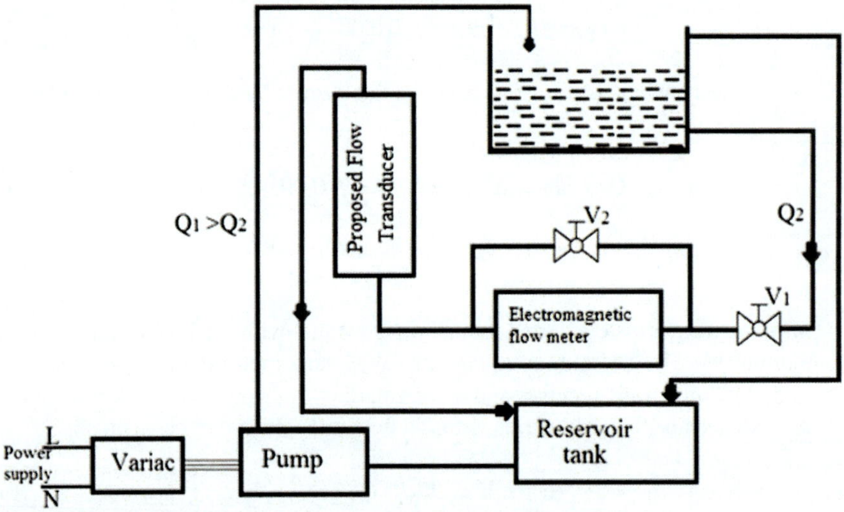

Fig. 2 Experimental set up

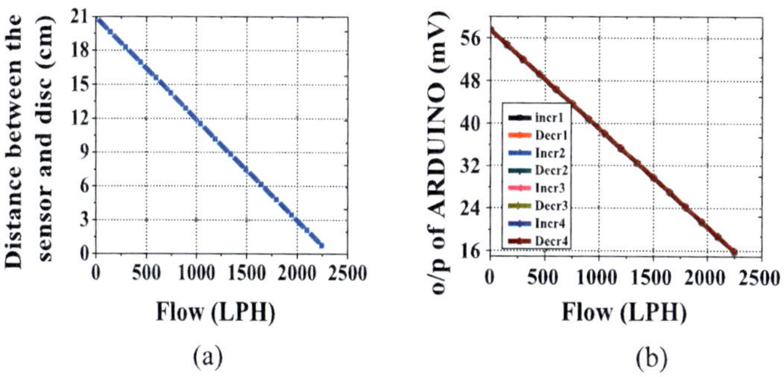

Fig. 3 **a** Disc position towards ultrasonic sensor versus flow rate. **b** The output voltage of Arduino (mV) w.r.t. the flow rate (LPH)

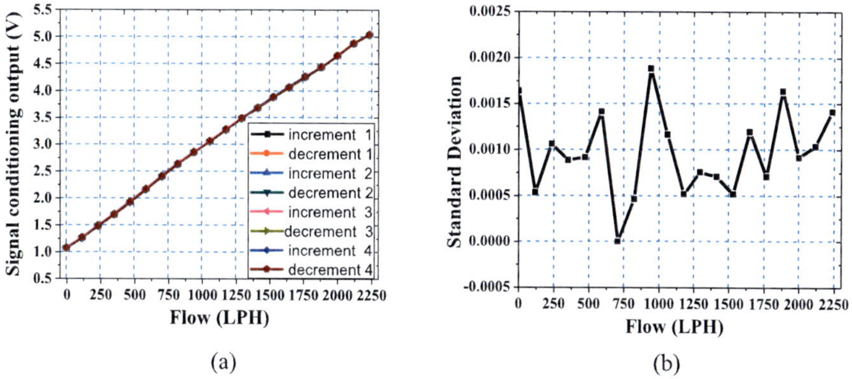

Fig. 4 The static characteristics graph of signal condition circuit. **a** Flow transducer characteristic. **b** Standard deviation curve for eight repeated experiments

circuit is shown in Fig. 4a. The standard deviation has been drawn and shown in Fig. 4b range from 0 to 0.0020.

In the third step, for the transmission of the measured information to the remote area or process location through a conductive wire, a *V–I* circuit is required. The characteristic of current with respect to the flow rate increases linearly with the flow as shown in Fig. 5a. The standard deviation has been shown in Fig. 5b range from 0 to 0.0013.

4 Discussions and Conclusion

From Fig. 3a, it has been observed that the distance between circular disc and ultrasonic sensor linearly decreases as the flow rate increases. It shows the upliftment of float in a rotameter due to increase in flow rate. The output of Arduino board in mV

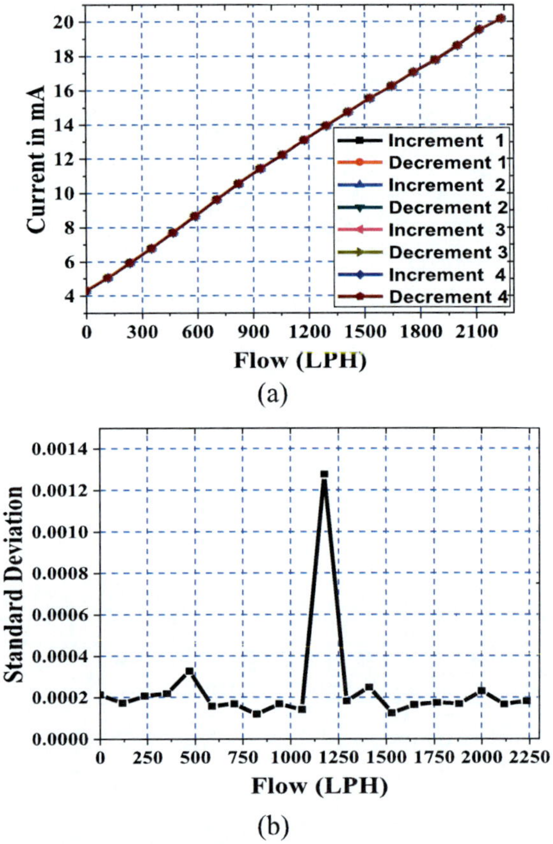

Fig. 5 Static characteristic graph of proposed flow transmitter. **a** Flow transmitter characteristic. **b** Standard deviation curve for eight repeated experiments

is inversely proportional to the flow rate shown in Fig. 3b. Signal conditioning circuit has been used for converting the voltage signal obtained from Arduino board in a standard voltage 1–5 V form which is shown in Fig. 4. The output of Arduino board and signal conditioning circuit give, respectively, linearly decreasing and increasing relationship with flow rate. For transmitting measured information to the control room, the current conversion is required which has been done by *V–I* converter circuit. The characteristic of this circuit is also increased linearly with respect to flow rate as shown in Fig. 5. The whole system is designed for long distance transmission purpose from field to control room and not affected by electromagnetic radiation. The measurement of the flow rate of the fluids which are not transparent is also possible with the help of the proposed technique.

References

1. Baker, R.: Flow Measurement Handbook: Industrial Designs, Operating Principles, Performance, and Applications. Cambridge University Press, Cambridge, UK (2000)
2. Doeblin, E.O.: Measurement System Application and Design, 4th edn. McGraw-Hill, New York (1990)
3. Dunn, W.C.: Fundamentals of Industrial Instrumentation and Process Control. Tata McGraw Hill, New York (2009)
4. Willatzen, M.: Flow acoustics modelling and implications for ultrasonic flow measurement based on the transit-time method. Ultrasonic **41**, 805–810 (2004)
5. Tanyildizi, V., Eren, H.: A new production technique for rotameters and venturimeters. Measurement **39**, 674–679 (2006)
6. Zhu, W., Xu, K., Fang, M., Wang, W., Shen, Z.: Mathematical modeling of ultrasonic gas flow meter based on experimental data in three steps. IEEE Trans. Instrum. Meas. **65**(8), 1726–1738 (2016)
7. Fang, Z., Hu, L., Qin, L., Mao, K., Chen, W., Fu, X.: Estimation of ultrasonic signal onset for flow measurement. Flow Meas. Instrum. **55**, 1–12 (2017)
8. Goh, C., Ruzairi, A., Hafiz, F., Tee, Z.: Ultrasonic tomography system for flow monitoring: a review. IEEE Sens. J. **1**(17), 5382–5390 (2017)

Design and Development of Bending Sensor-Based Pressure Transducer

Anamika Lata and Nirupama Mandal

1 Introduction

Accurate pressure measurement is one of the vital and required parameters for any process variable based industry. The pressure sensors are categories into two classes one is mechanical, and the other is electromechanical. In both the pressure sensors modules, the applied pressure is transformed into the displacement of the sensing part [1]. There are numerous methods available for transferring the pressure reading to a remote site, and various researchers still work on it. Bakhoum et al. [2] developed a capacitive pressure sensor with an enormous dynamic range. In this technique, the electrode's surface area changes with the application of the pressure. A pressure sensor using a Fibre Bragg grating and primary sensor as a metal diaphragm was developed in [3]. The proposed sensor has good repeatability and having a negligible hysteresis error. Bera et al. [4] have designed a pressure transmitter by an upgraded inductance bridge and Bourdon tube as a primary sensor. In this method, an inductive pick-up type sensor is attached to the tip of the Bourdon tube. By applying the pressure across the one end of the Bourdon tube, the ferromagnetic wire joined to the one tip of the Bourdon tube move inside the inductive pick up, which alters the self-inductance of the coil. The variation in self-inductance of coil is converted into voltage form with the support of enhanced type inductance measuring bridge. Chattapadhyay et al. [5] proposed and developed a pressure transmitter using a Bourdon tube as a primary sensor, and reluctive pick-up has been used as a secondary sensor for changing the displacement of primary pressure sensor into electrical output. A pressure transmitter using a bellow as a primary sensor has been designed by Kumar et al. [6]. The inductive pick-up has been used to change the bellow's mechanical movement into an electrical signal in this technique. To improve the linearity of the proposed pressure transmitter, artificial neural network (ANN) has been used in this method. Singh et al.

A. Lata (✉) · N. Mandal
Department of Electronics Engineering, Indian Institute of Technology (ISM), Dhanbad, India

J. K. Mandal and J. K. Roy (eds.), *Proceedings of International Conference on Computational Intelligence and Computing*, Algorithms for Intelligent Systems,
https://doi.org/10.1007/978-981-16-3368-3_14

[7] have developed the pressure transmitter using a bellow and capacitive sensor, and its sensing results gets transmitted to the remote region by FSK transmitter.

In this article, the bellow, which is a mechanical type pressure sensor, is used as the main sensor and the bending sensor is used as a secondary one. It is one of the simplest and cost-effective methods for transmitting the mechanical-based pressure sensor readings to remote locations. One end of the rectangular-shaped PVC flexible cantilever beam is attached below, and the other end of flexible support is fixed. The adjustable support is connected with the bellow and fixed support in a circular form with specific curvature and angle. The bending flex sensor is attached at the top of the cantilever beam. With the application of pressure, the bellow movement changes the curvature of a flexible cantilever beam, which in turn changes the flex sensor bending angle. With the change in the flex sensor bending angle, the resistance of the flex sensor gets changes. The resistance of the flex sensor gets converted into a voltage signal using resistance to voltage converter. The theoretical equations completely follow the experimental result.

2 Method of Approach

The application of applied pressure across the bellow causes the cantilever beam's movement whose one end is fixed and shaped deformed. The beam's action changes the resistance of the flex sensor, which is placed above the beam. So, the applied pressure is directly related to the change in resistance.

$$P_1 = k_1 R_{\text{Flex\_shaped}}(\alpha) \tag{1}$$

For the flat, ($\alpha = 0°$) the rectangular sensor of ($A \times T$).

Where A is the length of the rectangular sensor, and T is the active layer width. The total resistance of the flat flex sensor $R^{0°}_{\text{Flex}}$ is calculated by sheet resistance. $R^{0°}_{\text{Flex\_ sheet}}$, and it is expressed as,

$$R^{0°}_{\text{Flex}} = R^{0°}_{\text{Flex\_ sheet}} \frac{A}{T}$$

Let us consider that there is a change in the shape of the active layer of flex sensor, and it is described with its width $T(x)$ and it is expressed as

$$R^{0°}_{\text{Flex\_ shaped}} = R^{0°}_{\text{Flex\_ sheet}} \int_0^A \frac{1}{T(x)} \mathrm{d}x \tag{3}$$

When the substrate of the flex sensor gets to bend, the active layer gets stretched accordingly. The sheet resistance of the flex sensor rises around the bend axis. Let consider that the flex sensor sheet resistance increases around the rotational axis A_M. A_M is occurred in the middle of the sheet strip.

With the application of pressure, the bellow moves, and it causes the bending of the flexible cantilever beam and its shape changes, and it expressed through a Gaussian function,

$$G(x,\sigma) = \frac{1}{\sqrt{2\Pi\sigma^2}} \mathrm{e}^{\frac{-(x-A_M)^2}{2\sigma^2}} \tag{4}$$

The rotational axis position moving away from the half of rotation arc, the shift in axis on the resistive layer of flex sensor is calculated as,

$$C(\alpha) = \frac{1}{2}\frac{\alpha}{180^\circ} \times \frac{\Pi_S}{2} \tag{5}$$

Then, the total sheet resistance of the flex sensor is expressed as,

$$R_{\text{Flex\_ sheet}}(x,\alpha) = R^{0^\circ}_{\text{Flex\_ sheet}} + y(\alpha)G[x - A_M - C(\alpha), \sigma(\alpha)] \tag{6}$$

Let us assumed that the variance of the Gaussian function is equal to the shift axis, then the resistance of a rectangular flex sensor is expressed in terms of bending angle and is expressed as,

$$\begin{aligned} R_{\text{Flex\_shaped}}(\alpha) &= \frac{1}{T}\int_0^A R_{\text{Flex\_sheet}}(x,\alpha)\mathrm{d}x \\ &= \frac{R^{0^\circ}_{\text{Flex\_sheet}}}{T}(A) + \frac{y(\alpha)}{T}\int_0^A G[x - A_M - C(\alpha)]\mathrm{d}x \end{aligned} \tag{7}$$

We have,

$$\int_0^A G[x - A_M - C(\alpha)]\mathrm{d}x \cong 1 \tag{8}$$

$$R_{\text{Flex\_ shaped}}(\alpha) = R^{0^\circ}_{\text{Flex}} + \frac{y(\alpha)}{T} \tag{9}$$

The response of the flex sensor with the movement of bellow in respect of the applied pressure is non-uniform and is expressed as,

$$R_{\text{Flex\_ shaped}}(\alpha) = \int_0^A R_{\text{Flex\_ sheet}}(x,\alpha)\frac{1}{T(x)}\mathrm{d}x \tag{10}$$

$$R_{\text{Flex\_ shaped}}(\alpha) = R^{0^\circ}_{\text{Flex\_ shaped}} + y(\alpha)E_{\text{geom}}(\alpha) \tag{11}$$

From the above equation, it has been concluded that the deformed shape of the flexible cantilever beam on which the flex sensor is placed causes the change in resistance, and the bending resistance is a function of the bending angle. The movement causes the bending of the cantilever beam, which causes alterations in the flex sensor resistance. The variation in flex sensor resistance is transformed into a voltage signal using resistance to voltage converter and is expressed as:

$$V_{\text{Out\_ shaped}}(\alpha) = \left(g_1 - g_2 R_{\text{Flex\_ shaped}}(\alpha)\right) \tag{12}$$

The output voltage of resistance to voltage converter is related to the alteration in the flex resistance is examined from Eq. (12).

3 Experimental Result

The experiment is executed and conducted in two stages in the experimental arrangement, and it is presented in Fig. 1a, b. In the initial phase of the investigation, the variable pressure is applied across the bellow, which is joined to the deadweight tester. The displacement of bellow makes the deformation in the shape of a flexible cantilever beam; with the alteration in the cantilever beam's form, the flex sensor changes the bending angle, which changes the flex sensor's resistance. The variation in resistance with the deviation in pressure is measured with the help of 4 and 1/2 digital multimeter. This step is repeated six times. The characteristics of

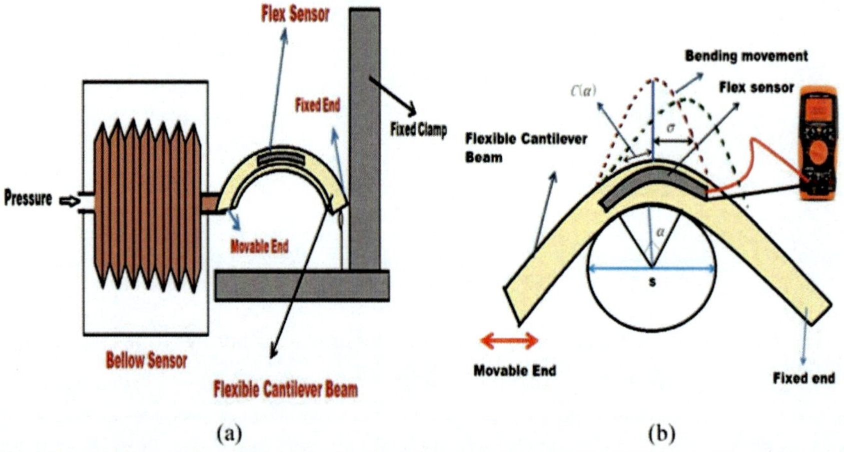

Fig. 1 **a** Graphical diagram of proposed pressure transducer. **b** Arrangement of flex sensor on the flexible cantilever beam

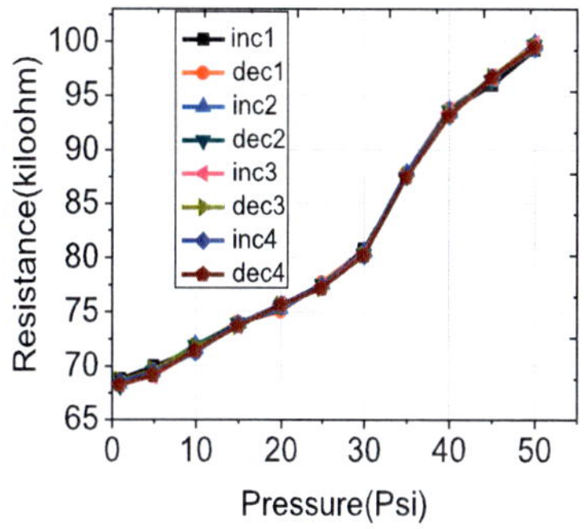

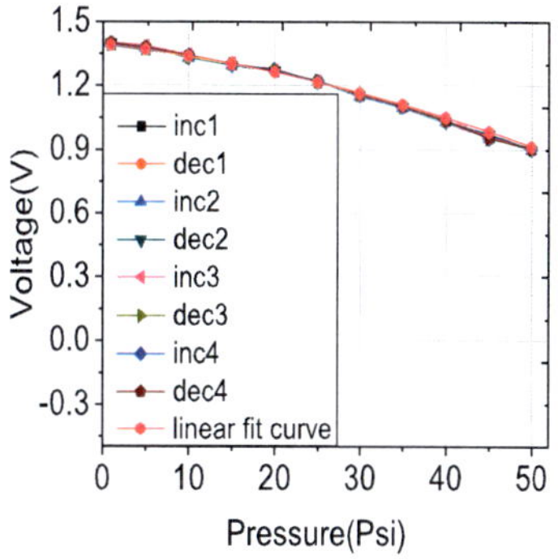

Fig. 2 **a** Characteristics curve of the proposed pressure sensor. **b** Characteristics curve of the proposed pressure transducer

the proposed pressure sensor have been obtained by plotting the curve between the change in flex sensor resistance against the change in applied pressure, and it is displayed in Fig. 2a. In the second stage of the experiment, the flex sensor terminal is joined to the inverting station of resistance to voltage converter is measured by 4 and 1/2 multimeter. The graph between the measured output voltages of resistance to voltage converter against the applied pressure has been drawn to obtain the proposed pressure transducer's characteristics, and it is displayed in Fig. 2b.

4 Discussion

Figure 2a shows the characteristics of the proposed pressure sensor. It shows that the proposed pressure sensor output rises with the increment in the pressure. With the increment in the applied pressure, the flexible cantilever beam gets bend from its central axis, leading to the alteration in the flex sensor bending angle placed above the beam. The flex sensor resistance increases with the increase in the bending angle. The characteristic of the proposed pressure sensor is nonlinear, and the percentage deviation lies between ±6%. The output of the proposed pressure sensor is fed to the *R* to *V* converter. The flex sensor is coupled at the inverting terminal of the *R* to *V* converter in series with the linearization resistance. The curve is shown in Fig. 2b, and it concludes that the output voltage of the proposed pressure transducer decreases with the increase of applied pressure across the bellow.

5 Conclusion

The proposed system is a simple, cost-effective type of pressure transducer. The displacement of pressure-sensitive bellows has been easily transformed into electrical signals using a flex sensor as a secondary sensor. The proposed pressure transducer has a linearity of $\pm 2.25\%$ with acceptable repeatability of 1.12%. The percentage of non-linearity may be decreased by using an artificial neural network, microcontroller and a piecewise linearization technique.

References

1. Liptak, B.G.: Process Measurement and Analysis, 3rd edn. Butterworth-Heinemaan, London, UK (1999)
2. Bakhoum, E.G., Cheng, M.H.M.: Capacitive pressure sensor with very large dynamic range. IEEE Trans. Compon. Packag. Technol. **33**(1), 79–83 (2010)
3. Ahmad, H., Chong, W.Y., Thambiratnam, K., Zulklifi, M.Z., Poopalan, P., Thant, M.M.M., Harun, S.W.: High sensitivity fiber Bragg grating pressure sensor using thin metal diaphragm. IEEE Sens. J. **9**(12), 1654–1659 (2009)
4. Bera, S.C., Mandal, N., Sarkar, R.: Study of a pressure transmitter using an improved inductance bridge network and bourdon tube as transducer. IEEE Trans. Instrum. Meas. **60**(4), 1453–1460 (2011)
5. Chattopadhyay, S., Sarkar, J.: Design and development of a reluctance type pressure transmitter. In: 2012 7th International Conference on Electrical and Computer Engineering, pp. 70–73. Dhaka, Bangladesh, 20–22 Dec 2012
6. Kumar, V.N., Narayana, K.V.L.: Development of an ANN-based pressure transducer. IEEE Sens. J. **16**(1), 53–60 (2016)
7. Sinha, S., Kachhap, R.V., Mandal, N.: Design and development of a capacitance based wireless pressure transmitter. IET Sci. Meas. Technol. **12**(7), 858–864 (2018)

Adaptive Time Duration Computation for Parallel Arc Fault in Wind-Solar Hybrid System

Debopoma Kar Ray, Tamal Roy, and Surajit Chattopadhyay

1 Introduction

Renewable power generations are of great concern in recent days due to the reduced future of conventional energy resources in the coming 50 years. Thus the condition monitoring of these systems is of utmost importance in recent days to ensure quality supply of renewable power in industrial, commercial, and domestic sector requirements. A recent study has shown the effect on solar modules at different inclination angles, fill factors, and seasonal effects, wherein it has been seen that 33.74° title angle of modules receives higher solar radiation [1]. Also, the study depicts the occurrence of decreased module efficiency and increased module temperature at increased solar radiances. Phase change material (PCM) integrated solar collectors have been seen with an increased contact surface for enhancing outlet temperature [2]. A comprehensive review has been seen to discuss the performance, fabrication, and characterization of a solar energy system based on mono nano-fluids with reports of increasing solar cell efficiency in due course. B ut it has also been seen that instability, increased friction factor, rheological issues, and increased pumping factor have restrained the use of these systems in large scale commercial applications [3]. A comprehensive study of thermoelectric generator (TEG) configuration for various type of solar thermal systems have been seen, wherein technologies have been discussed for improving the efficiency of TEGs [4]. The design of a 1-D high-temperature solar receiver has been seen for application in micro gas turbines with integrated process control monitoring (PCM), enabling variation of different geometrical factors for calculation of heat losses in PCM [5]. A vivid study has been seen to justify the effectiveness of nanofluids in efficient solar systems, wherein the flow

D. K. Ray (✉) · T. Roy
Department of Electrical Engineering, MCKV Institute of Engineering, Howrah, WB, India

S. Chattopadhyay
Department of Electrical Engineering, GKCIET, Malda, WB, India

J. K. Mandal and J. K. Roy (eds.), *Proceedings of International Conference on Computational Intelligence and Computing*, Algorithms for Intelligent Systems,
https://doi.org/10.1007/978-981-16-3368-3_15

rate, concentration, size, and type of nanofluids have been investigated to ascertain the efficiency of solar systems. Also, the effect of nanoparticles on various factors has been seen to be discussed in the study [6]. Performance and efficiency analysis of photovoltaic modules have been seen to be improved using PV modules, wherein compared to conventional modules, the cooling modules have been seen to show 6–7.2% higher electrical efficiency and performance ratio [7]. Performance analysis of a polished parabolic trough stainless steel solar cooker has been seen, wherein theoretical efficiency ranges between 50 and 30% and experimental efficiency between 38 and 5% with a maximum achieved water temperature of 53.6 °C at stagnated condition [8]. Performance analysis of a solar air heater encompassing paraffin wax and aluminum has been observed at different surface temperature conditions [9]. A cost-based allocation model has been seen for a solar-wind-thermal generation system to cater to load demands in a robust way [10]. A constraint-based iterative search algorithm has been observed for optimal sizing of wind-PV-battery energy storage systems. The source sizing part of the algorithm has been seen to determine the optimized sizes of renewable energy sources, and the battery sizing part of the algorithm has been seen to determine the optimal size of the battery unit [11]. Series arc fault detection has been seen to be done in a PV-wind hybrid micro-grid using Stockwell transform-based Skewness, Kurtosis value covariance analysis. The analysis done has been observed to be authenticated for data collected from industry [12, 13]. Analysis has been seen in PV-wind hybrid micro-grid in various domains, but none of the assessments done have been seen to deal with the parallel arc fault assessment in PV-wind hybrid micro-grid monitoring the micro-grid current Stockwell transform-based parametric analysis. Also, considering motivation from [12], the present work has been extended for more critical fault diagnosis in solar wind hybrid micro-grid using statistical parametric optimization.

Thus, the motivation of this paper is to assess parallel arc fault in the solar cell in a hybrid micro-grid system. Firstly, a prototype system of a hybrid system has been built in simulation environment. The grid current has been captured at healthy and for different cycles of parallel arc faults in the PV array and analyzed using Stockwell transform. The statistical values have been assessed computing the Skewness values, and the standard deviation, covariance analysis of the data at fault with respect to normal has been done. Features have been extracted at the end, and depending on the extracted features a technique has been proposed for effective parallel arc fault identification in the system.

2 System Data Acquisition

This work deals with the development of a solar-wind hybrid micro-grid, wherein data acquisition has been done at the grid end for normal and for the presence of parallel arc fault in the system. An electric arc is a current that flows from one conductor to the other with a gap between these two conductors. A parallel arc fault occurs when the insulation of two conductors of opposite polarity is damaged due to

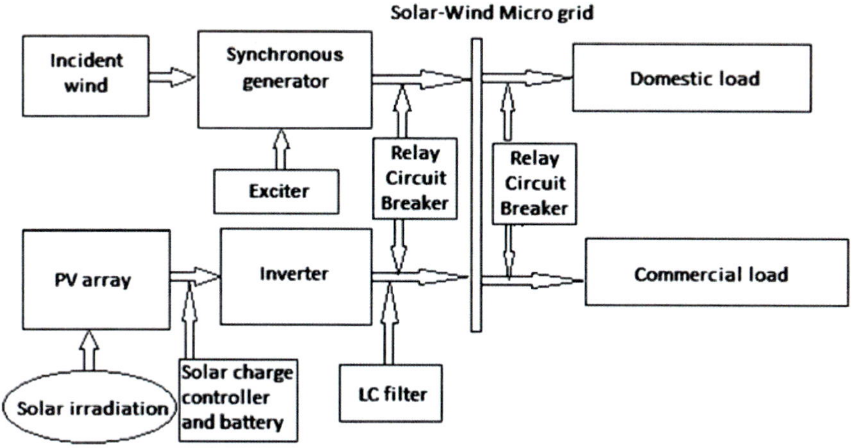

Fig. 1 Developed solar-wind micro-grid

hotspot generation due to external extremities resulting in the flow of high current in between the conductors [14]. The block diagram of the system under study has been presented in Fig. 1.

Here a 25 kVA, 440 V, 50 Hz synchronous generator with static exciter and a relay with an operating time of 10–60 ms has been used. The wind generator is connected with solar PV generation by the common grid.

The grid current has been assessed using Stockwell transform (ST) [12], a time–frequency distribution. The ST is defined as

$$s(\tau, f) = \mathrm{STFT}(\tau, f) = \int_{-\infty}^{\infty} h(t) \frac{|f|}{\sqrt{2\pi}} \mathrm{e} \frac{(\tau - t)^2}{2} \mathrm{e}^{-j2\pi f t} \mathrm{d}t \quad (1)$$

The ST coefficients have been assessed, computing the Skewness, Kurtosis, Mean, Mode, and Median values. Skewness is a measure of symmetry in any dataset. Kurtosis is a parameter that describes the shape of a random variable's probability distribution. The standard deviation and covariance of the data obtained at fault from normal have been computed to extract features for percentage of parallel arc fault occurring in the system. Standard deviation is a quantity expressing by how much a value deviates from the mean in any data distribution, and covariance is the property of a function of retaining its form when the variables are linearly transformed.

3 Fault Analysis Using S-Transform Based Parametric Values

The grid current has been recorded for normal and for parallel arc faults of different cycles (1–3) in the PV array, as presented in Figs. 2, 3, 4 and 5. To accomplish the percentage parallel arc fault, firstly, the healthy cross string resistance from simulation has been seen, and 1–3 cycle parallel arc faults in the photovoltaic module.

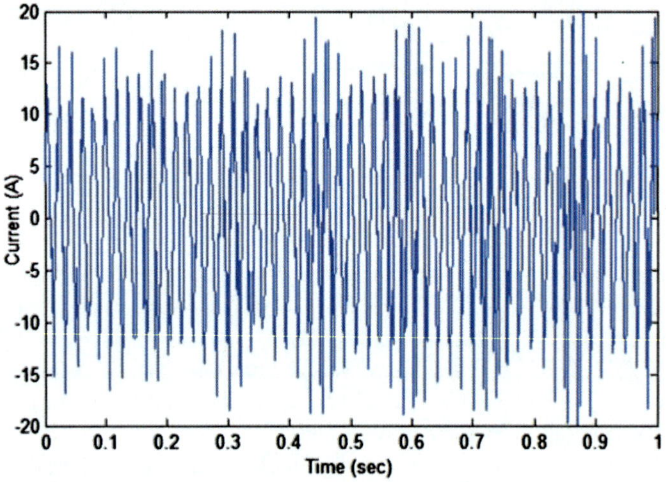

Fig. 2 Wind generator output current at healthy condition

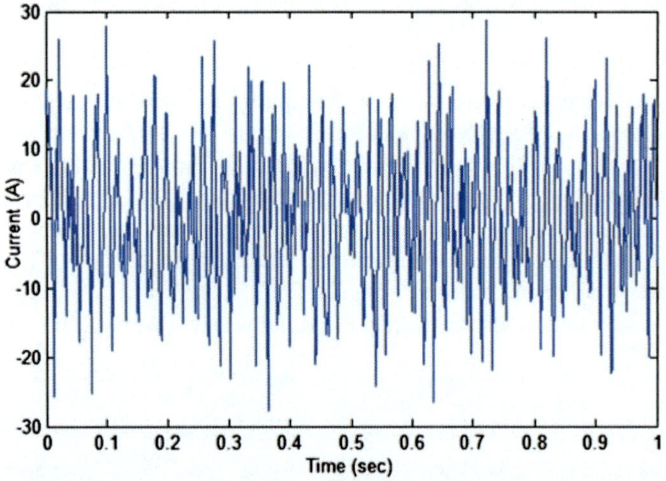

Fig. 3 Wind generator output current at parallel arc fault of 1 cycle duration

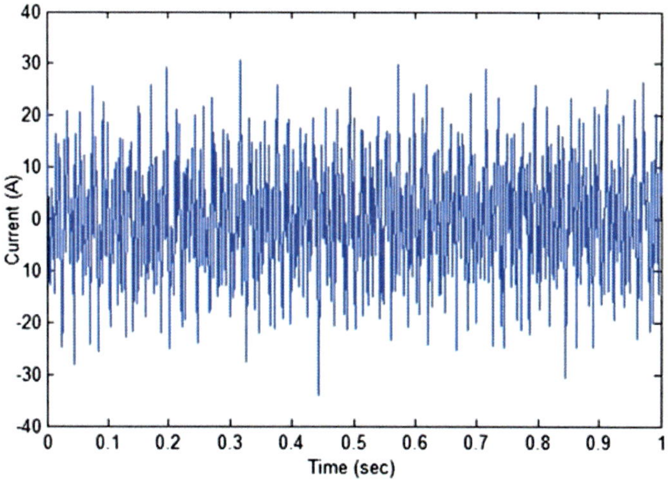

Fig. 4 Wind generator output current at parallel arc fault of 2 cycle duration

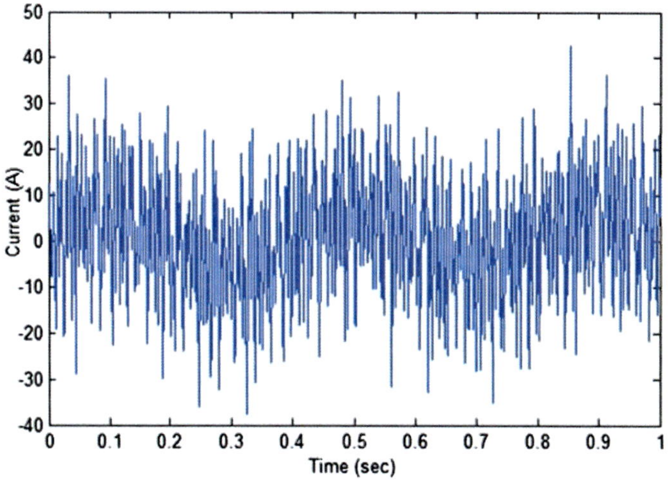

Fig. 5 Wind generator output current at parallel arc fault of 3 cycle duration

Current signature analysis has been done using the ST program with the calculation of Skewness values. Depending on the variation in standard deviation and covariance values of the statistical parameters, optimization has been done for fault detection in the system. The 2-D output frames have been presented in Figs. 6, 7, 8 and 9. The variation in the Skewness value has been presented in Table 1.

However, monitoring Figs. 2, 3, 4 and 5, it is very difficult to come to a definite conclusion. Thus ST algorithm has been applied to extract features. Outcomes have been shown in Figs. 6, 7, 8 and 9.

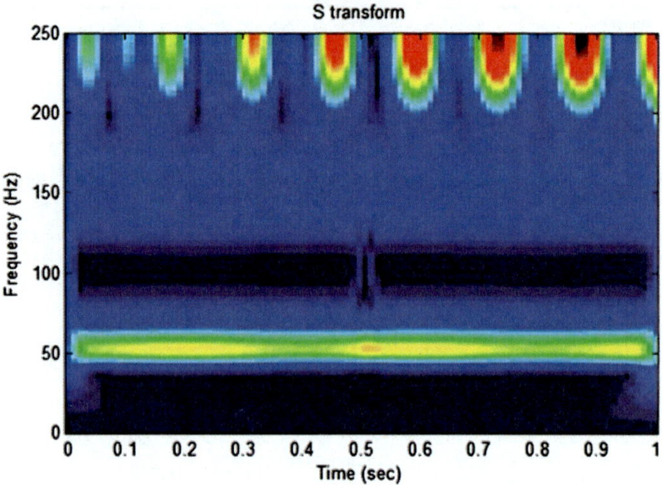

Fig. 6 Healthy system ST output frame

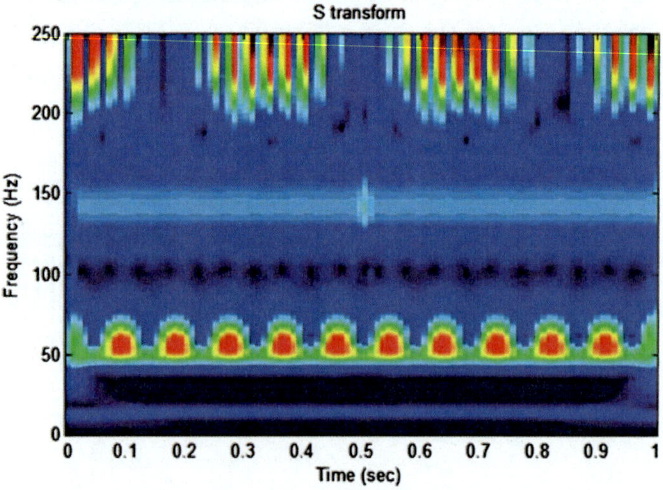

Fig. 7 ST output frame for parallel arc fault of 1 cycle duration

Monitoring Figs. 6, 7, 8 and 9, it has been seen that the time–frequency distribution of the grid current for different cycles of parallel arc faults in the PV array is different. However assessing only the nature of the ST distribution, fault assessment is very difficult. Thus statistical parameter monitoring has been done to extract features for effective fault estimation in the micro-grid. The computed Skewness values have been presented in Table 1, and the standard deviation and covariance analysis of the data at fault from normal has been presented in Table 2.

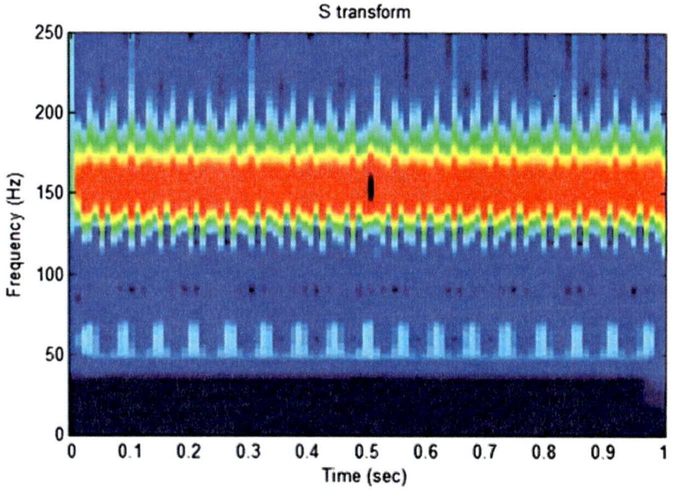

Fig. 8 ST output frame for parallel arc fault of 2 cycle duration

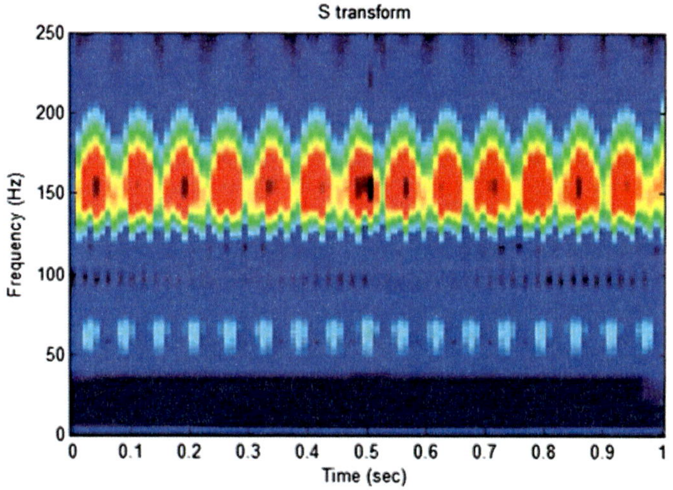

Fig. 9 ST output frame for parallel arc fault of 3 cycle duration

Table 1 Statistical value assessment from S-Transform

Parameters	Healthy	System condition		
		Arc fault of 1 cycle duration	Arc fault of 2 cycle duration	Arc fault of 3 cycle duration
Skewness	2.16	1.58	6.53	8.901

Table 2 Standard deviation and covariance analysis

Analyzing tool	System condition		
	Arc fault of 1 cycle duration	Arc fault of 2 cycle duration	Arc fault of 3 cycle duration
Standard Deviation	0.58	−4.37	−6.74
Covariance	1.1	1.15	0.08

4 Feature Extraction

The standard deviation and covariance of the statistical values obtained from Table 2 have been analyzed. Standard deviation and covariance calculated after arc fault of different duration have been calculated and shown in Fig. 10. They decrease with the increase of duration.

Depending on the extracted features as in Fig. 10, it has been observed that the slope for Skewness value standard deviation (SSD) has a decreasing nature with increased cycles of fault current. Thus standard deviation of the Skewness value has been considered as the best fit parameter in this case study. An algorithm has been developed for assessing the percentage parallel arc fault in the system:

- **Step 1**: Collect grid current
- **Step 2**: Calculate ST coefficients
- **Step 3**: Compute Skewness value of ST coefficients.
- **Step 4**: Compute *SSD*.
- **Step 5**: Extract features as in Fig. 10.
- **Step 6**: Compute fault duration.

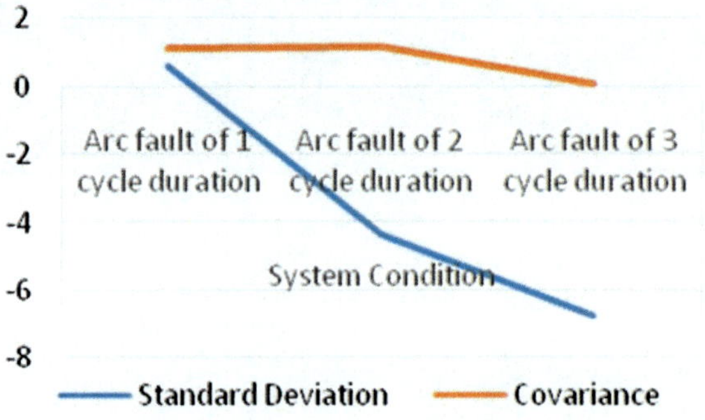

Fig. 10 Standard deviation and covariance calculated after arc fault of different duration

5 Conclusion

Solar-powered systems are considered to be the most viable alternative sources of energy in recent days. Thus the concern for effective condition monitoring of these types of systems has increased day by day. In this context, the work has been advanced with the detection of parallel arc fault in a solar-wind hybrid micro-grid. Here, a prototype system has been built in the software interface, wherein 1–3 cycle(s) parallel arc fault has been created wherefrom grid current has been recorded. Current signature analysis has been done using Stockwell transform-based algorithm where Skewness values have been computed from the Stockwell transform coefficients at normal and fault. The standard deviation and the covariance in the statistical values have been calculated, and depending on the linear optimization technique, the work has been advanced.

Thereafter, an algorithm has been developed for parallel arc fault detection in the solar wind hybrid micro-grid. The operating time of the algorithm from data acquisition to decision making has been, recorded and it has been observed to be satisfactorily less.

The future scope of this work aspires to deal with parallel arc fault detection in hybrid micro-grid systems using S-Transform-based Kurtosis, Mean, Mode, and Median value-based best fit optimization techniques. A comparative study between all of the statistical tools will be done to extract the mostly suitable parameter for this type of fault detection, followed by practical validation of the case concerned using real field data based analysis.

References

1. Chatta, M.B., Ali, H.M., Ali, M., Bashir, M.A.: Experimental investigation of mono- crystalline and polycrystalline solar modules at different inclination angles. J. Thermal Eng. **4**(8), 2137–2148 (2018).
2. Khan, M.M.A., Ibrahim, N.I., Mahbubul, I.M., Ali, H.M., Saidur, R., Sulaiman, F.A.A.: Evaluation of solar collector designs with integrated latent heat thermal energy storage: a review. Sol. Energy **166**(15), 334–350 (2018)
3. Shah, T.R., Ali, H.M.: Applications of hybrisnanofluids in solar energy, practical limitations and challenges: a critical review. Sol. Energy **183**(1), 173–203 (2019)
4. Karthick, K., Suresh, S., Hussain, M.M.M.D., Ali, H. M., Kumar, C.S.S.: Evaluation of solar thermal system configurations for thermoelectric generator applications: a critical review. Solar Energy **188**, 111–142
5. Bashir, M.A., Giovannelli, A., Ali, H.M.: Design of high-temperature solar receiver integrated with short term thermal storage for Dish-Micro Gas Turbine systems. Solar Energy **190**(15), 156–166
6. Wahab, A., Hassan, A., Qasim, M.A., Sajid, H.U.: Solar energy systems-Potential of nanofluids. J. Mol. Liquids **289**(1), 111049
7. Sajjad, U., Amer, M., Ali, H.M., Dahiya, A., Abbas, N.: Cost effective cooling of photovoltaic modules to improve efficiency. Case Stud. Thermal Eng. **14**, 100420 (2019)

8. Noman, M., Wasim, A., Ali, M., Jahanzaib, M., Hussain, S., Ali, H.M.K., Ali, H.M.: An investigation of a solar cooker with parabolic trough concentrator. Case Stud. Thermal Eng. **14**, 100436 (2019)
9. Baig, W., Ali, H.M.: An experimental investigation of performance of a double pass solar air heater with foam aluminum thermal stirage medium. Case Stud. Thermal Eng. **14**, 100440 (2019).
10. Bhowmick, D., Sinha, A.K.: Cost-based allocation model for hybrid power system considering solar, wind and thermal generations separately. IET Gener. Transmission Distrib. **11**(18), 4576–4587 (2017). https://doi.org/10.1049/iet-gtd.2017.0305
11. Akram, U., Khalid, M., Shafiq, S.: Optimal sizing of wind/solar/battery hybrid grid- connected micro-grid system. IET Renew. Power Gener. **12**(1):72–80 (2018). https://doi.org/10.1049/iet-rpg.2017.0010
12. Kar Ray, D., Das, D., Shah, O.P., Singh, S.K., Chattopadhyay, S.: Condition Monitor- ing of solar-wind hybrid micro-grid using Stockwell Transform based parametric values. In: Proceeding of MCCS-2019, May 2019. https://doi.org/10.1007/978-981-15-5546-6_23
13. Datta, S., Chattopadhyaya, A., Chattopadhyay, S., Das, A.: Line to ground and line to line fault analysis in IEEE standard 9 bus system. Model Measure. Control A **93**(1–4), 10–18 (2020). https://doi.org/10.18280/mmc_a.931-402 (2020)
14. https://support.industry.siemens.com/cs/attachments/109476961/manual_5SM_ADU_for_photovoltaic_en_en-US.pdf?download=true

A Study of Phonocardiography (PCG) Signal Analysis by K-Mean Clustering

Tanmay Sinha Roy, Joyanta Kumar Roy, and Nirupama Mandal

1 Introduction

Human heart makes certain sounds. The sounds come from the bicuspid/mitral valves, the tricuspid valves, and the aortic valves. As these valves open and close, it allows blood flow to and from the heart and thus they produce the heartbeat sound.

Feature Extraction [1, 2] and Classification proves to be very useful when it comes to Phonocardiography (PCG) signal analysis. Many advances have been made towards automated heart sound segmentation, heart pathology detection, and classification. However, an efficient method for noise handling method still would come as an important breakthrough for further development in this field, especially when it comes to working with PCGs collected in realistic environments such as hospitals and clinics. The feature extraction of Phonocardiography (PCG) signals have been made at different levels using Discrete Wavelet Transformation techniques and then the PCG signals have been analyzed by calculating different parameters at these levels. The parameters are Average Energy, Average Power, Standard deviation, Variance, Mean square error (MSE), Peak Signal to Noise Ratio (PSNR), Root Mean Square Error (RMSE), and Maximum Entropy (ME) values of human heart signal extracted from the Phonocardiogram. Then a classification made based on different Machine Learning algorithms as explained in Fig. 1(a).

T. S. Roy (✉)
Department of Instrumentation and Control Engineering, Haldia Institute of Technology, Purba Medinipur, WB, India

J. K. Roy
Eureka Scientech Research Foundation, Kolkata, India

N. Mandal
Department of Electronics Engineering, Indian Institute of Technology (ISM), Dhanbad, India

J. K. Mandal and J. K. Roy (eds.), *Proceedings of International Conference on Computational Intelligence and Computing*, Algorithms for Intelligent Systems,
https://doi.org/10.1007/978-981-16-3368-3_16

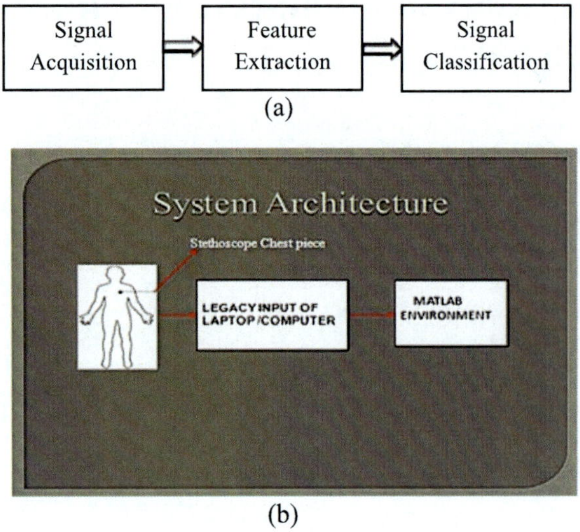

Fig. 1 **a** Various steps involved in phonocardiography (PCG) signal analysis [3]. **b** system architecture [3]

2 Objective

The whole work is focused on the detection, analysis of cardiac sound and diagnose different types of diseases related with heart using DWT used for feature extraction and K-mean Clustering as a classification tool with the help of MATLAB and PYTHON software as shown in Fig. 1b. To start with, a few words should be discussed about the physiology of heart sounds, which includes its cause, nature, and many other parameters. A healthy heartbeat is a continuous activity. There are varieties of abnormal heart sounds, some of which are harmless, while others can indicate serious heart related problems. The heart problems are typically generated from defects in the heart valves. The defects may be the aortic stenosis, aortic regurgitation, mitral stenosis, or mitral regurgitation of diseased heart valves.

Heart sounds [4, 5] are the disturbances generated by the beating heart and the resultant flow of blood through it. Accurately, the sounds reflect the turbulence created when the heart valves expand and contract. In cardiac auscultation, a doctor uses a stethoscope to hear for these heart sounds that provide important auditory information regarding the condition of the heart valves.

The Relationship of heart sound with the cardiovascular system is given in Fig. 2.
In healthy adults, there is two normal heart sounds often described as:

1. *lub*2. *Dub*(or *dup*)

References [7, 8] LUB and DUB occur in sequence with each heartbeat. These are the **first heart sound** (S1) and **second heart sound** (S2), produced by the closing of the AV valves and semilunar valves, respectively. In addition to these normal sounds, a variety of other sounds may be present in the abnormal heart, which is:

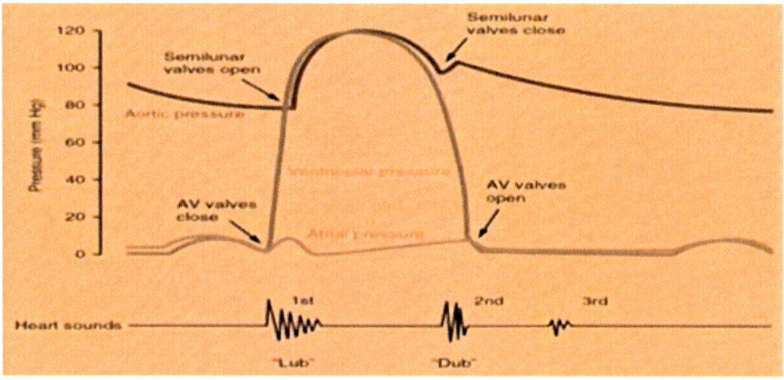

Fig. 2 Relationship of heart sound with the cardiovascular system [4]

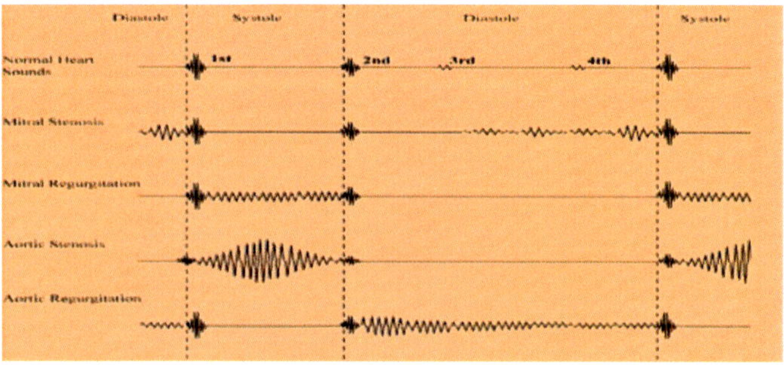

Fig. 3 Systolic and diastolic murmurs [6]

Cardiac murmurs: Heart murmurs [1, 6, 9] as given in Fig. 3 are caused by the turbulent flow of blood occurs inside the heart. Murmurs may be benign or malignant (Abnormal). Abnormal murmurs can be caused by narrowing of the heart valve opening resulting in disturbance as blood flows through it which is known as Stenosis. The abnormal murmur sound may also occur with valvular insufficiency which is known as Regurgitation. It allows the backflow of blood when a diseased heart valve closes only partially. Different types of murmurs [9, 10] are heard in different parts of the heart's cardiac cycle, depending on the cause of the murmur which is related to various heart valve disorders. **A Third Heart Sound (S3)** is sometimes heard, especially in teenagers. This sound occurs from 0.1 to 0.2 s after the second heart sound is contributed to the rush of blood flow from the atria into the ventricles, which causes disturbance and some vibrations of the inner ventricular walls.

The fourth heart sound (S4) is aa low-intensity sound and late diastolic [11, 12] that corresponds to late ventricular filling through an active atrial contraction.

S1 has a frequency range of 30–45 Hz. S2 [9, 12] has a frequency range of 50–70 Hz. S3 (Third heart sound) is an extremely weak vibration ranges below 30 Hz as given in Fig. 4.

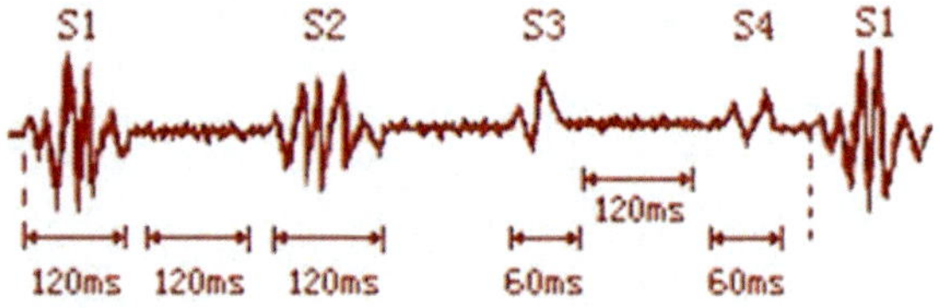

Fig. 4 Different segments of heart sound [3]

3 Methods and Materials

PCG Signal Analysis using Discrete Wavelet Transform

A Normal PCG signal is represented interms of Detailed Coefficients and Approximate coefficients using DWT.A 6-Level DWT of a Normal Heart sound representation has been shown in the below Fig. 5.

In this method [2, 12] PCG signals are decomposed up to eight levels, and the variance of detail coefficients obtained at level 5, 6 and 7 are shown to contain classification information to discriminate normal, aortic stenosis, aortic regurgitation, mitral stenosis, and mitral regurgitation PCG signals.

Figure 6 shows the various PCG signals taken for the proposed work [8, 14]. Each PCG signal is decomposed up to level eight by selecting 'db6' as a mother wavelet. Then the variance of details coefficients is calculated and is shown in Table 1. One can observe from Table 1 that variance of detail coefficients at level 5, 6, and 7 can be taken as a set of features to classify PCG signals. These values are plotted in Fig. 7 to discriminate Aortic Stenosis (AS), Aortic Regurgitation (AR), Mitral Stenosis (MS), Mitral Regurgitation (MR), and Normal (N) PCG signals.

Detailed Coefficients of DWT at Level 5(CD5) has also been used as a feature for classification of heart sounds. RMS Value, Avg. Power and Energy of Level-5

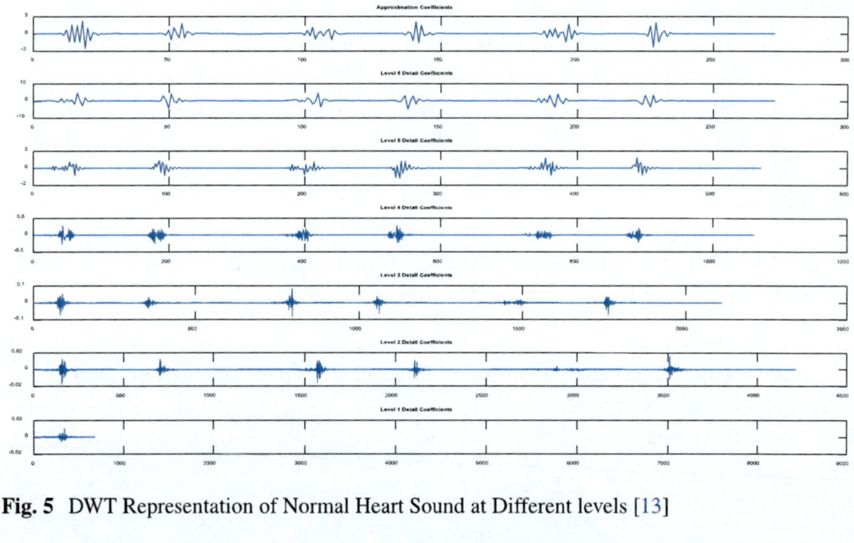

Fig. 5 DWT Representation of Normal Heart Sound at Different levels [13]

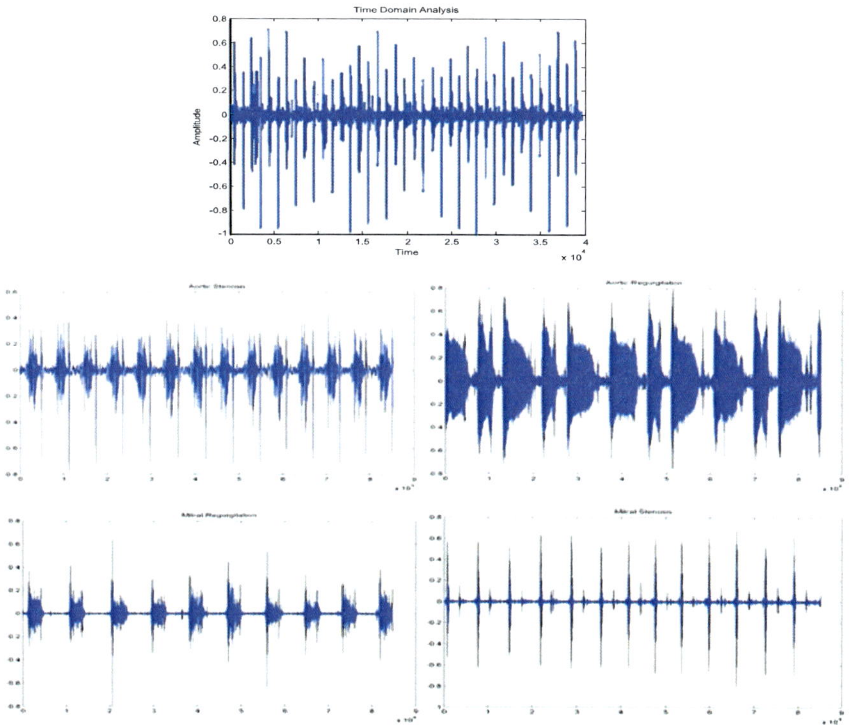

Fig. 6 N, AS, AR, MS and MR Phonocardiography (PCG) signals [13]

Table 1 Features extracted for classification

Variance of detail coefficient	Normal	Aortic stenosis	Aortic regurgitation	Mitral stenosis	Mitral regurgitation
d1variance	6.98E–06	4.97E–08	2.04E–06	1.56E–09	7.98E–10
d2variance	1.32E-05	9.51E–08	5.33E–06	4.88E–07	3.16E–08
d3variance	3.53E–05	1.12E–06	6.25E–05	2.55E–05	7.89E–06
d4variance	8.49E–04	4.06E–04	0.006	4.82E–04	5.41E–05
d5variance	1.01E–02	0.0252	0.11	0.0413	0.0011
d6variance	0.0147	0.1073	0.0712	0.1983	0.0057
d7variance	1.70E–03	0.577	0.0022	0.0823	0.0958
d8variance	3.04E–05	0.8519	8.49E–04	0.0012	1.8891

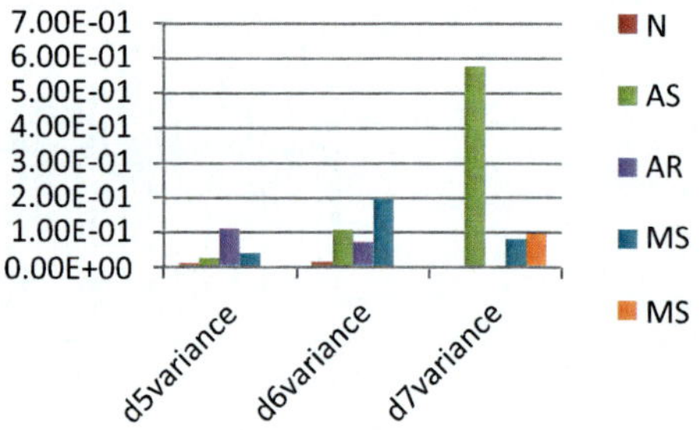

Fig. 7 Comparison of different heart sounds [15]

detailed coefficients have been evaluated and used as a feature extraction technique given below.

From the Comparative study of 1000 odd samples of different kind of heart sounds (Normal, Aortic Stenosis, Mitral Regurgitation, Mitral Stenosis, Mitral Valve Prolapse), it can be concluded that Normal heart sound can easily be identified and distinguished. As RMS Value of Normal Heart sound is the lowest given in Fig. 8.

From the Comparative study for 1000 odd samples of different kind of heart sounds (N, AS, MR, MS, MVP), it can be concluded that Normal heart sound can easily be identified and distinguished. As Average Energy for Normal Heart sound is the highest given in Fig. 9.

From the Comparative study for 1000 odd samples of different kind of heart sounds (N, AS, MR, MS, MVP), it can be concluded that Normal heart sound can easily be identified and distinguished. As Average Power of Normal Heart sound is the lowest given in Fig. 10.

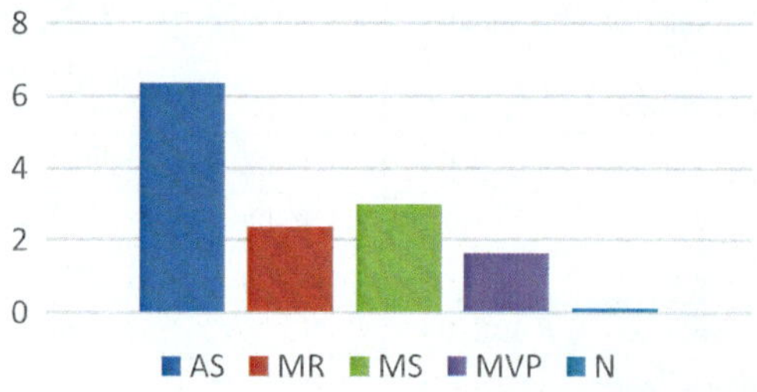

Fig. 8 RMS of different heart sounds at level 5 (CD5)

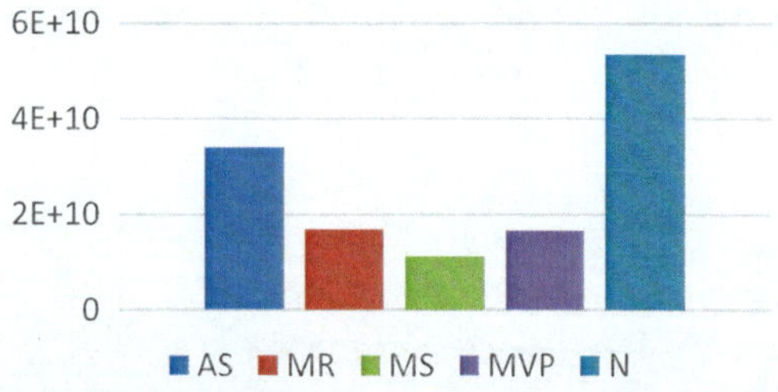

Fig. 9 Average energy of detailed coefficients at level 5 (CD5)

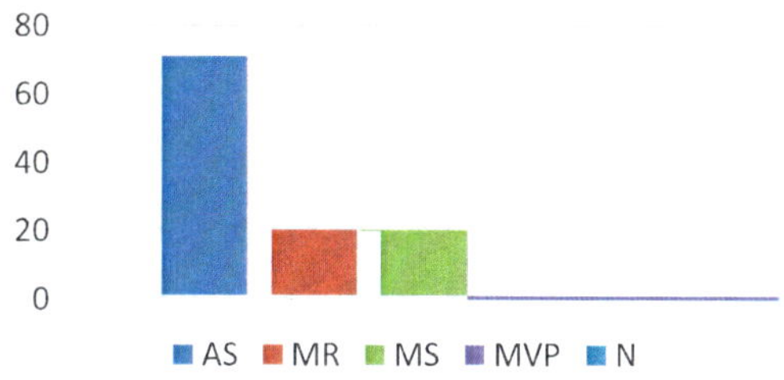

Fig. 10 Average power of detailed coefficients at level 5 (CD5)

4 Time Domain Feature Extraction

4.1 Root Mean Square (RMS)

The RMS value used in the time domain is given below [16,17]

$$\mathrm{RMS}\{x(n)\} = \left(1/N \sum |x(n)|^2\right)^{1/2} \tag{1}$$

The RMS equation used in the frequency domain, using Parsevals Theorem.

$$\sum x^2(n) = 1/N \sum |x(m)|^2 \tag{2}$$

where, $x(m) = \mathrm{FFT}\{x(n)\}$

$$\begin{aligned}\mathrm{RMS}\{x(n)\} &= \left(1/N \sum |x(n)|^2\right)^{1/2} \\ &= \left(\sum |x(m)/N|^2\right)^{1/2}\end{aligned}$$

Average Power of a signal $x(n)$ used in the time domain can be written in terms of RMS value as

$$Average - Power = (\mathrm{RMS}\{x(n)\})^2 \tag{3}$$

From the Comparative study for 1000 odd samples of different kind of heart sounds (N, AS, MR, MS, MVP), it can be concluded that Normal heart sound can easily be identified and distinguished. As RMS value for Normal Heart sound is the highest given in Fig. 11.

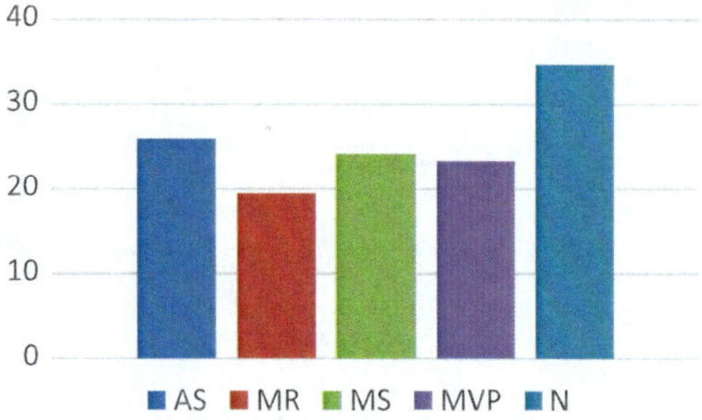

Fig. 11 Comparison of RMS value of different heart sound

4.2 *Signal Energy and Power*

Average Energy and Average Power for Continuous-Time Signals

The terms signal energy and signal power are used to describe a signal. They are not the measures of energy and power of any continuous signal. The explaination of signal energy and power of any signal $x(t)$ is given below.

The Average Energy of signal $x(t)$ in time domain is given by

$$E = \int_{+\infty}^{-\infty} |x(t)|^2 \mathrm{d}t \tag{4}$$

The Average- Power of signal $x(t)$ in time domain is given by

$$P = \lim_{T \to \infty} 1/2T \int_{+\infty}^{-\infty} |x(t)|^2 \mathrm{d}t \tag{5}$$

If $0 < E < \infty$, is satisfied then the signal $x(t)$ is called an energy signal. However, there are some signals where this condition is not satisfied. For such signals, power of the signal is considered. If $0 < P < \infty$, is satisfied then the signal is called a power signal. Remember that the power for an energy signal is zero ($P = 0$) and that the energy for a power signal is infinite ($E = \infty$). There exists some signals which are neither energy nor power.

Energy and Power for Discrete-Time Signals

The definition [18, 19] of signal energy and power for discrete signals are very much similar in definitions for continuous signals. A discrete-time energy signal is defined as one for which $0 < E < \infty$ and a discrete-time power signal is defined as one for which $0 < P < \infty$. A discrete-time signal can be neither an energy signal nor a power signal.

The Average Energy of the discrete-time signal $x(n)$ is given as

$$E = \sum |x(n)|^2 \quad \text{where } n \text{ starts from } -\infty \text{ to } +\infty \tag{6}$$

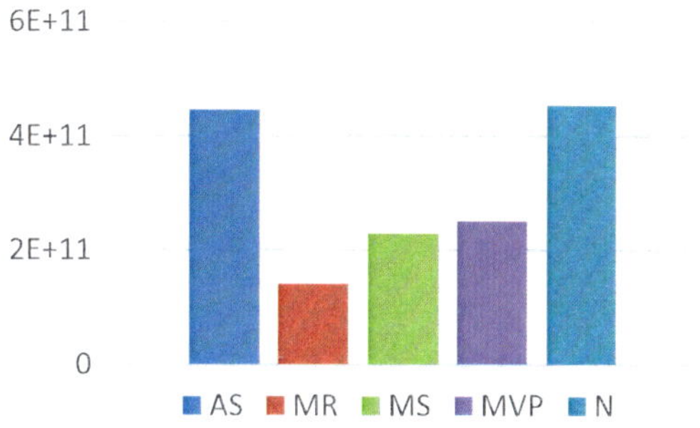

Fig. 12 Comparison of average energy of different heart sounds

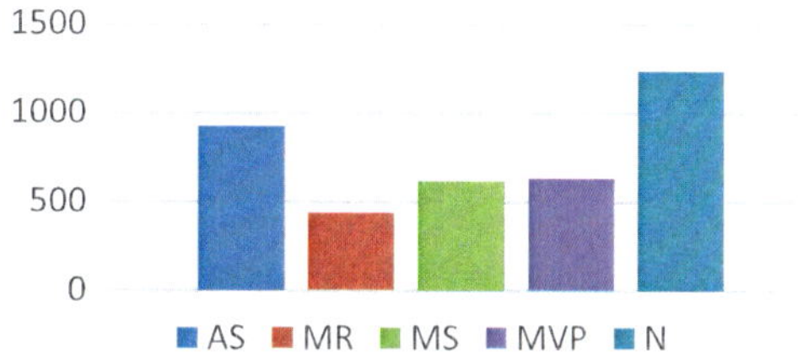

Fig. 13 Comparison of average power of different heart sound

The Average signal power of the signal $x(n)$ is given as

$$P = \lim_{N \to \infty} 1/N \sum |x(n)|^2 \tag{7}$$

where n starts from $-N$ to $+N$.

From the Comparative study for 1000 odd samples of different kind of heart sounds (N, AS, MR, MS, MVP), it can be concluded that Normal heart sound can easily be identified and distinguished. As Avg. Energy for Normal Heart sound is the highest given in Fig. 12.

From the Comparative study for 1000 odd samples of different kind of heart sounds (N, AS, MR, MS, MVP), it can be concluded that Normal heart sound can easily be identified and distinguished. As Avg. Power for Normal Heart sound is the highest given in Fig. 13.

5 Classification Methods Using Machine Learning

Machine Learning [2, 20] is the branch of science that gives computers the ability to learn without being getting programmed. Machine Learning is one of the most interesting and challenging technologies in todays world. As it is quite obvious from the name itself, it makes the computer and laptops more similar to human being. Machine learning is actively being used today in most of the real world applications.

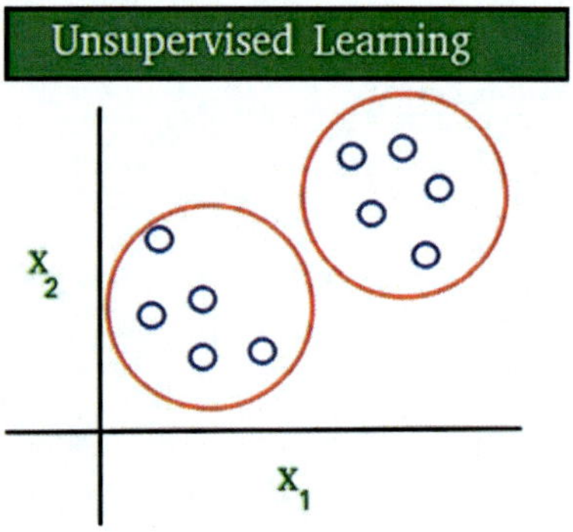

Fig. 14 Unsupervised learning [16]

5.1 K Means Clustering Algorithm

It is seen from Fig. 14 clearly, the data in unsupervised learning is unlabelled.

K-means clustering [5, 21] is a clustering algorithm that aims to partition n observations into k clusters. There are primarily three steps involved in it:

1. Initialization Step—K initial "means" (centroids) are initialized arbitrarily in the first step.
2. Assignment Step—K clusters are generated by associating each observation with the nearest centroid location.
3. Update Step—The centroid of the clusters in each iteration becomes the new mean location.

Assignment Step and Update Step are repeated iteratively until convergence occurs. The final result is that the sum of squared errors is minimized between points and their respective centroids.

6 Results and Discussions

We have considered 1000 heart sounds of 5 different kinds (Normal, Aortic Stenosis, Mitral Regurgitation, Mitral Stenosis, Mitral Valve Prolapse), each having 200 heart sound samples. With the help of K-Mean Clustering, we can be able to Classify different heart sounds. Features like Energy, Avg-Power, and RMS values are considered, and then K-Mean Clustering has been applied for different values of k.

Figure 15 shows the clustered data set after the application of K-Mean Clustering, where black dots represent the Cluster Centroids. In this diagram, number of clusters (k) have been choosen as 5.

Finally we have applied different values of k and obtained various plots for $k =$ 1 to 5 as shown in Fig. 16.

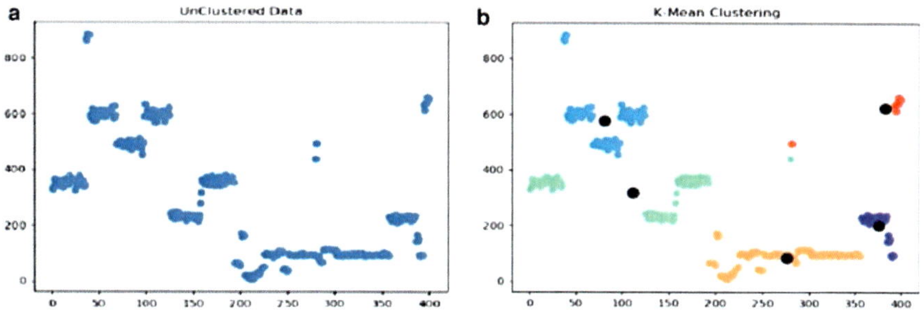

Fig. 15 **a** Unclustered data, **b** clustered data

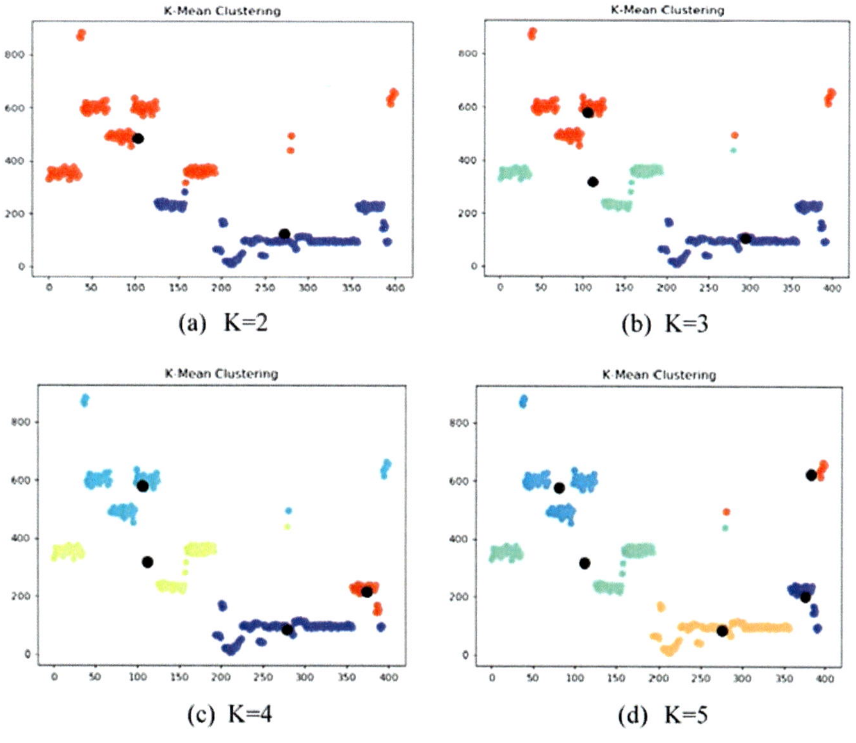

Fig. 16 K-mean clustering for different values of *K*

Figure 16 shows application of K-mean Clustering for different values of *k* applied on 1000 number of odd heart sound samples (N, AS, MR, MS, MVP) after feature extraction

As we can see from Table 2, WCSS(Within Cluster Sum of Square) is gradually decreasing with the increase in the number of clusters (K), and at the same time, the

Table 2 Within cluster sum of squares

	$K = 2$	$K = 3$	$K = 4$	$K = 5$	$K = 6$	$K = 7$	$K = 8$	$K = 9$	$K = 10$
WCSS	6,364,368	3,598,022.2	2,891,658	2,241,251.5	1,658,363.23	1,326,962	948,372.06	627,838.69	389,895.25

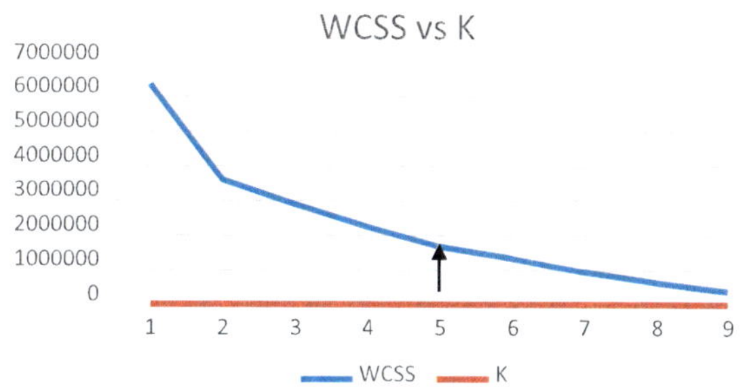

Fig. 17 Plot of WCSS versus K using elbow method

distance between the clusters is increasing with the increase in the number of clusters (Fig. 17).

With the help of elbow method in K-Mean clustering, we can easily determine the number of clusters, classes or groups we need to have while using a supervised machine learning algorithm. Here in our research work the correct number of clusters or classes happens to be at $k = 5$.

7 Conclusion

The PCG signal analysis is a trending and challenging research area, where the research is going on to design different techniques of PCG signal analysis, which can provide more accurate and efficient measurements to the doctors and at the same time, can help in the diagnosis of a diseased heart. Though a huge amount of research work has been carried out in PCG signal analysis in the biomedical engineering domain, the lack of a good research paper is often one of the main drawbacks of broader acceptance of the idea of PCG Signal analysis. In this research paper, various points of Feature-Extraction techniques have been covered. The basic idea about PCG Signal analysis has been provided, and along with it various signal processing algorithms used in PCG Signal analysis have also been discussed. Various applications of PCG Signal analysis in different domains have been summarized in this paper. In the signal processing part of PCG Signal analysis, DWT method has been used as a feature extraction method. Finally, the Classification of the PCG Signals and finding out the number of clusters in an unclustered data set using elbow method have also been discussed with the help of K-Mean clustering. Further the right number of clusters or classes can help in distinguishing a diseased heart with a normal heart using any supervised machine learning algorithm.

References

1. Randhawa, S.K., Singh, M.: Classification of heart sound signals using multimodal features. In: Second International Symposium on Computer Vision and the Internet, vol. 58, pp. 165–171. Elsevier, Amsterdam (2015)
2. Roy, J.K., Roy, T.S., Mandal, N., Postolache, O.A.: A Simple technique for heart sound detection and identification using Kalman filter in real-time analysis. In: ISSI 2018, First International Conference, 6–7 Sept 2018. 978-1-5386-5638-9/18/$31.00 ©2018 IEEE
3. Gupta, C.N., Palaniappan, R., Rajan, S., Swaminathan, S., Krishnan, S.M.: Segmentation and classification of heart sounds. In: International Conference: Canadian Conference on Electrical and Computer Engineering, June 2005. https://doi.org/10.1109/CCECE.2005.1557305
4. Roy, J.K., Roy, T.S.: A simple technique for heart sound detection and real-time analysis. In: Proceedings of ICST 2017 held at Macquarie University Sidney, Sensing Technology (ICST), 2017 Eleventh International Conference, 4–6 Dec 2017. https://doi.org/10.1109/ICSensT.2017.8304502
5. Roy, JU.K., Roy, T.S., Mukhopadhyay, S.C.: Heart sound: detection and analytical approach towards diseases. In: Mukhopadhyay, S.C. (eds.) Modern Sensing Technologies, pp. 103–145. Springer Nature, Switzerland. https://doi.org/10.1007/978-3-319-99540-3_7
6. Cardiac cycle, https://en.wikipedia.org/wiki/Cardiac_cycle
7. Amarnath, R.: Methods for classification of phonocardiogram. TENCON2003. In: Conference on Convergent Technologies for the Asia-Pacific Region 2003, vol. 4, pp. 1514–1515
8. Liang, H., Hartimo, I.: A heart sound feature extraction algorithm based on wavelet decomposition and reconstruction. In: Proceedings of 20th Annual International Conference on IEEE Engineering in Medicine and Biology Society, vol. 20, pp. 1539–1542 (1998)
9. Anju, Kumar, S.: Detection of Cardiac Murmur. Int J Comput Sci Mobile Comput **3**(7):81–87. ISSN 2320–088X
10. Heart sounds - Wikipedia. https://en.wikipedia.org/wiki/Heart_sounds
11. Heart murmur causes, https://www.nhlbi.nih.gov/health/health-topics/topics/heartmurmur/causes
12. Dewangan, N.K., Shukla, S.P., Dewangan, K.: PCG signal analysis using discrete wavelet transform. Int. J. Adv. Manag. Technol. Eng. Sci. **8**(III) (2018). ISSN NO: 2249–7455
13. Venkata Hari Prasad, G., Rajesh Kumar, P.: Analysis of various DWT methods for feature extracted PCG signals. Int. J. Eng. Res. Technol. (IJERT) **4**(04) (2015). ISSN: 2278-0181
14. Roy, A.K., Misal, A., Sinha, G.R.: Classification of PCG signals: a survey. Int. J. Comput. Appl. Recent Adv. Inform. Technol. (2014). ISSN NO: 0975-8887
15. Mishra, G., Biswal, K., Mishra, A.K.: Denoising of heart sound signal using wavelet transform. Int. J. Res. Eng. Technol. **02**(04) (2013). ISSN: 2319-1163
16. Singh, M., Cheema, A.: Heart sounds classification using feature extraction of phonocardiography signal. Int. J. Comput. Appl. **77** (4) (2013). ISSN NO:0975-8887
17. Liang, H., Lukkarinen, S., Hartimo, I.: Heart sound segmentation algorithm based on heart sound envelogram. Comput. Cardiol. **24**(7), 105–108 (1997)
18. Misal, A., Sinha, G.R.: Denoising of PCG signal by using wavelet transforms. J. Adv. Comput. Res. **4**(1), 46–49 (2012) ISSN: 0975-3273 & E-ISSN: 0975-9085
19. Muruganantham (2003) Methods for classification of phonocardiogram. TENCON (2003)
20. Debbal, S., Bereksi-Reguig, F.: Graphic representation and analysis of the PCG signal using the continuous wavelet transform. Internet J. Bioeng. **2**(2)
21. Javed, F., Venkatachalam, P.A., Ahmad Fadzil, M.H.: A signal processing module for the analysis of heart sounds and heart murmurs. J. Phys. Conf. Ser. **34**, 1098–1105 (2006)
22. Janse, V., Magre, S.B., Kurzekar, P.K., Deshmukh, R.R.: A comparative study between MFCC and DWT feature extraction technique, vol. 3(1), pp. 3124–3127

An Improved Decision Support System for Automated Sleep Stages Classification Based on Dual Channels of EEG Signals

Santosh Kumar Satapathy, D. Loganathan, Hari Kishan Kondaveeti, and Rama Krushna Rath

1 Introduction

Sleep stage classification is one of the primary steps for any type of sleep-related diseases. As per the current survey of Non-Communicable Diseases (NCDs) unacceptably increased day by day and as per NCDs 2018 report in global wise 57 million affected with this chronic neurological disorder and out of that 71% of 57 million has died globally with these diseases, mostly death has occurred with cardiovascular and hypertension diseases [1]. Nowadays sleep-related disorder has shown with any age groups of human beings. Currently, sleep-related diseases complaint is highly increased and it is one of the research topics for the health sector regarding how effectively handle the diagnosis and treatments for several types of neurological type of disorders like parasomnia, insomnia, sleep apnea, narcolepsy, bruxism and hypopnea. Now it's more important for clinicians that how to manage with suitable investigation in related to identifying abnormality and choose a certain diagnosis according to evaluation reports. One human being covered one–third of the time spends as asleep in his/her life. In human life sleep is a basic need to maintain the balance between mental and physical systems of our body and a proper sleep quality requires for accurate functioning of our immunity system. In many things of our body is interlinked with developing comfortable sleep during night hours such as memorization productivity, hormone growth, learning capability, and concentration.

S. K. Satapathy (✉) · D. Loganathan
Pondicherry Engineering College, Puducherry, India
e-mail: santosh.satapathy@pec.edu

H. K. Kondaveeti
VIT University, Andhra Pradesh, India

R. K. Rath
Anna University, Chennai, India

J. K. Mandal and J. K. Roy (eds.), *Proceedings of International Conference on Computational Intelligence and Computing*, Algorithms for Intelligent Systems,
https://doi.org/10.1007/978-981-16-3368-3_17

To detect the sleep disorder the first important step is to conduct a sleep study on patients, which is called a polysomnography test. In this test, generally neurophysiological and cardiorespiratory signal has to be collected from patients continuously during sleep hours in the night. In polysomnography (PSG), test is the basic tool for recorded and studied sleep disorder of patients, and during this test, multiple information required for proper analysis with regard to sleep abnormality. All these information has to be gathered from subject's body during sleeping time. Normally a PSG signal contains three basic physiological signals that is electroencephalogram (EEG), electromyogram (EMG), electrooculogram (EOG), and electrocardiogram (ECG).In the year 1968, the two researchers named Rechtschaffen and Kales have published a new sleep standards, which became more popular in the later time with subject to diagnosis of any type neurological-related disorders [2]. As per the R&K rules, the total sleep hours have segmented into six stages such as Wake (W), non-Rapid Eye Movement (NREM1, NREM2, NREM3, NREM4) stages and Rapid Eye Movement (REM) stage. After long years to gap in the year 2007, scientists from the American Academy of Sleep Medicine (AASM) has updated the earlier sleep manuals designed by R&k and according to the new AASM standards, both NREM3 AND NREM4 represented in to one stage that is called as NREM3 [3]. Henceforth as per AASM standards stage, N3 and N4 combined into one stage as N3.So that according to AASM manuals the total sleep hours have annotated in to five states of sleep such as Wake, non-Rapid Eye Movement (NREM1, NREM2, and NREM3), and REM. Generally, sleep stages repeated in the cyclic process during night sleep time. The cycle normally starts from NREM stages and ends into the REM stage and this cycle repeated 4–6 times in a whole night sleep time. Each individual has to be pass through all these mentioned sleep stages and each stage reflected different characteristics during sleep with range of frequency sub-bands. The behavior of EEG signals with different sleep stages with criteria to amplitude, frequency and signal patterns presents in Table 1. Sleep scoring through manual approach requires more sleep experts and its one of a time-consuming task. The major difficulty with manual diagnosis is that it has taken more hours to record the PSG signals from patient and ultimately it has very difficult for the clinician to monitor the signal patterns throughout the whole night and it leads some error in sleep score which affect the accuracy of classification in between different sleep stages. The complete sleep stages

Table 1 Behavior of EEG signals during sleep

Sleep stage	Chracters of wave patterns
Wake	The maximum activities are in nature of alpha-band(8-13 Hz),beta-band (12–30 Hz)
NREM1	Most of the wave patterns are in behavior of theta-band(4–8 Hz)
NREM2	K-complex(1 Hz), Spindle (12–14 Hz)
NREM3	The major activities are in delta-band wave forms
REM	Mixing of wave patterns are reflected like alpha-band, beta-band, theta-band, sawtooth wave

annotations approach has followed according to either AASM rules or R&K rules. As per AASM rules, the PSG records has segmented into different fixed time frame epochs but the standard epoch segmentation is the 30 s and allows for the discriminate of different sleep–wake stage. According to channelings that occurred in a manual approach for that reason the most of the researchers obtained an automatic sleep stage scoring approach which alternatively saves time and to reduces interpretations of experts during PSG recordings and scoring. This approach has achieved good accuracy in the case of identifying sleep disorder that happened in different stages of sleep (Tables 2 and 3).

Table 2 Sleep states EEG epochs representation of subject with no sleep problem

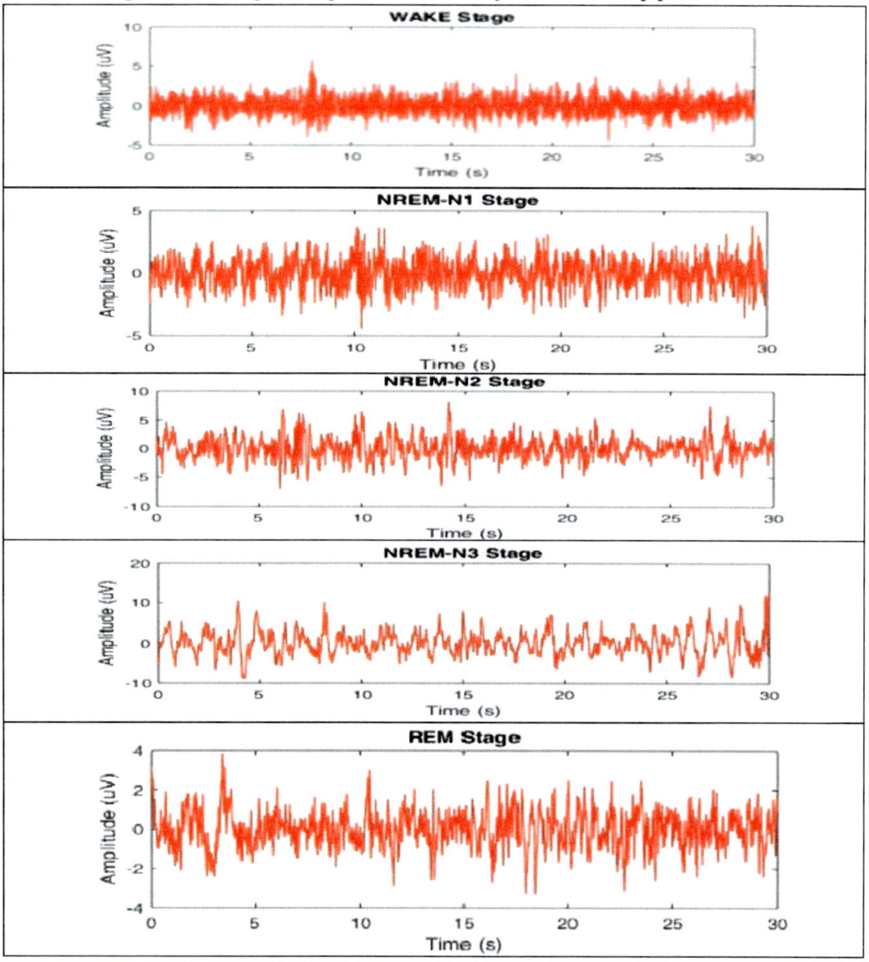

Table 3 Sleep states EEG epochs representation of subject with light sleep problem

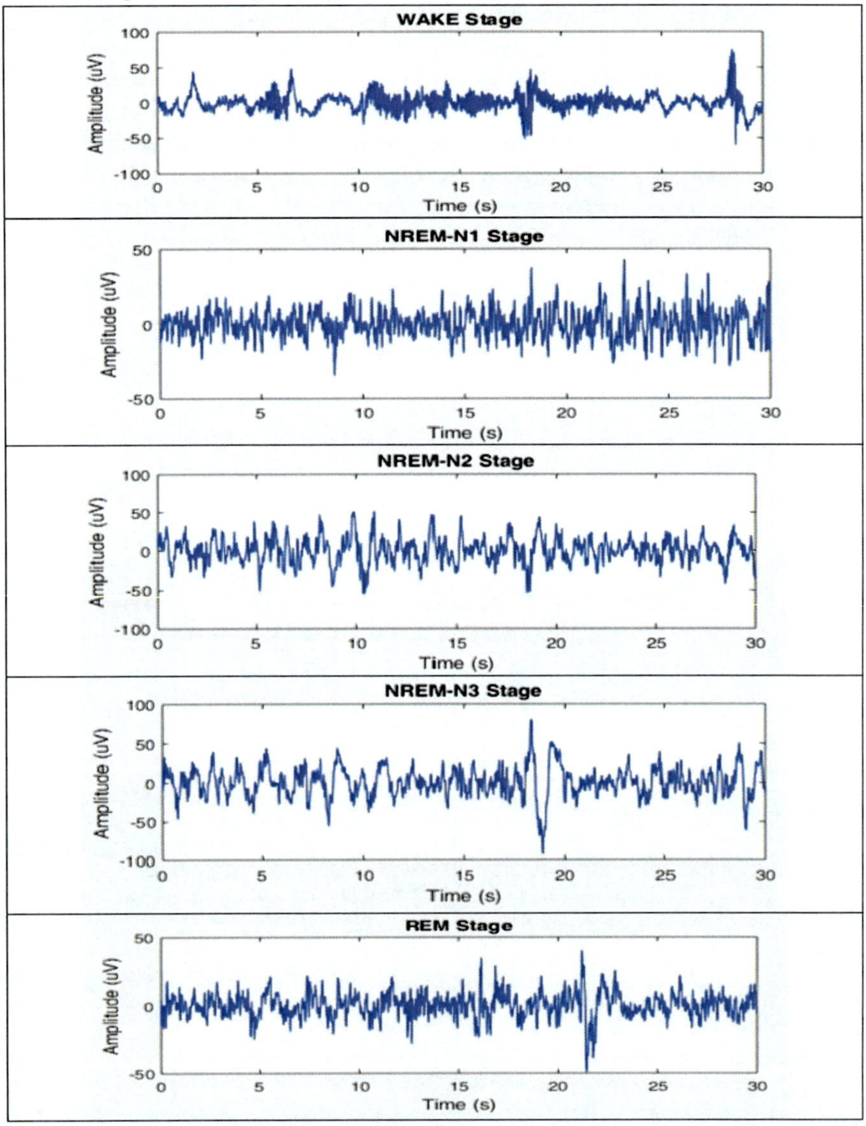

1.1 Literature Survey

Various prior studies have deliberated by different researchers to perform sleep stage scoring and with every experimental work; the authors have obtained different approaches, techniques, and methods subject to feature extraction, feature selection, and classification techniques. Besides, some experimental work has based upon different sleep states classification (Two, Three, Four, Five and Six Stages). Different authors have followed different rules of sleep staging such as R&K rules or AASM rules. Diykh et al. has proposed sleep stage classification based on structural graph similarity concept of feature extraction, only statically features are extracted and extracted features by utilizing the graph similarity concept and obtained K-means classifier to classify the stages [4]. Fraiwn et al. developed four sleep state classifications by considering 16 recordings of a single channel of EEG (C3-A1). He obtained random forest classification techniques for classifying the multiple sleep stages and it has found that the overall accuracy of classification reached to 84% [5]. In [6] the author has designed a sleep study based on single channel, only considered the energy features and obtained recurrent neural network classification techniques and achieved the overall classification accuracy rate as 81.8%. In [7] the sleep scoring evaluation is based upon multiple classifiers such as SVM and ANN for classification tasks. In this work, the authors have also used the concept of wavelet tree analysis and neighborhood component analysis concerning feature extraction and feature selection. The accuracy percentage was scored around 83%. Khald Aboalayon et al. have considered Sleep-EDF datasets used for experimental work. In this study features extracted from different frequency bands such as α, β, σ, δ and θ respectively and extracted features from different sub-band of EEG signal forwarded into SVM classifier for discrimination among different sleep stages. The overall accuracy results reached 92.5% [8]. Ahanf Rashik Hassan et al. has proposed a decision support system for automated investigation of sleep stage irregularities during sleep by considering EEG signal. Here, authors obtained data from single channel and have utilized the wavelet transform concept for signal segmentations. The proposed study experiments upon two datasets such as the Sleep-EDF database and DREAMS subjects' database. For two-state classification, it has reached an overall accuracy of 92.43% [9]. Md Hayat et al. were used to combine multi-channel of EEG (C4-A1 and C4-P4) signals for the detection of the sleep disorder. In this work, the author has focused two-state classifications in between S1 and rapid eye movement and utilized welch techniques for extracting the features from input channels. For classification model, here author has obtained decision tree classification algorithm. This experimental work carried through combinations of dual channels and achieved overall accuracy of 81.25%. [10].

In our literature review, we found that most of the research articles with subject to automated investigation of sleep stages irregularities based on EEG signal and also used one type of dataset frequently by researchers as the Sleep-EDF dataset from physio net. It has observed from existing contributed work that the researchers considered either fully affected with sleep problems or mild effect with sleep problems. Finally, we have considered these challenges for our proposed sleep study. We are trying to solve these basic shortfalls found from contributed research work in our proposed experimental work on SleepEEG Study. In the present research work, we present an application on automatic sleep stage classification from multiple channels of EEG signals. The main contribution of this research work is threefold.

1. The first objective of this work is to design a better decision model for investigating the sleep disorders detection through new ASSC methods. Our experimental work is obtained different gender health condition subjects for demonstration purpose.
2. The second objective of this study is to make a brief comparative study of different machine learning classifiers to decide which classification techniques are most suitable for sleep–wake detection.
3. To identify which channel is most effective for the diagnosis of wake-sleep detection.
4. In this research part, we have used the ISRUC-Sleep dataset, in which we consider both category subjects such as healthy subjects with no medication earlier with sleep problems and another one to be mild difficulty with sleep problems.

The rest of the research work is described as follows. The proposed methodology is described in Sect. 2. Section 3 explains the experiments conducted, present their outcomes and discussion about their significance with related to proposed objectives. Finally, the concluding of this paper and mention some points for the future scope of this proposed work is described in Sect. 4.

2 Proposed Work

This research work proposed an automated decision model with subject to investigation of sleep-related irregularities. The proposed scheme includes processes such as raw data collection from respective channels, preprocessing raw signals followed by feature extraction and finally obtained a classification model. The complete framework of the proposed decision model is shown in Fig. 1. Here, we have obtained the subject behavior during sleep and those recordings are extracted from two different fixed electrodes such as F3-A2 and C3-A2 and those channel recording are obtained from public repository named as ISRUC-SLEEP database. The channel recordings of 30 s time window are picturized in Table 4. In the process of preprocessing, we applied bandpass filter techniques to reduce the artifacts, irrelevant noise from recorded channels. Afterward extracted statistical features from considered channels. Next to this step, we have adopted feature selection techniques to select most relevant features for classification task. At last, we have deployed two machine learning classifiers such as DT and KNN to classify the human sleep stages.

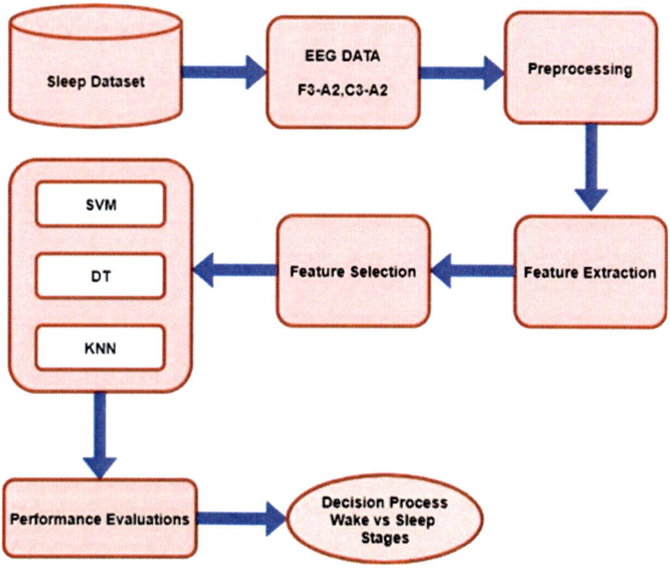

Fig. 1 A schema of our proposed automated sleep stage classification framework

2.1 *Subjects and Signals Acquisition Under Study*

In this sleep study, we have obtained data from all-night PSG records; each of recording time was 8 h. Here we have considered recordings from one of the effective public datasets named ISRUC-Sleep. This dataset completely prepared for sleep disorder analysis and all those recordings done by the Sleep Medicine Centre of the Hospital of Coimbra University (CHUC) [11]. This dataset segmented into three groups of data. This database contained three different set of recordings from different category of patients such as healthy subjects with no sleep symptoms, subjects who has effected with sleep-related disorders and subjects under sleep medication. The recorded data was interpreted through two sleep experts. Out of these three subsection data, only in subsection-II contained two sets of PSG recordings with different dates under supervision of two sleep experts. This dataset exclusively used by many sleep researchers to investigate the sleep disorder. In this proposed work we have used two subjects for our experimental work one from a subject with a sleep disorder (subsection-II) and other from a subject with under sleep medication (subsection-III). In subsection-II, out of 8 subjects recordings, 6 of them are male category and 2 of them are belongs to the female category and age 46 ± 18 years. In subsection-III, out of 10 subjects, 9 subjects under the male category and one belong to the female category and age 40 ± 10 years. All the acquired channel signal data were sampled at 200 MHz and all these subject recordings have been annotated accordingly with AASM manuals on 30 s epochs wise. This dataset contained bio-signals of EEG, EOG and, EMG from 11 electrodes as per 10–20 international electrode placement

Table 4 A 30 s segment of the EEG recording analysis of original and preprocessed channels of F3-A2 and C3-A2

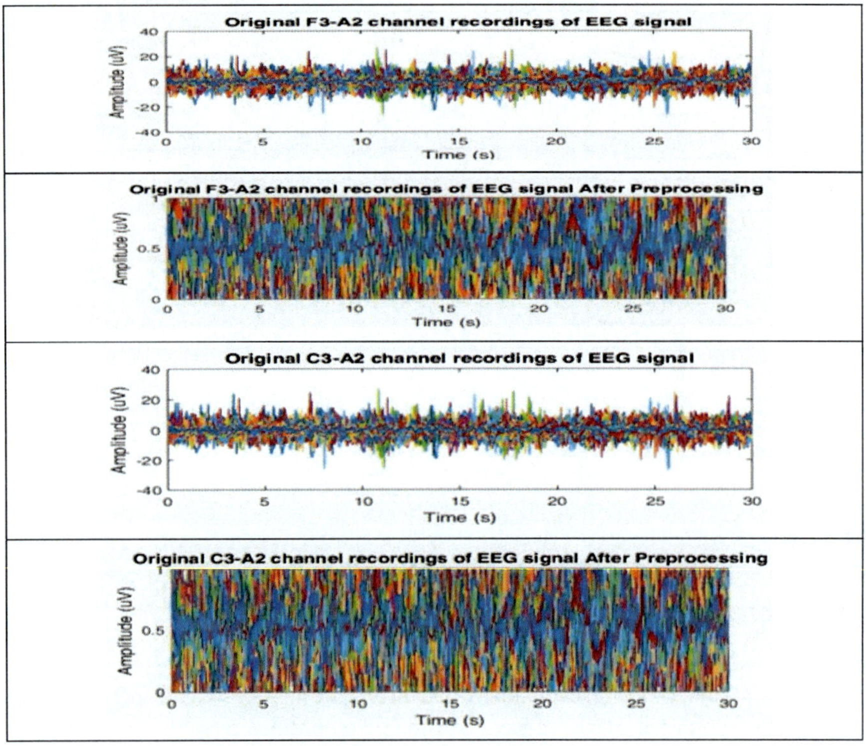

standards. In this study, we have selected only two subjects with dual channels of EEG signals such as F3-A2 and C3-A2. Each obtained subjects five sleep stage of EEG signal segments (healthy and mild sleep problem) are shown in Tables 2 and 3. The EEG dual channels 30 s recordings for both subjects is shown in Table 4. The brief information regarding channel acquisition of this dataset shown in Table 5. The epochs distribution with each sleep stages for both obtained subjects is presented in Table 6.

Table 5 Structure of dataset of ISRUC-Sleep Subsection-II and subsection-III

Bio-signal Information	Acquired Channels
EEG	C3-A2, C4-A1, F3-A2,F4-A1,O1-A2,O2-A1
EOG	LOC-A2,ROC-A1
EMG	Chin EMG,Leg-1 EMG,Leg-2 EMG

Table 6 Description of sleep stages for obtained subjects

Subject label	Total epochs	Awake stage	NREM1 stage	NREM2 stage	NREM3 stage	REM stage
Subject-3 (Subsection-II)	750	67	127	272	174	110
Subject-1 (Subsection-III)	750	149	91	267	158	85

2.2 *Signal Preprocessing*

In the data preprocessing step we have applied some data preprocessing techniques upon acquired raw signals for reducing the noise factor and some irrelevant signal characteristics. Normally the acquisition EEG signals are composed of several signal patterns such as alpha-band (α), beta-band (β), theta-band (θ), delta-band (δ), sawtooth, spindle and K-complex. The characteristics of different signal wave patterns are shown in Table 1. Our data preprocessing method was applied filtering mechanism for removing inconsistency signal compositions through obtained band-pass butter worth filter techniques, with a lower cut-off frequency of 0.5 Hz and a higher cut-off frequency of 45 Hz.

2.3 *Feature Extraction*

Next to data preprocessing, we have obtained the feature extraction step and to discriminate the behavior of among sleep stages, we have extracted certain properties from input channels. Here we have extracted the statistical features from respective input channels to distinguish between the different behaviors that occurred during sleep hours and which ultimately help to investigate the irregularities that happened between different transition states of sleep. The detailed explanations of the extracted features of this proposed study is shown in Table 7.

2.4 *Feature Selection*

Feature selection is one of the important parts of sleep study experimental work for attribute selection. It is the process to reduce the input features into most suitable informative features for the classification model construct. In this study, we have obtained online streaming feature selection (OSFS). These techniques help for selecting the strongly relevant and non-redundant features from features extracted vector. These OSFS techniques categorized the extracted features into three section strongly and weakly features and add in to with one more set called Best Candidate

Table 7 Extracted features

Features extracted	Equation	Features extracted	Equation
Mean (Fe1)	$\frac{1}{n}\sum_{1}^{N} Xi$	Zero crossing rate (Fe7)	$\frac{1}{N-1}\sum_{i=1}^{N-1} Xi - X_{i-1}$
Maximum (Fe2)	max[Xn]	Percentile (Fe8)	$\frac{P}{100} \times (N+1)$
Minimum (Fe3)	min[Xn]	Skewness (Fe9)	$\sum (Xi - \bar{X})^2$
Standard deviation (Fe4)	$\frac{\sum_{i=1}^{N}(Xi-\bar{X})}{N-1}$	Kurtosis (Fe10)	$\sum_{i=1}^{N} \frac{(Xi-\bar{X})/N}{\mathrm{SD}^4}$
Median (Fe5)	$\frac{(N+1)}{2}$ th	Energy (Fe11)	$\sum_{i=0}^{N-1} \lvert X[i] \rvert^2$
Variance (Fe6) $\frac{\sum x^2 - \frac{(\sum x^2)}{N}}{N-1}$			

Table 8 Best combinations of features for classification

Participants name/gender	Channel	Best feature combinations
Subject-3 (with light sleep problem)	F3-A2	Fe1, Fe2, Fe7
	C3-A2	Fe1, Fe8, Fe10
Subject-1 (healthy subject)	F3-A2	Fe1, Fe2, Fe4, Fe5, Fe7, Fe10
	C3-A2	Fe1, Fe2, Fe3, Fe4, Fe7, Fe9, Fe10, Fe11

Feature (BCF). Before any new features considered for classification model first of all OSFS has scrutinized either it is strongly relevant according to class label or not. If it is suitable as per class label then add into BCF else discarded. In this manner, OSFS is finalized the feature vectors for the classification model [12]. Here we have also selected the best feature set for each input channel through this algorithm and the details is shown in Table 8.

2.5 *Classification*

Classification process in this research work carried through two machine learning classifiers such as Decision Tree (DT) and K-Nearest Neighbor (KNN). Among the various classification techniques, these two classifiers have the most effective in terms of the sleep staging problem of human beings because of their good learning capability and robustness property. Table 9 indicates proposed two-class sleep problem.

Decision Tree (DT): It is mainly used to solve classification problems based on the machine learning approach. This algorithm is simple, very fast and handles multi-dimensional data. Generally, the DT concept completely based upon the algorithmic

Table 9 Two-state classification problem considered in this paper

Classification	Sleep stages
2-Class Problem	Wake stage versus sleep stage (N1 + N2 + N3 + REM)

approach, here the whole dataset divided into the number of branches and sub-branches depend upon certain conditions. The main goal of this approach to create a model that predicts the target variables by inferring some decision rules and the rules are followed through if-then-else statements. The more depth of tree-level the rules are more complex and that model is more fit to the objective [13].

K-Nearest Neighbor (KNN): It is one of the non-parametric oriented algorithms. This classifies the different classes based on feature similarity. KNN measures closeness through some distance metric equations such as Euclidean and Mahalanobis. KNN compares the test samples into training instances presented in N-dimensional spaces by measuring their closeness and assigns those test samples to appropriate class labels based on the majority of the vote of its k-nearest samples [14].

3 Result Analysis and Discussion

Here we explain details about our proposed experimental process and performances of the classification model, here we have considered two subjects from different health condition categories. One subject we have selected for our experimental work with no early medication related to sleep problem or alternatively we called as healthy subject and the other subject chosen from that subject which has faced a light sleep problem during night hours. Here we have considered 750 epochs from each subject with time periods of individual epochs is 30 s. In this research study; we have extracted brain signals from dual channels (F3-A2 and C3-A2) of EEG signals. We have also selected different gender subjects for this analysis work. The entire work was carried through the MAT LAB 2017a platform. This study attempts two class (Wake vs. Sleep) classification problem to classify the sleep stages and different machine learning classification techniques were applied to discrimination in between irregularities occurred during sleep. In this research work, we have obtained evaluation metrics that is accuracy, sensitivity (also known as recall), specificity, precision (positive predictive value), and *F*-Score for make a comparison in between different classification model [15–19].

$$\begin{aligned}
\text{Accuracy} &= (\text{TP} + \text{TN})/(\text{TP} + \text{FN} + \text{TN} + \text{FP})\% \\
\text{Sensitivity} &= \text{TP}/(\text{TP} + \text{FN})\% \\
\text{Specificity} &= \text{TN}/(\text{TN} + \text{FP})\% \\
\text{Precision} &= \text{TP}/(\text{TP} + \text{FP})\% \\
\text{Fscore} &= (2 * \text{Sen} * \text{Pre})/(\text{Sen} + \text{Pre})\%
\end{aligned}$$

where TP is the number of true positives, TN is the number of true negatives, FP is the number of false positives, and FN is the number of false negatives. Additionally to evaluate the system performances, we used k-fold cross-validation techniques. In this process, the whole dataset is divided randomly into k equal-size subsets. In individual fold, the $(k-1)$ subsets used as the training and validating data and one subset are used as testing subset. This approach continued k time periods and in this approach, each subset used exactly once for testing called an outer fold. Each outer fold, k inner fold has further segmented in such a manner that both training and validating part is equally segmented into k equal-size attributes. For each inner fold, the $(k-1)$ subsets are considered as training part and 1 subset is considered as validating part. This process continued until each individual subset is used once as a validating part. Finally, the results from inner folds are considered as system parameters and averaged outcomes of k-fold are treated as performance of the proposed system. The experimental performances of the proposed sleep stage scoring study is shown in Tables 10 and 11 for a healthy subject and similarly, for mild sleep problem the performance metric is presented in Tables 12 and 13 for both the two channels.

For the 2-class problem, using ISRUC-SLEEP healthy subject dataset the overall accuracy rate achieved 81.6% through the DT classifier and it is reached 73.7% in case KNN classifier for channel F3-A2.In a similar manner for the C3-A2 channel, DT

Table 10 Performance parameter result of healthy subject (F3-A2 Channel)

F3A2	Accuracy (%)	Sensitivity (%)	Specificity (%)	Precision (%)	*F*-score (%)
DT	81.6	95.6	24.8	83.6	89.1
KNN	73.7	84.1	31.5	83.2	83.6

Table 11 Performance parameter result of healthy subject (C3-A2 Channel)

C3A2	Accuracy (%)	Sensitivity (%)	Specificity (%)	Precision (%)	*F* score (%)
DT	79.2	92.3	26.1	83.4	87.6
KNN	77.7	87.6	37.5	85	86.2

Table 12 Performance parameter result of mild sleep subject (F3-A2 Channel)

F3A2	Accuracy (%)	Sensitivity (%)	Specificity (%)	Precision (%)	*F* score (%)
DT	91.1	99.8	14.9	91.1	95.3
KNN	84.8	91.6	14.9	91.6	91.6

Table 13 Performance Parameter Result of Mild Sleep Subject (C3-A2 Channel)

C3A2	Accuracy (%)	Sensitivity (%)	Specificity (%)	Precision (%)	Fscore (%)
DT	90.9	99.8	0	91	95.2
KNN	85.9	92.8	14.9	91.7	92.2

and KNN classifiers achieved an overall accuracy of 79.2% and 77.7% respectively. In other parts, for mild effected light sleep subject the performance of an overall accuracy achieved for classification through DT reached 91.1% and for KNN it is reported to 85.9%. As it can be observed from Tables 10, 11, 12, and 13 the proposed system performance is more suitable with respect to sensitivity for both healthy subject and mild sleep affected subject. It is reported for healthy subjects 95.6% for DT and 84.1 KNN classifier for channel F3-A2.In the same way for the C3-A2 channel, the sensitivity reached 92.3% and 87.6% for DT and KNN respectively. It was observed that the performance of the subject with suspected light sleep problems in terms of sensitivity is reported 99.8% and 91.6% with respective to DT and KNN classifier for the F3-A2 channel. With same subjects and classifiers for the C3-A2 channel, the sensitivity achieved to 99.8 and 92.8%.All the result statistics of the proposed evaluation metrics represented in Figs. 2, 3, 4 and 5 for both subjects and input channel. The performance comparison of our proposed system and one existing contribution of state-of-the-art work is summarized in Table 14.

The performance of earlier some contributed work on sleep stage classification are given and made a comparison with achieved performance of proposed work.

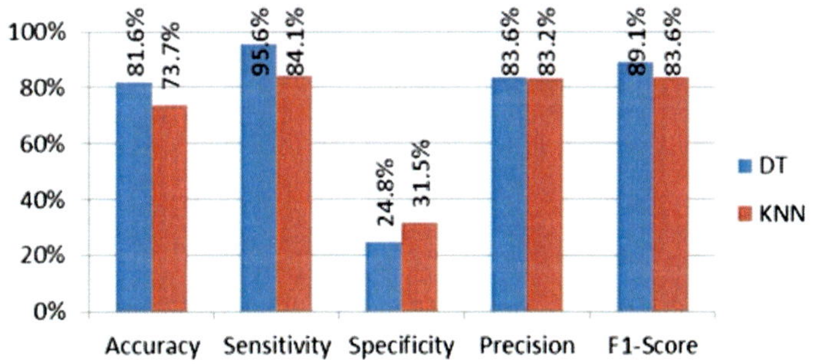

Fig. 2 Overall performance metric evaluation (healthy subject, F3-A2 Channel)

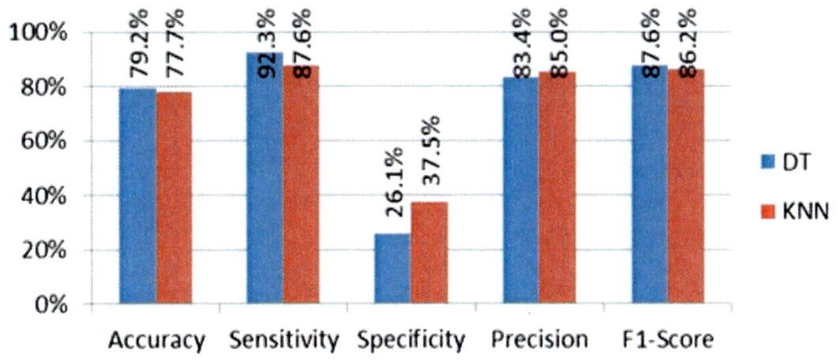

Fig. 3 Overall performance metric evaluation (healthy subject, C3-A2 Channel)

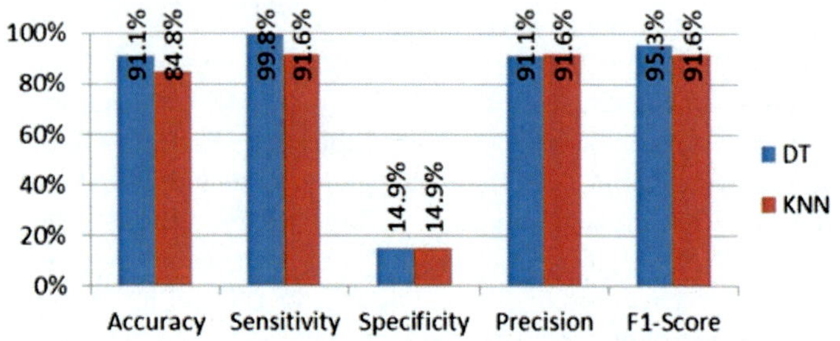

Fig. 4 Overall performance metric evaluation (suspected light sleep problem subject, F3-A2 Channel)

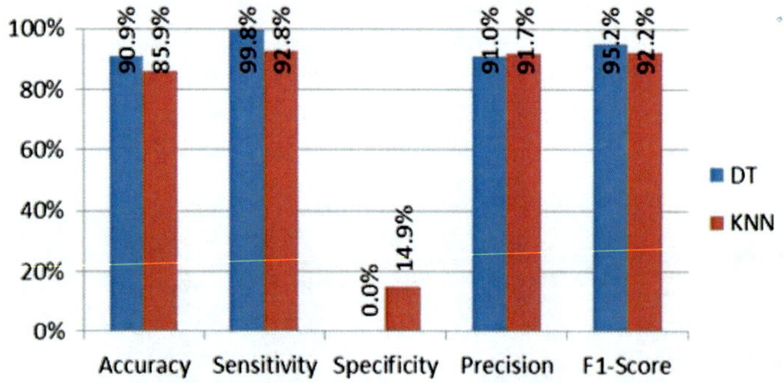

Fig. 5 Overall performance metric evaluation (suspected light sleep problem subject, C3-A2 Channel)

Table 14 Classification performance comparison of the proposed scheme with some existing schemes

System	Year	Accuracy (%)	Dataset Used
Khalighi et al. [11]	2016	88.87	ISRUC-Sleep
Sousa et al. [20]	2015	86.75	ISRUC-Sleep
Katerina et al. [21]	2018	75.29	ISRUC-Sleep
Alizadeh Savareh et al. [22]	2018	89.93	PhysioNet Sleep-EDF
Proposed Work	2020	91.1	ISRUC-sleep

4 Conclusion

In this research work, we proposed a 2-class sleep stage classification system using dual channels of EEG signal. In this sleep study, we have segmented each epoch recording into 30 s. Eleven statistical features were extracted from dual input channels from subjects. In this experimental work, we have considered two subjects with suspected light sleep problem and the other was healthy with no prior medication is related to sleep problems. The OSFS algorithm was applied for selecting the most effective features for the classification task. In this proposed study, classification is based on DT and KNN. The proposed work considering some important factors such as the datasets used the number and the type of channel selection and the type of cross-validation.

The proposed system has some advantages the first one it signifies the channel effectiveness and selection for data recordings which indirectly helps to sleep experts to take certain decision during the selection of input channel. Second, this work has experimented in between completely healthy and with suspected sleep problems, this comparison performance supports sleep experts to make certain decisions during treatment. Third one the system performance does not degrade as much even also we have dealt with suspected light sleep problem subjects. Some limitations also reported in the proposed study, i.e., the number of samples consideration is very less and the second one we have only recorded the recordings from two channels. In future, the model will be extended for other sleep stages classification problem and also we will consider the other bio-signals such as EOG and EMG signals for evaluation of sleep pattern irregularities. A combination of different domain feature extraction and different ensemble classification techniques employed in future research works.

References

1. No communicable diseases country profiles 2018. Geneva: World Health Organization; 2018. Licence: CC BY-NC-SA 3.0 IGO.Available from: https://www.who.int/ nmh/publications/ncd-profiles-2018/en/
2. Rechtschaffen, A., et al.: A manual of standardized terminology, techniques and scoring system for sleep stages of human subjects (1968)
3. Iber C, et al.: The AASM Manual for the Scoring of Sleep and Associated Events: Rules, Terminology and Technical Specifications. American Academy of Sleep Medicine, Westchester, IL (2007)
4. Diykh, M., Li, Y., Wen, P.: Eeg sleep stages classification based on time domain features and structural graph similarity. IEEE Trans. Neural Syst. Rehabil. Eng. **99**, 1–1 (2016). https://doi.org/10.1109/TNSRE.2016.2552539
5. Fraiwan, L., Lweesy, K., Khasawneh, N., Wenz, H., Dickhaus, H.: Automated sleep stage identification system based on time–frequency analysis of a single EEG channel and random forest classifier. Comput. Methods Programs Biomed. **108**(1), 10–19 (2012). https://doi.org/10.1016/j.cmpb.2011.11.005
6. Hsu, Y.-L., Yang, Y.-T., Wang, J.-S., Hsu, C.-Y.: Automatic sleep stage recurrent neural classifier using energy features of EEG signals. Neurocomputing **104**, 105114 (2013). https://doi.org/10.1016/j.neucom.2012.11.003
7. Savareh, B.A., Bashiri, A., Behmanesh, A., Meftahi, G.H., Hatef, B.: Performance comparison of machine learning techniques in sleep scoring based on wavelet features and neighboring component analysis. Peer J. **2018**(7), 1–23 (2018)
8. Aboalayon, K.A.I., Ocbagabir, H.T., Faezipour, M.: Efficient sleep stage classification based on EEG signals. IEEE Long Island Systems, Applications and Technology (LISAT) Conference 2014 (2014). https://doi.org/10.1109/lisat.2014.6845193

9. Hassan, A.R., Subasi, A.: A decision support system for automated identification of sleep stages from single-channel EEG signals. Knowl.-Based Syst. **128**, 115–124 (2017). https://doi.org/10.1016/j.knosys.2017.05.005
10. Heyat, M.B.B., Lai, D., Zhang, F.I.K.Y.: Sleep bruxism detection using decision tree method by the combination of C4-P4 and C4-A1 Channels of Scalp EEG. IEEE Access 1–1 (2019). https://doi.org/10.1109/access.2019.2928020
11. Khalighi, S., Sousa, T., Santos, J.M., Nunes, U.: ISRUC-Sleep: a comprehensive public dataset for sleep researchers. Comput. Methods Programs Biomed [Internet] **124**, 180–192 (2016). Available from: https://doi.org/10.1016/j.cmpb.2015.10.013
12. Xindong, Wu., Kui, Yu., Ding, W., Wang, H., Zhu, X.: Online feature selection with streaming features. IEEE Trans. Pattern Anal. Mach. Intell. **35**(5), 1178–1192 (2013). https://doi.org/10.1109/tpami.2012.197
13. Aboalayon, K.A.I., Almuhammadi, W.S., Faezipour, M.: A comparison of different machine learning algorithms using single channel EEG signal for classifying human sleep stages. In: 2015 Long Island Systems, Applications and Technology (2015)
14. İlhan, H.O., Bilgin, G.: Sleep stage classification via ensemble and conventional machine learning methods using single channel EEG signals. Int. J. Intell. Syst. Appl. Eng. **5**(4), 174–184
15. Sanders, T.H., McCurry, M., Clements, M.A.: Sleep stage classification with cross frequency coupling. In: Proceedings of 36th Annual International Conference on IEEE in Engineering and Medicine Biology (EMBC), Aug 2014, pp. 4579–4582
16. Bajaj, V., Pachori, R.B.: Automatic classification of sleep stages based on the time-frequency image of EEG signals. Comput. Methods Programs Biomed. **112**(3), 320–328 (2013)
17. Hsu, Y.-L., Yang, Y.-T., Wang, J.-S., Hsu, C.-Y.: Automatic sleep stage recurrent neural classifier using energy features of EEG signals. Neurocomputing **104**, 105–114 (2013)
18. Güneş, S., Polat, K., Yosunkaya, Ş: Efficient sleep stage recognition system based on EEG signal using k-means clustering based feature weighting. Expert Syst. Appl. **37**(12), 7922–7928 (2010). https://doi.org/10.1016/j.eswa.2010.04.043
19. Hassan, A.R., Bashar, S.K., Bhuiyan, M.I.H.: On the classification of sleep states by means of statistical and spectral features from single channel Electroencephalogram. In: 2015 International Conference on Advances in Computing, Communications and Informatics (ICACCI) (2015). https://doi.org/10.1109/icacci.2015.7275950
20. Sousa, T., Cruz, A., Khalighi, S., Pires, G., Nunes, U.: A two-step automatic sleep stage classification method with dubious range detection. Comput. Biol. Med. **59**, 42–53 (2015)
21. Tzimourta, K.D., Tsilimbaris, A., Tzioukalia, K., Tzallas, A.T., Tsipouras, M.G., Astrakas, L.G., Giannakeas, N.: EEG-based automatic sleep stage classification. Biomed. J. Sci. Tech. Res. **7**(4) (2018)
22. Alizadeh Savareh, B., Bashiri, A., Behmanesh, A., Meftahi, G.H., Hatef, B.: Performance comparison of machine learning techniques in sleeps scoring based on wavelet features and neighboring component analysis. Peer J. **6**, e5247 (2018)

Development of Interactive Smart Mirror for Implementation in College Environment

Bansari Deb Majumder, Raktim Pratihar, Ratul Saha, and Sourab Ghosh

1 Introduction

Nowadays, life is closely associated with smart devices all around. Smart devices are hugely popular in advancing various areas like smart homes, smart cities, and the internet. Both academia and industries are involved in the development of smart devices in the past recent years. Smart devices are designed to be intelligent and solving the critical problems of society more conveniently. Some of the intelligent devices are smartwatches, smart televisions, smartphones, and more. A smart mirror is one of the intelligent device introduced in the thirteen years before. The smart mirror provides touch-free user interaction for the user on the screen via widgets. Due to time, many types of smart mirrors are introduced by the researchers categorized based on applications. A review paper on smart mirrors based on IoT has been presented, which provides a comprehensive review of the various methodologies of development of smart mirrors [1]. Smart mirrors are very popular as a significant element in the development of smart homes. Smart mirrors are developed incorporating the microcontrollers, Raspberry Pi boards as the core processing elements [2–4]. Researchers have added the Internet of things with smart mirrors to control remotely [5] effectively. The designed smart mirrors have ease of use, simple construction, and prospects of broad applications.

The security aspect of the IoT-enabled smart mirror has also been discussed by the researchers [6]. Henriquez et al. have developed a voice-controlled smart mirror with intelligent and security [7]. Njaka et al. have also developed a voice-controlled smart mirror with multiple authentications [8]. An adequate time-saving smart mirror has been developed with an assistant which offers the news updates [9]. A smart magic mirror is trendy for monitoring the children's activities at home when they work at

B. D. Majumder (✉) · R. Pratihar · R. Saha · S. Ghosh
Department of Electronics and Instrumentation Engineering, Narula Institute of Technology, Kolkata 700109, West Bengal, India

J. K. Mandal and J. K. Roy (eds.), *Proceedings of International Conference on Computational Intelligence and Computing*, Algorithms for Intelligent Systems,
https://doi.org/10.1007/978-981-16-3368-3_18

different locations [10]. Another exciting and more frequently used application is using the smart mirrors in the salon and boutiques for dressing purposes [11]. Smart mirrors are an integral part of the home automation systems. In the past ten years, much work has been reported on smart mirrors for implementing home automation [12–15]. In [15], the authors have reported a smart mirror system as a part of office automation where the device saves time and acts as an interactive medium in the employees' free time.

This paper provides a low-cost design of a smart mirror to be implemented in a college environment. The smart mirror is designed to interact with the user with voice commands and display the information user's question. To implement the said purpose, an assistant has been developed in the Python environment for interaction with the voice commands. The user is benefited not only with the general information, but also with the specific information related to the college. The information comprises the college map, the department's schedule routines, and the college's ongoing events. The smart mirror will also provide live updates of the news, calendar, and weather reports. The assistant (software) is intelligent and can do voice interaction with pre-registered users. It provides information by searching Wikipedia both in display and voice mode. IoT is one of the significant features implemented to access the smart mirror remotely. It stores the user interaction data and provides it to the cloud platform. Similar work has been reported by Akshaya et al., where they developed a smart mirror for implementing a digital magazine in the university [16]. This work is novel in the perspective of the application of the smart mirror in the college environment.

The paper is organized as follows: Sect. 2 will describe the methods and materials of the work. The result analysis and discussion will be shown in Sect. 3, and a conclusion is presented in Sect. 4.

2 Methods and Materials

Figure 1 shows the system architecture of the smart mirror. The smart mirror development is classified into two parts, (a) hardware and (b) software development.

A. Construction of hardware of smart mirror

To construct the smart mirror, a thin two-way reflective sheet is considered with an O- LED screen monitor of dimension 120 cm × 50 cm × 8 cm as a display unit fitted to it. The two-way reflective sheet acts as a mirror for the user. The O-LED monitor is placed at the mirror's back to display the content on the reflective mirror. Raspberry Pi 3B+ board is mounted at the back of the screen and connected with an HDMI cable. The Raspberry Pi board is a credit card size embedded board, and it has 40 G.P.I.O pins for acquiring the analog signals [17]. The Raspberry Pi board is the most popularly used microcontroller for developing various projects worldwide [18, 19]. An auxiliary mic of a 3.5 mm jack and a 10 W stereo Bluetooth

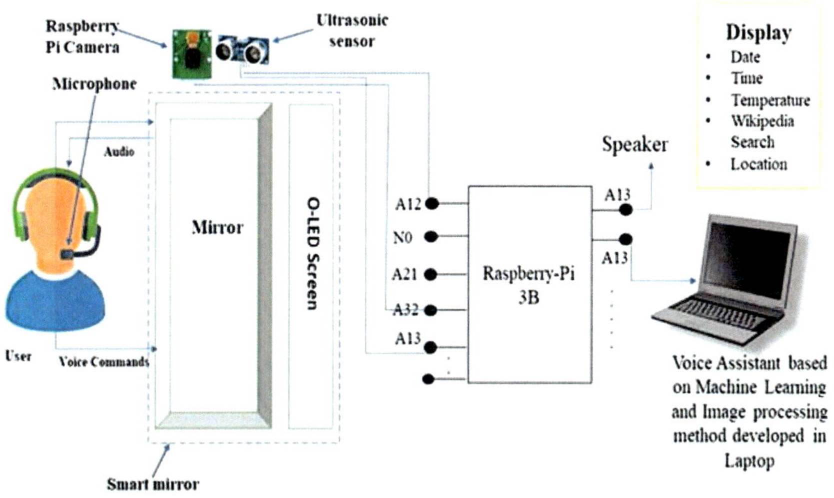

Fig. 1 System architecture of smart mirror

speaker is used as input and output devices. A voice assistant has been developed using machine learning, which takes the audio input from the microphone. Moreover, the audio output from the system is fed to the speaker for the user. Both the input and output devices are connected with the Raspberry Pi 3B+ board. An ultrasonic sensor is connected with the pi board to detect a user's presence before the smart mirror. For considering the smart sensor's power-saving factor, this feature is added to the smart mirror. A Raspberry Pi camera is also connected to capture the user's image and store the data for further processing. Once the image data is captured and processed, it saves the information and gives appropriate pre-determined voice commands like "Hey Ratul, You are looking handsome". Other auxiliary components like a breadboard, memory card of raspberry pi, and connecting wires are used and assembled inside the smart mirror's wooden structure, making it compact and easy to carry. Once the smart mirror's hardware structure is developed, then the software part needs to be configured for the smart mirror's operation.

B. Software development of smart mirror

The software development comprises of the following steps,

i. Installing the Debian-based operating system for the raspberry pi unit.
ii. All the codes and programs are developed in the python platform.
iii. For graphical user interface (G.U.I.), Tkinter has been used.
iv. All the necessary information about the college and general information are displayed when the user is in front of the mirror.
v. Image processing of the captured image of the user has been carried out and recorded.

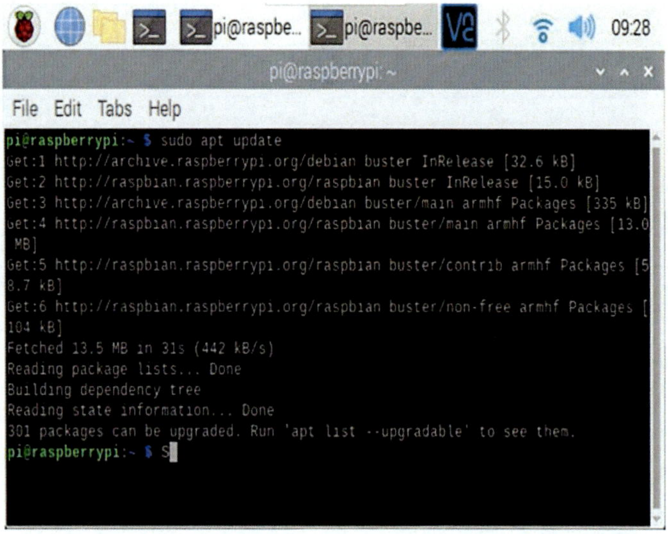

Fig. 2 Installation of Raspbian operating system

vi. A machine learning (ML)-based voice assistant has been developed for interaction with the user.

Now, each of the steps of developing the software of the smart mirror is presented elaborately.

At the initial stage, the Debian operating system for the Raspberry Pi unit installed in the 32 GB memory card with S.S.H enabled on Raspberry Pi, which allows connecting to Linux shell remotely. Now, the system is ready to run the commands and administer the system. A Putty has been used as a S.S.H. client application. The TightVNC has been used as an V.N.C. client, installed on a Pi board to access the laptop/P.C. remotely. Finally, the "Sudo apt-get update" code is given to update the system. Figure 2 shows the screen after the completion of the said installations.

Once the platform is ready, now, different modules of the programs have been implemented. To configure each of the modules, it requires the position values. A module position is assigned to each of the modules, and it also provides the best position values to the developer. Some of the pre-installed modules are news, calendar, weather, and clock. The google calendar has been configured for display at the smart mirror. The calendar can show the holidays, and it has been personalized by adding the upcoming events of the college. The modules are implemented by importing the URL from google in iCAL format. More details can be obtained by clicking the info icons for documentation. Using the news ticker for information can be used to feed the news updates to the display unit. The current API key is needed to retrieve current and forecast weather data from open weather map to show the current weather conditions. The developer has to sign up with an open weather map's website with a free API for getting the weather data. After obtaining the API

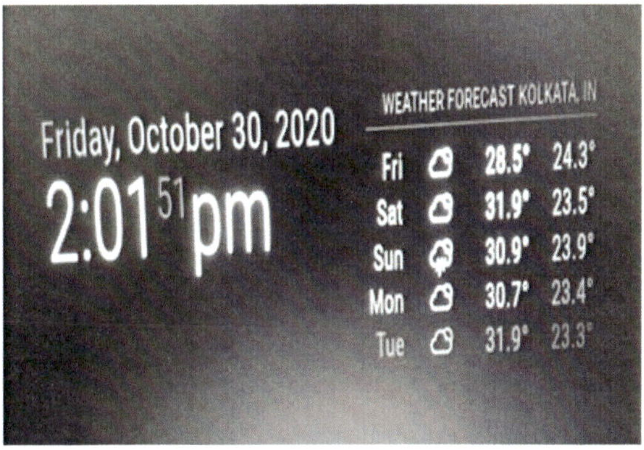

Fig. 3 Installation of modules of weather, date, clock, and news

key, the location I.D. of the open weather (https://openweathermap.org/find) need to be searched. Finally, the current weather conditions are displayed on the O-LED screen, and the developer gets the location I.I.D. and API key. Figure 3 shows the display unit after installing the news, weather, calendar, and clock modules.

Now, to identify the user, a predefined set of images has been given to the processor. When a user comes in front of the smart mirror, the fitted Raspberry pi camera clicks the user's image. The new image is compared with the reference images and generates the output in terms of the predefined compliments of the day. The unique feature incorporated in the mirror and adding the feature of smartness to it. The methodology of implementation is as follows, (i) Capturing the image as input, (ii) Dividing the images into different sections and each section is considered as a separate image, (iii) Now, all the sections of images are passed to CNN, and it is classified, and (iv) Finally, we can re-construct the image and get the original image with detected objects. The said method of processing of images has been implemented in the smart mirror. The image detection and identification of the user are shown in Fig. 4.

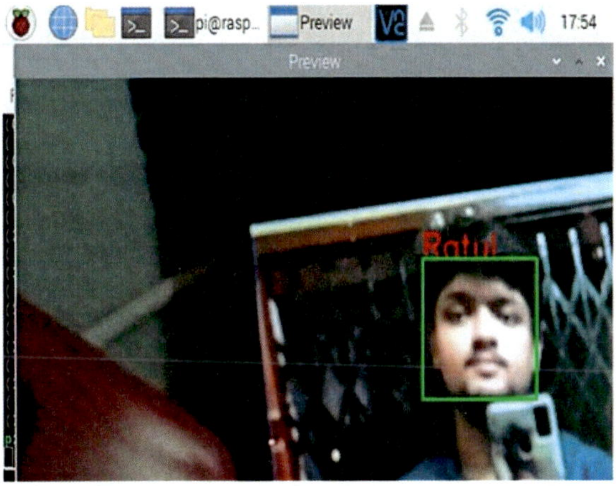

Fig. 4 User identification by image processing in python platform

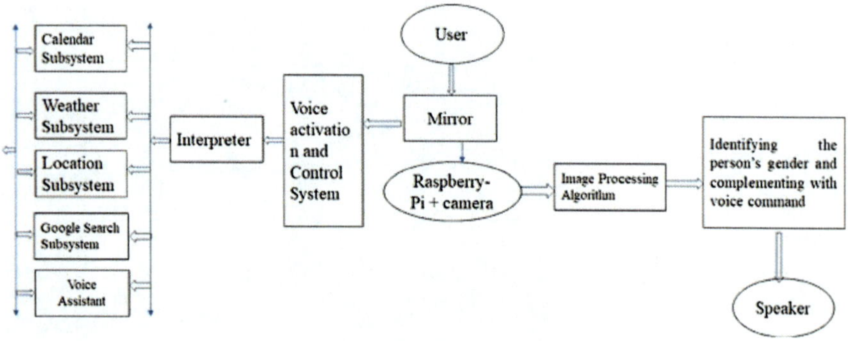

Fig. 5 Dataflow in the smart mirror

The dataflow of the smart mirror is shown in Fig. 5. The dataflow firstly starts with identifying the user with the voice activation and activations of the control systems. As the logic goes well, the modules of the raspbian are executed with the display of date, time, weather, and location. The program also takes the voice command, and the voice assistant is activated. The voice assistant answers the questions by searching Wikipedia and generates its output as voice commands. The second part of the program is the image processing part, where the user's identification is made by comparing with a stored predefined set of data. Finally, the smart mirror displays the user's name with some of the predefined compliments and quotations.

C. **Development of the voice assistant**

The voice assistant for the smart mirror is developed using machine learning. In this method, the voice command is taken as the input signal. Then, the algorithm of machine learning has been developed in a python environment. For developing the algorithm, the following library functions are used: speech_recognition module is required for voice recognition. Then, the pyttsx3 module is used to convert the text to speech data. The machine has then been trained with a logic database or set comprising the output voice commands matching the input voice commands.

This training set data is developed using IF-ELSE statements. Finally, the output signal as voice commands is generated by the program. This voice assistant can also search the data from Wikipedia to answer the questions of the user. As the assistant's application is specifically for the information of the college, the authors have not used google assistant or Alexa as the assistant of the smart mirror. Here, the authors can customize the voice assistant's data from time to time, which is not possible for other assistants.

3 Results Analysis and Discussion

Tables 1 and 2 show the results of identifying the images by the smart mirror, considering the sample of 50 images. In Table 1, 10, images are considered from the sample of 50 images. Then their visibility and non-visibility of the no. of images are indicated. In Table 2, 5, images are considered from the sample of the same 50 images. Finally, the visibility and non-visibility of the 10 samples are indicated.

Both the table shows that the visibility of the images depends on the image quality given to the system's database. The graphical view of the observation data is shown in Figs. 6 and 7.

Figure 8a, b shows the display unit of the smart mirror. When the smart mirror identifies the user, Fig. 8a is displayed on the O-LED screen. Further, Fig. 8b indicates the interaction between the user and the smart mirror. Finally, the smart mirror has been tested successfully and installed in the college premises. It has been beneficial to the user with time saving and getting different informations at a glance.

Table 1 Visibility of the images considering 10 samples

No. of trials	visibility (no. of images)	Non-visibility (no. of non-images)
1	8	2
2	5	5
3	7	3
4	8	2
5	6	4

Table 2 Visibility of the images considering 5 samples

No. of trials	visibility (no. of images)	Non-visibility (no. of images)
1	4	1
2	3	2
3	2	3
4	4	1
5	4	1
6	3	2
7	1	4
8	3	2
9	4	1
10	4	1

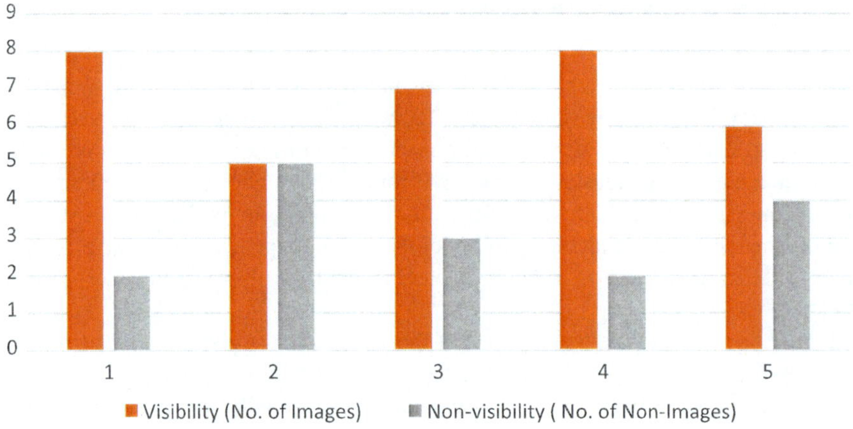

Fig. 6 Visibility of the images considering 10 samples

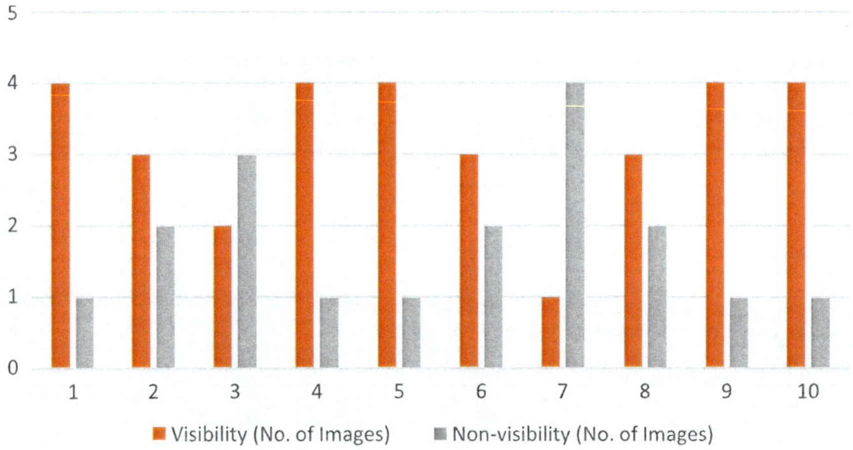

Fig. 7 Visibility of the images considering 5 samples

4 Conclusion

The designed smart mirror can prove an interactive experience to the user with the smart mirror, displaying the general information and specific information related to the institution by integrating computing and communication technologies. The system is providing a faster response and also conserves the traditional operation of a mirror or can be a security system in at an emergency. However, the system has some disadvantages to be overcome, (a) If the system loss Internet connection, then none of the data will be saved in the database, (b) data transaction gets affected if the server slows down, and (c) If the Internet connection gets off, then the voice assistant

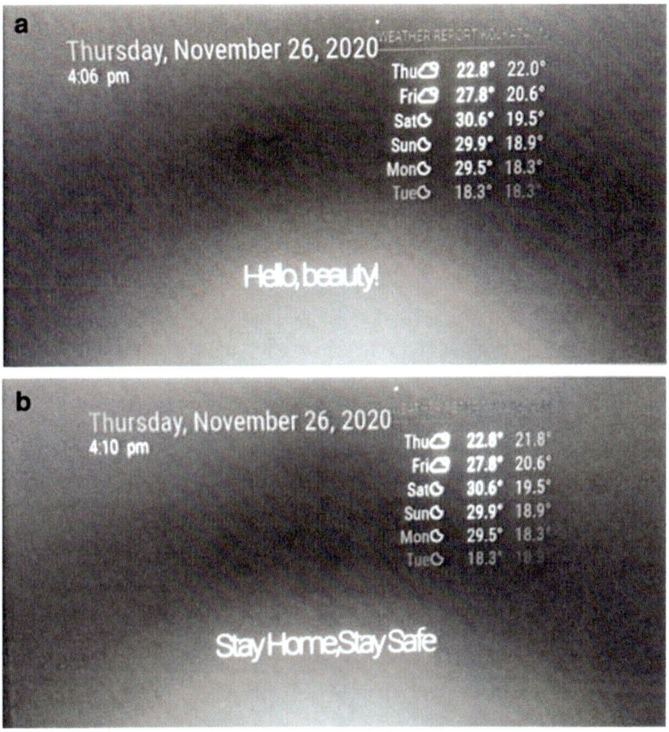

Fig. 8 **a** Display unit of smart mirror with comment "Hello Beauty". **b** The display unit of smart mirror with comment "Stay Home Stay Safe"

fails to provide the searching information to the user. With the rise in technological advancement, the designed smart mirror is superior to others in terms of the availability of the feature of voice and image identification methods. In the future, the authors will not restrict the application of the smart mirror to the college environment's perspective instead focusing on other environments as well. The addition of a touch pad with the smart mirror will also enhance the interactive feature to the user.

References

1. Alboaneen, D.A., Alsaffar, D., Alateeq, A., Alqahtani, A., Alfahhad, A., Alqahtani, B., Alamri, R., Alamri, L.: Internet of things based smart mirrors: a literature review. In: 3rd International Conference on Computer Applications and Information Security (I.C.C.I.C.C.A.I.S.), Mar 2020. https://doi.org/10.1109/ICCAIS48893.2020.9096719
2. Sun, Y., Geng, L., Dan, K.: Design of smart mirror based on Raspberry Pi. In: International Conference on Intelligent Transportation, Big Data and Smart City (I.C.I.I.C.I.T.B.S.), Xiamen, China, Jan 2018. https://doi.org/10.1109/ICITBS.2018.00028

3. Jin, K., Deng, X., Huang, Z., Chen, S.: Design of the smart mirror based on Raspberry PI. In: 2018 2nd IEEE Advanced Information Management, Communicates, Electronic and Automation Control Conference (I.M.CI.M.C.E.C.), Xian China, May 2018. https://doi.org/10.1109/IMCEC.2018.8469570
4. Singh, V., Singh, D.: Smart interactive mirror display. In: International Conference on Machine Learning, Big Data, Cloud and Parallel Computing (COMITCon), Faridabad, India, Feb 2019. https://doi.org/10.1109/COMITCon.2019.8862180
5. Wani, M., Ahire, P.: Real-time smart mirror system using internet on things. In: 5th International Conference on Computing, Communication, Control and Automation (I.C.C.UI.C.C.U.B.E.A.), Pune, India, Sept 2019. https://doi.org/10.1109/ICCUBEA47591.2019.9129315
6. Nadaf, R., Bonal, V.: Smart mirror using Raspberry Pi as a security and vigilance system. In: 2019 3rd International Conference on Trends in Electronics and Informatics (I.C.OI.C.O.E.I.), Apr 2019. https://doi.org/10.1109/ICOEI.2019.8862537
7. Henriquez, P., Matuszewski, B.J., Andreu, Y., Bastiani, L., Colantonio, S., Coppini, G., D'Acunto, M., Favilla, R., Germanese, D., Giorgi, D., Marraccini, P., Martinelli, M., Morales, M.-A., Pascali, M.A., Righi, M., Salvetti, O., Larsson, M., Stromberg, T., Randeberg, L., Bjorgan, A., Giannakakis, G., Pediaditis, M., Chiarugi, F., Christinaki, E., Marias, K., Tsiknakis, M.: Mirror mirror on the wall… an unobtrusive, intelligent multisensory mirror for well-being status self-assessment and visualization. IEEE Trans. Multimedia **19**(7), 1467–1481 (2017). https://doi.org/10.1109/TMM.2017.2666545
8. Njaka, A.C., Li, N., Li, L.: Voice controlled smart mirror with multifactor authentication. In: 2018 IEEE International Smart Cities Conference (ISC2), Sept 2018. https://doi.org/10.1109/ISC2.2018.8656932
9. Johri, A., Jafri, S., Wahi, R.N., Jafri, S., Pandey, D.: Smart mirror: a time-saving and affordable assistant. In: 2018 4th International Conference on Computing Communication and Automation (I.C.CI.C.C.C.A.), Dec 2018. https://doi.org/10.1109/CCAA.2018.8777554
10. Siripala1, R.M.B.N., Nirosha, M., Jayaweera, P.A.D.A., Dananjaya, N.D.A.S., Fernando, S.G.S.: Raspbian magic mirror—a smart mirror to monitor children by using Raspberry Pi technology. Int. J. Sci. Res. Publ. **7**(12) (2017). ISSN 2250-3153
11. Gao, G., Bai, C., Zheng, W., Liu, C.H.: The future of smart dressing mirror: an open innovation concept video. In: 2015 IEEE 12th International Conference on Ubiquitous Intelligence and Computing, Beijing China, Aug 2015. https://doi.org/10.1109/UIC-ATC-ScalCom-CBDCom-IoP.2015.82
12. Hossain, M.A., Trey, P.K., El Saddik, A.: Smart mirror for ambient home environment. In: 3rd I.EI.E.T. International Conference on Intelligent Environments (I.I.E., 07), pp. 589–596 (2007)
13. Yusri, M.M., Kasim, S., Hassan, R., Ruslai, Z.A.H., Jahidin, K., Arshad, M.S.: Smart mirror for smart life. In: 2017 6th I.CI.C.T. International Student Project Conference (ICT-ISPC), May 2017. https://doi.org/10.1109/ICT-ISPC.2017.8075339
14. Mathivanan, P., Anbarasan,G., Sakthivel, A., Selvam, G.: Home automation using smart mirror. In: IEEE International Conference on System, Computation, Automation and Networking (I.C.S.I.C.S.C.A.N.), Mar 2019. https://doi.org/10.1109/ICSCAN.2019.8878799
15. Hamza, M., Lohar, S.A., Ghulamani, S., Shah, A.: Smart mirror for home and work environment. IEEE 6th International Conference on Engineering Technologies and Applied Sciences (I.C.E.I.C.E.T.A.S.), Dec 2019. https://doi.org/10.1109/ICETAS48360.2019.9117296
16. Akshaya, R., Niroshma Raj, N., Gowri, S.: Smart mirror-digital magazine for University implemented using Raspberry Pi. In: International Conference on Emerging Trends and Innovations in Engineering and Technological Research (ICETIETR). Ernakulam, India, July 2018. https://doi.org/10.1109/ICETIETR.2018.8529005
17. Foundation, R.P.I.: Teach, learn, and make with raspberry pi. Raspberry Pi [Online]. Available: https://www.raspberrypi.org/

18. Web enabled pool control—raspberry pi (2016) [Online]. Available: http://www.instructables.com/id/Web-Enabled-Pool-Control-Raspberry-Pi/
19. Patkar, M.: 6 best raspberry pi smart mirror projects we've seen so far (2016) [Online]. Available: https://www.makeuseof.com/tag/6-best-raspberry-pi-smart-mirror-projects-weve-seen-far/

Review of Segmentation and Classification Techniques in Computer-Aided Detection of Brain Tumor from MRI

Sucharita Jena, Mamata Panigrahy, and Jitendra Kumar Das

1 Introduction

Brain tumor is collection of abnormal cells. The abnormal cells can be cancerous (malignant), premalignant, or non-cancerous (benignant). Benignant tumors occur rarely and are homogenous in nature. They do not contain cancerous cells so are surgically cured [1]. Types of Benignant tumors are Meningiomas and Gliomas. Premalignant tumors are not cancerous but they have the tendency to become cancerous. Malignant tumors are heterogenous in nature as it contains cancerous cells which are cured by radiotherapy or chemotherapy. Types of malignant tumors are Glioblastoma and Astrocytoma. Glioma is the maximum occurring brain tumor which is found in the glial cells. Our nervous system comprises of supportive cells knows as glial cells. These cells provide nutrients, cleans cellular waste, and breaks down dead neurons. The tumor present in these cells are astrocytoma, oligodendroglia, and glioblastomas which originate in the cerebrum, frontal-temporal lobes, and in supportive brain tissues respectively [2]. Out of these three, the most aggressive type is glioblastomas. According to the survey of world health organization (WHO), brain tumor is the second major issue of death occurring worldwide [3]. There are four different types of brain tumors such as (grade I to grade IV) according to their level of severity. Diagnosis of a brain tumor is carried out by neurologists. The neurologist performs extensive medical imaging at various stages of the treatment. The medical imaging modalities used for brain tumor detection are Positron emission tomography

S. Jena (✉) · M. Panigrahy · J. K. Das
School of Electronics Engineering, KIIT University, Bhubaneswar, India

M. Panigrahy
e-mail: mamta.panigrahyfet@kiit.ac.in

J. K. Das
e-mail: jkdasfet@kiit.ac.in

J. K. Mandal and J. K. Roy (eds.), *Proceedings of International Conference on Computational Intelligence and Computing*, Algorithms for Intelligent Systems,
https://doi.org/10.1007/978-981-16-3368-3_19

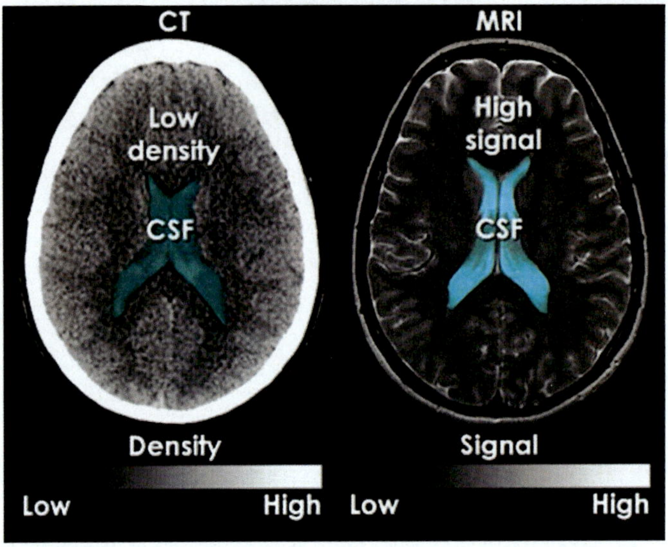

Fig. 1 CT and MRI

(PET), single-photon emission computed tomography (SPECT), computed tomography (CT), and magnetic resonance spectroscopy (MRS) [4]. Out of all modalities, MRI is preferred due to its high resolution and accuracy. MRI provides noninvasive, quantitative measurement of structural, anatomical, and functional information about human tissue. MRI uses strong radio frequency (RF) waves to detect the area and dimensions of the tumor tissues. Brain tissues are defined by two settling times namely vertical settling time (T1) and transverse settling time (T2) [3]. T1 and T2 are the time taken by the excited protons to be in phase and out of phase with the magnetic field respectively. The healthy tissues of the brain are differentiated in T1-images whereas swelling areas in T2-images. Figure 1 presents the difference between CT and MRI whereas Fig. 2 illustrates different types of brain tumors.

With the help of sophisticated medical imaging techniques, a large volume of quality imaging is coming out for interpretation and investigation. Manual investigation of imaging slides and proper interpretation and diagnosis of the probable causes of the disease requires an expert knowledge which is not possible in every hospital. Therefore, a hybrid approach is required where some form of automation (by the use of state-of-art image processing algorithms) is required. This paper provides an overall analysis of computer-aided detection of brain tumor from MRI image focusing on image segmentation and classification. Different limitations of image segmentation and ways to overcome the issues have been discussed.

Section 2 provides an overall description of computer-aided detection of brain tumor from MRI image and Sect. 3 provides conclusive remarks.

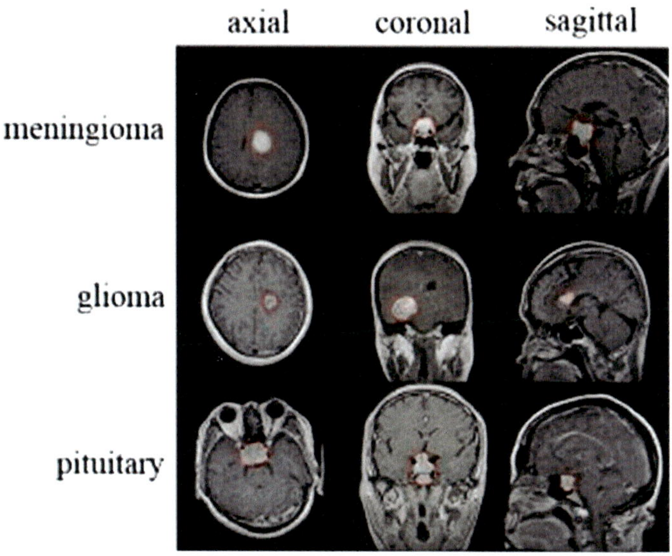

Fig. 2 Different types of brain tumor

2 Computer-Aided Detection of Brain Tumor from MRI Image

The initial step in computer-aided brain tumor detection starts with acquisition of brain MRI in image format followed by pre-processing, segmentation, feature extraction, feature selection, and finally classification. Figure 3 depicts the processing steps of computer aided detection of MRI image.

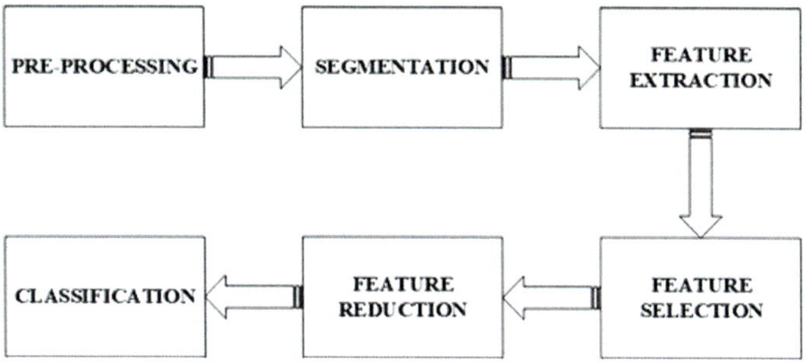

Fig. 3 Computer-aided detection of brain tumor from MRI image

Table 1 Different Dataset Available for MRI Image

Dataset	Samples	Age	Gender
IXI	562	46.85 ± 16.4	249 male 313 female
COBRE	146	36.97 ± 12.78	37 male 109 female
ABIDE	1102	17.08 ± 8.06	163 male 939 female

2.1 MRI Database

Table 1 provides different datasets available in research community where MRI images can be acquired 1.

MICCAI-BraTS is one of the benchmark image segmentation databases used for benchmark results for brain MRI analysis.

Synthetic Brain MR Simulator BrainWeb is an online interface to a 3D MRI simulated brain database (SBD) available at https://brainweb.bic.mni.mcgill.ca/brainweb/. SBD comprises of normal as well as multiple sclerosis (MS) anatomical model of the brain and these are simulated using three different sequences. Variety of slice thickness (in mm), noise level, and level of intensity non-uniformity can be simulated. Different orthogonal views such as (transversal, sagittal and coronal) also available.

2.2 Pre-processing

Despite rapid advances made in the field of MRI, different undesired artifacts are present in the MRI images. The presence of artifacts hamper the image processing steps and therefore, these are required to be removed before proceeding to computer-based segmentation of image.

- Noise
- Intensity inhomogeneity (IIH) or Bias field (smoothly varying intensity inside tissues)
- Partial volume effect (voxel contributing multiple types of tissue).

These limitations of the images are corrected using different mathematical formulations such as de-noising of image, IIH correction, contrast enhancement, or edge-preserving method.

Once the said corrections are made, then computerized image segmentation is carried out. Figure 4 depicts the brain tumor segmentation techniques.

De-Noising MRI images are corrupted by noise during capturing, transmission and digitization. The noises prevalent in MRI images are Salt and pepper (impulse), Gaussian, and Speckle noise which are removed by mean, median, and Gaussian

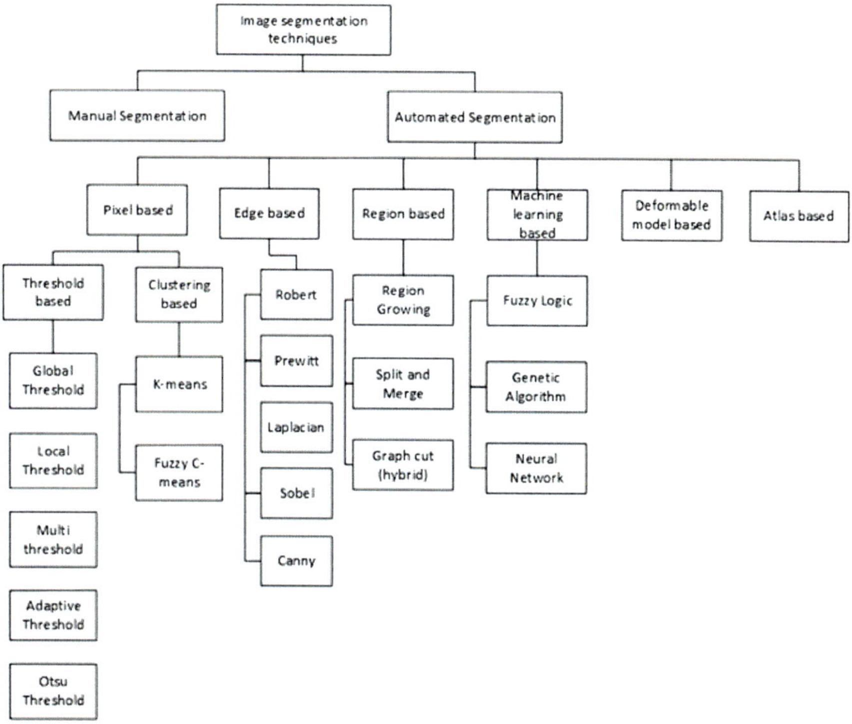

Fig. 4 Different types of image segmentation

filters. De-Noising is done to free MRI images from such noises. Table 2 shows the types of median filters proposed till date as cited in the paper [5, 6].

In [13], the authors present standard as well as conventional filters which distort the fine details of the image and smoothens the image continuously. De-noising and skull stripping algorithm is used as pre-processing technique to improve the image quality. In [14] improves the signal-to-noise ratio of MR images by modified sigmoid function which uses adaptive contrast enhancement technique.

The author Fernandes et al. [15] proposes image enhancement, fusion, and thresholding as the required scheme in pre-processing. The author Gilanie et al. [16] employed Gaussian filter and local histogram equalization for noise removal and contrast enhancement respectively. The paper [17] proposes median filter which works only with distorted pixels to derive the edges by separating the normal and distorted pixels. It removes the extreme values and maintains the edges securing the image boundaries. MRI images have high resolution and quality but leaves behind traces in the scans, which have to be handled. The paper [18] proposes three filters namely anistrophic, gradient, and threshold(sigmoid) filter along with skull stripping as pre-processing steps of MRI image. The paper [19] deals with image preprocessing which includes noise filtering, skull detection, etc.

Table 2 Merits and demerits of different median filters

Median filter types	Paper	Advantages	Disadvantages
Standard median filter (SMF)	[7]	Removes low-density noises without distorting the edges	At high densities, the edges are not preserved
Adaptive median filter (AMF)	[8]	Effective both in high and low noise densities	Blurs the image when window size is increased at high noise densities
Adaptive switching median filter (ASMF)	[9, 10]	It is based on predefined threshold value	The details and edges are not defined properly at high noise level
Decision-based algorithm (DBA)	[11, 12]	If the pixel is 0 or 255, it is processed else left unaltered. At high noise level, the neighboring pixel is used as replacement	Continuous replacement of neighboring pixel produces streaking effect
Decision-based unsymmetric trimmed median filter (DBUTMF)	[12]	Replaces the noisy pixel by trimmed median value	At high noise level, when the pixels are 0 or 255 or both, trimmed median value is not obtained
Modified decision-based unsymmetric trimmed median filter (MDBUTMF)	[5]	At high noise densities, trimmed value is obtained	Do not work on variation of window sizes
Adaptive switching modified decision based unsymmetric trimmed median filter (ASMDBUTMF)	[6]	It works with multiple window sizes which are switched in an adaptive way to generate blur-free noise reduction	It removes noise at more than 90% density level

Anisotropic diffusion process is a non-linear diffusion method which reduces noise from the flat surface in a more effective manner also reduces the diffusivity at the edge of the images. The author [20] proposes segmentation algorithm in four steps namely finding gradient image using Sobel Operator, calculation of image dependent threshold iteratively, applying Closed-Contour Algorithm and finally image segmentation based on pixel intensity within closed contour. The paper [14] proposes Berkeley wavelet transformation (BWT) to enhance the operation and decrease the complexity involved in medical image segmentation. The author [21] describes segmentation process using weighted aggregation (SWA) algorithm. To enhance the speed of segmentation in 2D images with accuracy [22] proposed memory-based learning with two-fold histogram stretching. The paper [23] uses mathematical morphological operations for image segmentation. Usually, medical images have low contrast. So paper [24] employed contrast stretching to enhance the quality of the image. After enhancement, subtractive clustering algorithm is used for cluster center detection, taking potential value of the image into account. The MRI images posses high divergence in the appearance of tumor and unstructured boundaries. To solve this issue, the author Huang et al. [25] proposes a method which exhibit the segmentation of tumor as a classification problem such as local independent projection-based classification (LIPC) method. It classifies each voxel into different classes. The author Kanmani et al. [17] proposes a two-stage segmentation approach using Threshold Based Region Optimization (TBRO). The paper [18] proposes segmentation of MRI images based on combination of region growing and geodesic level set. The author [26] implements an improved region growing algorithm which uses a threshold segmentation algorithm to help in the automatic selection of seed points. The author [19] segments the tumor part from the brain using fast bounding box (FBB). In order to segment MRI images automatically, the author [27] proposed a multiobjective semi-supervised clustering technique. This method uses AMOSA search engine, an optimization technique based on simulated annealing to automatically determine the exact division of MR brain image dataset. The paper [28] developed an idea based on texture features and sparse kernel coding for FLAIR sequence MRI images. Sparse coding is used for noise reduction and compression sensing.

The techniques involved in segmentation process are broadly divided into five categories. They are Manual segmentation, Pixelwise, Region, Edge-based, Deformable Model, Machine learning, and Atlas as given by author in [29].

Manual Segmentation Manual segmentation is time-consuming, expensive, and difficult as tumor regions are to be individually located on all the affected parts of the brain. This process is subjected to manual interruption or variations as each individual will draw different conclusions regarding the tumor. So to avoid these problems an automated segmentation technique is required.

Pixelwise Segmentation It differentiates each pixel of the image based on its intensity and leave the neighbor pixels. The MRI of brain consists of three tissues namely grey matter, white matter, and cerebrospinal fluid which are differentiated using techniques. The histogram analysis of an image determines the threshold in this technique. Thresholding can be of two types: Global thresholding, Local thresholding, multi-thresholding, adaptive thresholding, and Otsu thresholding. Global thresholding uses a single threshold value for segmenting an image having similar intensities. Global thresholding fails when tissues have overlapping intensities. Local Thresholding works when image gradient is smaller than its size. Adaptive thresholding is used to separate the foreground and background image objects depending on the variations in the pixel intensities. However, it works only in smooth background. Last but not least, if segmentation is done by incorporating two threshold limits from histogram analysis of an image, it is called otsu thresholding as cited in paper [30]. The drawback of pixel-based segmentation is that it fails to preserve the edges as it considers only intensity information which affects the bonding between the pixels.

Region-Based Segmentation This technique divides the image into parts based on its equality criteria. It enables the original image to be assembled exactly by combining all regions together without any overlap between two regions. It consists of region growing, region splitting, region merging and region split and merge as cited in the paper [31]. Region growing (seeded segmentation) starts with a single pixel (seed) and slowly add on pixels. The algorithm of region growing starts initially by choosing the seed pixel, secondly adds the neighboring pixels to the region if they resemble the same criteria to the seed chosen. Region merging, region splitting and region split and merge are the higher approach of region growing method. The main feature of this technique is to segment and create similar regions. Region growing suffers from the following drawbacks: it is easily prone to noise, if the seed pixel is not chosen properly, the region extends beyond the image and combines with another image that is not a part of the desired image. Watershed segmentation is a part of region growing which is a common method for MRI brain segmentation. It incorporates erosion dilation, closing, and opening of an image based on the structuring element. The disadvantage of watershed is that it suffers from oversegmentation where the image is divided into greater number of regions. Figure 5 illustrates the algorithm of region growing segmentation.

Edge Based Segmentation This technique is used to find the edges or boundaries of an image. It is fast computationally and does not require any initial information about the image data. It is highly sensitive to acute changes, so it can identify minor edges of individual pixels as in paper [32]. It overcomes the drawback of threshold-based segmentation by determining the proper edges of the segmented image.

Deformable Model-Based Segmentation Deformable model-based segmentation is basically suitable for segmenting images with missing edges, contrast mismatch between background and foreground objects of an image. This segmentation technique is the most recommended as it has the ability to adapt the changes of life structures of various patients. Due to this application it is widely used in a variety of applications specially in medical imaging. It is mainly used for 3D images. This is the best of all methods as it do not require training data. Moreover, it defines the feature that is nearer to the targeted image. The advantage of deformable model is its capability to acquire features of the edges for similar regions.

Machine Learning Based Segmentation This approach is effective for automated analysis of medical images. It is divided into three types: supervised, unsupervised and semi-supervised segmentation. If the training data is labeled manually then it is called supervised segmentation. The advantage is that it can be used to perform other tasks by just altering the training data. If the data is labelled automatically by grouping similar pixels, then it is called unsupervised segmentation. To achieve accurate results is a tedious task especially when the tumors have varying intensities and blurry edges.

Atlas-Based Segmentation When large tumors change shape and place of brain structures, it is very difficult to analyze and segment it visually. At this point of time, atlas-based segmentation provides the details of brain anatomy (location, shape, size) of brain structures without any prior knowledge between regions and pixel intensities. Thus guiding as a reference to segment other images. It has the advantage of segmenting any brain tumor available in the atlas without extra cost. It is really challenging and further research is going on to segment neonatal brains using this

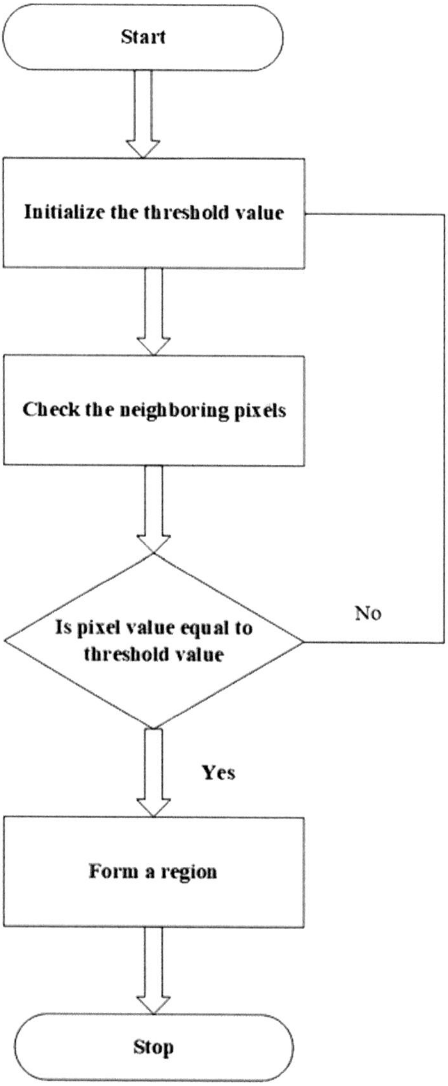

Fig. 5 Algorithm of region-based segmentation

segmentation. As MRI segmentation of such brain is tidious job as it is in the growing stage and reflects poor quality MRI. Therefore, a probabilistic atlas with varying structures is used to segment such brains. The time consumption is very high in atlas methods which accounts for its drawback.

Feature Extraction The next stage in image processing is feature extraction. It extracts features such as size, area, texture, and binary patterns. These features characterize the brain tumor for classification purposes. The brain is a complicated structure so extraction is a challenging task. The methods involved in feature extraction are texture features, wavelet transform, decision boundary, nonparametric weighted, gabor features, minimum noise fraction and Spectral analysis (Fig. 6).

The author [13] proposes discrete wavelet transform (DWT) coefficients as feature vectors for extraction of features. The author [14] follows two steps for feature extraction from the medical images. Initially, gray level co-occurrence matrix (GLCM) is computed, and second the texture features based on the GLCM are calculated. The author [21] proposes two extraction algorithms. First, he discusses an approach that uses saliency-based extraction. Second, he presents a new extraction algorithm that is based on tracing a voxels model details. The author [23] uses GLCM to extract first-order statistical features. The author [16] uses Gabor filter to extract texture features of both normal and abnormal brain MR slices. The author [25] presents a patch-based technique for extracting the image feature. The author [17] uses Harris Laplace detection for extraction of seed points. The author [33] proposes DWT to extract features. The author [19] deals with feature extraction of MR brain images

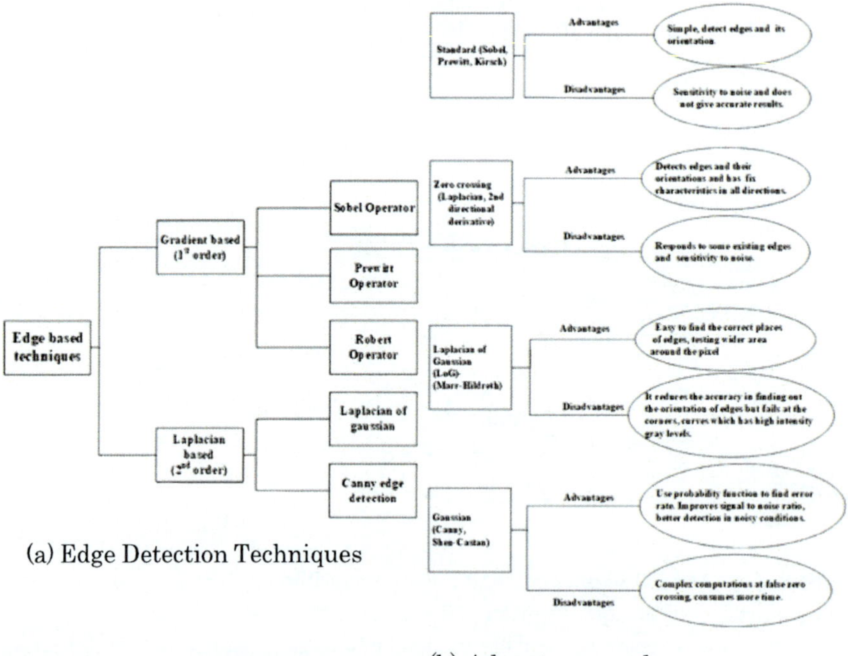

(a) Edge Detection Techniques

(b) Advantages and Disadvantages of Edge Detection Methods

Fig. 6 Edge detection techniques

using gray level co-occurrence matrix. The author [34] here, uses DWT for extracting GLCM for statistical feature extraction.

2.3 Feature Selection and Reduction

If feature vector space increases, there is a chance of considerable decrease in the system accuracy. Therefore, feature selection is used to select the most important features. It increases efficiency of learning models, improves generalization capability, and enhances model interpretability. The methods employed for feature selection are Genetic algorithm (GA), Sequential forward selection (SFS), Particle swarm optimization (PSO), and Sequential backwards selection (SBS). The author [35] proposes hybrid feature selection approach which uses a globally optimized Artificial Neural Network Input Gain Measurement Approximation (GANNIGMA). There are chances of selecting multiple or same features during feature selection. Thus, in order to reduce the dimensionality of features, Feature reduction is done. The methods used are Independent component analysis (ICA), Principal component analysis (PCA), and kernel PCA. It increases the accuracy of classifying input data. The author [23] uses principal component analysis (PCM) for reduction of similar features. The author [33] suggests Genetic algorithm (GA) and Principal Component Analysis (PCA) technique for reducing the number of features.

2.4 Classification

Classification of normal and abnormal brain is the last step in the process of image processing. The classifiers used are Feedforward ANN, Support vector machine (SVM), Self-organizing mapping neural network (SOMNN), Kernel SVM (KSVM), Learning Vector quantization (LVQ), Multilayer Perceptron (MLP), *K*-nearest Neighbor (*K*-NN), Normalized Cross-correlation (NCC), Least Square Feature Transformation (LSFT), Fuzzy *C*-means (FCM), Unsupervised Linear discriminant analysis (ULDA), and Convolutional neural network (CNN) [36]. Out of all these algorithms, effective classification is achieved by deep learning algorithms. The author [13] deploys self organizing map (SOM) neural network and *K*-nearest neighbor (KNN) classification algorithm. The author [14] employs Support vector machine (SVM) classifier for classification purposes. He used Gaussian kernel function for transformation. By this, the nonlinear samples are converted into a high-dimensional future space where the separation of nonlinear samples are possible, for proper classification. The author [23]. Classifies the images using Support Vector Machines (SVM) with GRB kernel. SVM with cubic kernel is a nonlinear, nonparametric classification technique and is best suited for classification of binary data. The paper shows that SVM has comparatively better performance for binary classification. Therefore, the author [16] proposes SVM with cubic kernel as a classifier.

Table 3 Different quantitative measures for MRI image

Metric	Description
True positive rate TPR	$\text{TPR} = \frac{\text{TP}}{\text{TP+FN}}$
Positive predictive rate PPV	$\text{PPV} = \frac{\text{TP}}{\text{TP+FP}}$
Dice similarity coefficient DSC	$\text{DSC} = \frac{2\text{TP}}{2\text{TP+FP+FN}}$
Volume difference rate VDR	$\text{VDR} = \frac{\lvert\text{FP}-\text{FN}\rvert}{\text{TP+FN}}$
Lesion wise true positive rate LTPR	$\text{LTPR} = \frac{\text{TPL}}{\text{TPL+FNL}}$
Lesion wise positive predictive value LPPV	$\text{LPPV} = \frac{\text{TPL}}{\text{TPL+FPL}}$

To improve robustness and reduce computational cost, the author [25] suggested a multiresolution framework. For a multiresolution framework with P levels, trilinear interpolation method is used. The author [35] classifies the test samples using decision tree in combination with Bootstrap aggregating or bagging machine learning algorithms. The author [17] proposes the concept of MP-KDD algorithm by using the SIFT detector and SURF descriptor is deployed to make efficient classification between normal and abnormal images. The author [33] proposes support vector machine (SVM) for brain tumor classification. The author [19] deals with classification of inputs into normal or abnormal using Least Squares Support Vector Machine classifier with Multilayer perceptron kernel. The author [34] uses probabilistic neural network (PNN) classifier for normal and abnormal brain classification.

2.5 Performance Evaluation

Table 3 refers to the different quantitative measures of MRI images.

3 Conclusion

A concise review of various techniques used in medical MR images is provided in this paper. This gives us information regarding the detection of brain tumors at the earliest with maximum accuracy. In this paper, various methods are reviewed which are used in image processing techniques. The advantages and disadvantages of various methods are illustrated. The motive behind this review is to develop advanced techniques in the field of medical image processing. Further, if we combine the best methods of each technique into one, then an automated system will be developed for brain tumor diagnosis which will be accurate and clinically useful.

References

1. Despotović, I., Goossens, B., Philips, W.: MRI segmentation of the human brain: challenges, methods, and applications. Comput. Math. Methods Med. (2015)
2. Desai, A., Yan, Y., Gerson, S.L.: Concise reviews: cancer stem cell targeted therapies: toward clinical success. Stem Cells Trans. Med. **8**(1), 75–81 (2019)
3. Saman, S., Narayanan, S.J.: Survey on brain tumor segmentation and feature extraction of MR images. Int. J. Multimed. Inf. Retrieval **8**(2), 79–99 (2019)
4. Abd-Ellah, M.K., Awad, A.I., Khalaf, A.A.M., Hamed, H.F.A.: A review on brain tumor diagnosis from MRI images: practical implications, key achievements, and lessons learned. In: Magnetic Resonance Imaging (2019)
5. Esakkirajan, S., Veerakumar, T., Subramanyam, A.N., Chand, C.H.P.: Removal of high density salt and pepper noise through modified decision based unsymmetric trimmed median filter. IEEE Sig. Process. Lett. **18**(5), 287–290 (2011)
6. Sheela, C.J.J., Suganthi, G.: An efficient denoising of impulse noise from MRI using adaptive switching modified decision based unsymmetric trimmed median filter. Biomed. Signal Processing and Control **55**, 101657 (2020)
7. Astola, J., Kuosmanen, P.: Fundamentals of Nonlinear Digital Filtering, vol. 8. CRC Press (1997)
8. Hwang, H., Haddad, R.A.: Adaptive median filters: new algorithms and results. IEEE Trans. Image Process. **4**(4), 499–502 (1995)
9. Zhang, S., Karim, M.A.: A new impulse detector for switching median filters. IEEE Sig. Process. Lett. **9**(11), 360–363 (2002)
10. Ng, P.-E., Ma, K.-K.: A switching median filter with boundary discriminative noise detection for extremely corrupted images. IEEE Trans. Image Process. **15**(6), 1506–1516 (2006)
11. Srinivasan, K.S., Ebenezer, D.: A new fast and efficient decision-based algorithm for removal of high-density impulse noises. IEEE Sig. Process. Lett. **14**(3), 189–192 (2007)
12. Jayaraj, V., Ebenezer, D.: A new switching-based median filtering scheme and algorithm for removal of high-density salt and pepper noise in images. EURASIP J. Adv. Sig. Process. **1**, 690218 (2010)
13. Anitha, V., Murugavalli, S.: Brain tumour classification using two-tier classifier with adaptive segmentation technique. IET Comput. Vis. **10**(1), 9–17 (2016)
14. Bahadure, N.B., Ray, A.K., Thethi, H.P.: Image analysis for MRI based brain tumor detection and feature extraction using biologically inspired BWT and SVM. Int. J. Biomed. Imaging
15. Fernandes, S.L., Tanik, U.J., Rajinikanth, V., Karthik, K.A.: A reliable framework for accurate brain image examination and treatment planning based on early diagnosis support for clinicians. Neural Comput. Appl. 1–12
16. Gilanie, G., Bajwa, U.I., Waraich, M.M., Habib, Z., Ullah, H., Nasir, M.: Classification of normal and abnormal brain MRI slices using gabor texture and support vector machines. Sig. Image Video Process. **12**(3), 479–487 (2018)
17. Kanmani, P., Marikkannu, P.: MRI brain images classification: a multi-level threshold based region optimization technique. J. Med. Syst. **42**(4), 62 (2018)
18. Kermi, A., Andjouh, K., Zidane, F.: Fully automated brain tumour segmentation system in 3d-MRI using symmetry analysis of brain and level sets. IET Image Proc. **12**(11), 1964–1971 (2018)
19. Praveen, G.B., Agrawal, A.: Hybrid approach for brain tumor detection and classification in magnetic resonance images. In: 2015 Communication, Control and Intelligent Systems (CCIS), pp. 162–166. IEEE (2015)
20. Aslam, A., Khan, E., Beg, M.M.: Improved edge detection algorithm for brain tumor segmentation. Proc. Comput. Sci. **58**, 430–437 (2015)
21. Corso, J.J., Sharon, E., Dube, S., El-Saden, S., Sinha, U., Yuille, A.: Efficient multilevel brain tumor segmentation with integrated Bayesian model classification. IEEE Trans. Med. Imaging **27**(5), 629–640 (2008)

22. Debnath, S., Talukdar, F.A.: Brain tumour segmentation using memory based learning method. In: Multimedia Tools and Applications, pp. 1–18 (2019)
23. Devkota, B., Alsadoon, A., Prasad, P.W.C., Singh, A.K., Elchouemi, A.: Image segmentation for early stage brain tumor detection using mathematical morphological reconstruction. Proc. Comput. Sci. **125**, 115–123 (2018)
24. Dhanachandra, N., Manglem, K., Chanu, Y.J.: Image segmentation using *k*-means clustering algorithm and subtractive clustering algorithm. Proc. Comput. Sci. **54**, 764–771 (2015)
25. Huang, M., Yang, W., Yao, Wu., Jiang, J., Chen, W., Feng, Q.: Brain tumor segmentation based on local independent projection-based classification. IEEE Trans. Biomed. Eng. **61**(10), 2633–2645 (2014)
26. Li, G., Jiang, D., Zhou, Y., Jiang, G., Kong, J., Manogaran, G.: Human lesion detection method based on image information and brain signal. IEEE Access **7**, 11533–11542 (2019)
27. Saha, S., Alok, A.K., Ekbal, A.: Brain image segmentation using semi-supervised clustering. Exp. Syst. Appl. **52**, 50–63 (2016)
28. Tong, J., Zhao, Y., Zhang, P., Chen, L., Jiang, L.: MRI brain tumor segmentation based on texture features and kernel sparse coding. Biomed. Signal Process. Control **47**, 387–392 (2019)
29. Jaglan, P., Dass, R., Duhan, M.: A comparative analysis of various image segmentation techniques. In: Proceedings of 2nd International Conference on Communication, Computing and Networking, pp. 359–374. Springer (2019)
30. Bousselham, A., Bouattane, O., Youssfi, M., Raihani, A.: Towards reinforced brain tumor segmentation on MRI images based on temperature changes on pathologic area. Int. J. Biomed. Imaging
31. Gordillo, N., Montseny, E., Sobrevilla, P.: State of the art survey on MRI brain tumor segmentation. Magn. Reson. Imaging **31**(8), 1426–1438 (2013)
32. Ba¸stan, M., Bukhari, S.S., Breuel, T.: Active canny: edge detection and recovery with open active contour models. IET Image Process. **11**(12), 1325–1332
33. Kumar, S., Dabas, C., Godara, S.: Classification of brain MRI tumor images: a hybrid approach. Proc. Comput. Sci. **122**, 510–517 (2017)
34. Shree, N.V., Kumar, T.N.R.: Identification and classification of brain tumor MRI images with feature extraction using dwt and probabilistic neural network. Brain Inf. **5**(1), 23–30 (2018)
35. Huda, S., Yearwood, J., Jelinek, H.F., Hassan, M.M., Fortino, G., Buckland, M.: A hybrid feature selection with ensemble classification for imbalanced healthcare data: a case study for brain tumor diagnosis. IEEE Access **4**, 9145–9154 (2016)
36. Mohan, G., Subashini, M.M.: MRI based medical image analysis: survey on brain tumor grade classification. Biomed. Sig. Process. Control **39**, 139–161 (2018)

Estimation of Sundarban Reserve Forest Using Self-organizing Maps and Remote Sensing Data

Krishan Kundu, Prasun Halder, and Jyotsna Kumar Mandal

1 Introduction

Forest is an important valuable natural resource in the world. Mangrove forests are detected in the intertidal regions along with the coastal province in most tropical and semi-tropical zones [1, 2]. It creates a natural wall to defend the erosion of soil, protection of settlement area, equilibrium of Earth temperature, manages the biodiversity, climate control, rainfall quantity control and it is also reduced the natural adversity [3–5] includes cyclones, storms, tsunamis, hurricanes, etc. Besides, it is utilized to the growth of economy for the nation through collection of timber wood and the source of medicine. In the past, many authors have reported that forest is decreasing over the world due to the expansion of agricultural land, industrialization, increased population, urbanization, change of climate [6], large number of cutting trees, increased built-up area, increased road, bridge, environmental pollution [7], etc. Forest degradation impacts on the environment such as reduction of biodiversity, reduction of an important sink for atmospheric carbon dioxide, impact on local and regional climate, and negative effects on the livelihoods of people [8, 9].

In the present scenario, many authors are trying to develop new approaches which will be capable of detecting the changes of land-use land cover particularly with the Artificial Neural Networks (ANN) [10, 11]. Change detection has been used in many

K. Kundu (✉)
Department of Computer Science and Engineering, Government College of Engineering and Textile Technology, Serampore, Hooghly, West Bengal, India

P. Halder
Department of Computer Science and Engineering, Ramkrishna Mahato Government Engineering College, Purulia, West Bengal, India

J. K. Mandal
Department of Computer Science and Engineering, University of Kalyani, Kalyani, Nadia, West Bengal, India

J. K. Mandal and J. K. Roy (eds.), *Proceedings of International Conference on Computational Intelligence and Computing*, Algorithms for Intelligent Systems,
https://doi.org/10.1007/978-981-16-3368-3_20

areas like remote sensing, motion detection, and medical diagnosis. An application of remote sensing such as change detection of urban building is determined using morphological building index which is recently developed. While in this way estimation of change detection is too much computationally expansive, time-consuming, and complex process.

Another approach to change detection operation has been performed using the grouping of similar patterns in the image and this process is done with the help of self-organizing maps which is a subset of artificial neural network [12]. Generally, there are two types of learning methods are used for classification of satellite image such as supervised and unsupervised learning. In the supervised classification technique supervision is required to classify the image that means according to their learning and get a response in accordance with this learning. Therefore, need a training sample to classify the image. In the unsupervised classification method classify the pattern without a prior learning, only according to the similar properties. Self-organizing Maps (SOM) is an unsupervised learning artificial neural network (ANN) to produce a low-dimensional, non-continuous representation of input space of the training samples. It is distinguished from other artificial neural networks as they use competitive learning as opposed to error-correction learning, and in the sense that they use a neighborhood function to preserve the topological properties of the input space [13, 14]. The main objective is that satellite images have been classified using self-organizing maps (SOM) and result is also compared with the other existing classification methods in terms of accuracy. Another intention of this study is to finding the present distribution of various features and identified the dynamic changes of different features in Sundarban reserve forest.

2 Materials and Methods

2.1 Study Site

Sundarban is the largest mangrove forest delta in the world. It is placed on the eastern side of India which extends from Hooghly River to the border line of India and Bangladesh. It lies between 21° 30′ 00″–22° 40′ 48″ N latitude and 88° 10′ 48″–89° 04′ 48″ E longitude and created by the influence of the Ganges, Brahmaputra, and Meghna River in the Bay of Bengal. Figure 1 illustrated the geographical location of present study area. The average altitude of this island is 7.5 m above the sea level and temperature varies from 20 to 34 °C. In this region, heavy rainfall happened during the monsoon season. Weather of this region is almost always moist and continuously blowing the humid air from the Bay of Bengal. The total forest cover area (India and Bangladesh) is around 10,000 km^2 (40% in India and 60% in Bangladesh). Most of the region is reserved for forest. There are many tree species and several biodiversities that exists in this region. The most abundant tree species are Sundari (Heritiera fomes), Gewa (Excoecaria agaallocha), Hetal (Phoenix paludosa), and etc.

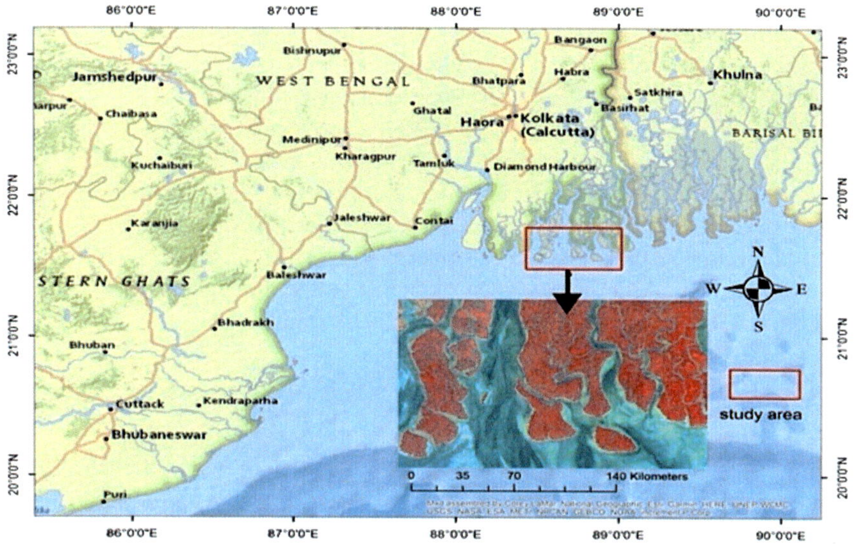

Fig. 1 Geographical position of the present study area

2.2 *Collection of Data*

Multispectral satellite images were collected from United States Geological Survey (USGS) which are freely available on the Internet. Five Landsat satellite data (1975, 1990, 2000, 2010, and 2018) were used to extract the significant information from the images. Table 1 displayed the detailed summary of Landsat satellite data such as satellite name, sensor name, date of acquisition, path\row, no of bands, and resolution. From this table, it is visibly observed that all the images were acquired in the winter season due to their availability on the web. Therefore, temporal change errors may be minimized. The spatial resolution of each image is 30 m except the 1975 image (57 m).

Table 1 Summary of Landsat satellite images

Satellite name	Sensor name	Date of acquisition	Path/row	No of bands	Resolution (m)
Landsat 3	MSS	05-12-1975	148/45	4	57 × 57
Landsat 5	TM	03-01-1990	138/45	7	28.5 × 28.5
Landsat 5	ETM	17-11-2000	138/45	8	28.5 × 2.5
Landsat 7	ETM+	24-01-2010	138/45	8	30 × 30
Landsat 8	OLI	28-02-2018	138/45	11	30 × 30

2.3 Image Preprocessing

Image preprocessing operation is needed before using the image. This preprocessing operation was performed by using the GIS-based software. If image contains any noise such as drop lines or stripping lines, clouds, etc. these were reduced using the geometric correction and radiometric calibration operation. If any dark is recorded on the image due to the atmospheric path radiance which can be minimized using dark object subtraction method (DOS). Post atmospheric correction, images were normalized with the help of linear regression estimated between images. Then employed the normalized functions and generate the normalized images.

2.4 Methods

Post collection of satellite images were preprocessed using the GIS-based image processing software. Then, crop the study area to carry out the present work. A layer stacking tool is used to integrate the three bands into one layer. Then, input images were projected to the UTM zone and the nearest neighbor algorithm is applied to resample the data. Each image was resampled to 30 m ground resolution. In this study, four features are considered such as healthy vegetation, unhealthy vegetation, water bodies, and wet land. The characteristic of healthy vegetation is that it absorbs most of the visible lights and reflects huge quantity of near-infrared light where the tree canopy density is above 50%. The property of unhealthy vegetation is that it reflects more visible light as compared with the near-infrared light where the tree canopy density is less than 50%. Wet land coverage areas are low land area, swamp land, marsh land, etc. Water body includes area covered by ponds, aquatic farms, small channel, etc. Satellite images were classified using the self-organizing maps. The SOM algorithm is relatively simple. It is composed of weights initialization, iterative over the input data, finding the winning neuron for each input, and adjusts the weights based on the location of that winning neuron. The flowchart of the SOM algorithm is presented in Fig. 2.

The SOM algorithm procedure is given below: initially assign the weight vectors of each node. From these weight vectors randomly one input vector is selected and the map of weight vectors is searched to find which weight best represents that sample. Each weight vector has neighboring weights that are near to it. The weight that is chosen is rewarded by being able to become more like that randomly selected sample vector. The neighbors of that weight are also rewarded by being able to become more like the chosen sample vector. From this step the number of neighbors and how much each weight can learn decreases over time. This entire process is continued until reached specific number of iterations. After classification of each image, two classified raster images are merged, and then converted into the vector images. Afterwards, two vector images are compared and evaluated the changes of each feature area. Then changes of each feature are reflected on the map and change

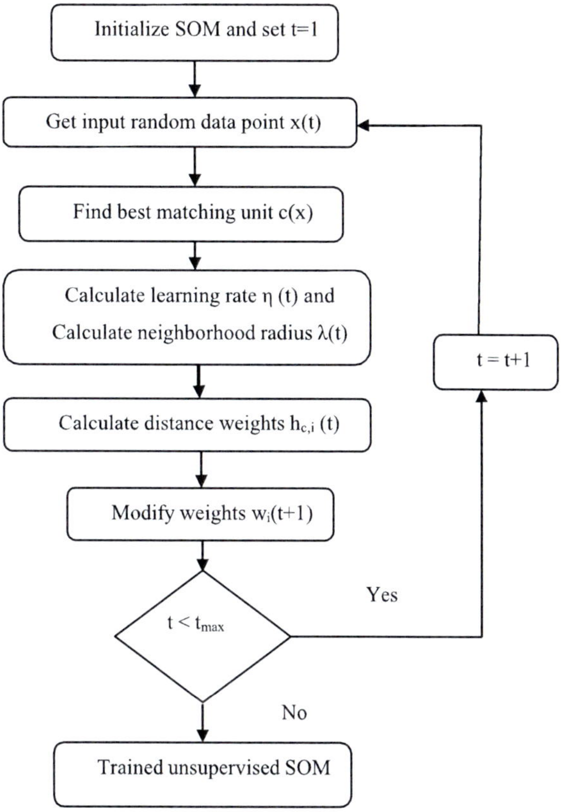

Fig. 2 Flowchart of the SOM algorithm

detection maps are generated. Figure 3 displayed the step by steps procedure for the present work.

3 Results and Discussion

3.1 Image Classification

The classification results are displayed in Table 2. From the table, it is clearly seen that healthy vegetation areas were gradually declined from 39.75% in 1975 to 23.74% in 2010, while the same were rapidly increased to 33.23% in 2018. However, net turnover of the healthy vegetation were declined during the study period. The unhealthy vegetation areas were progressively increased from 7.39% in 1975 to 18.51% in 2010, but during the period of 2010–2018 the same were decreased. The overall vegetation areas (healthy and unhealthy vegetation) were declined from 47.14% in 1975 to 44.67% in 2018. The causes of forest degradation are natural

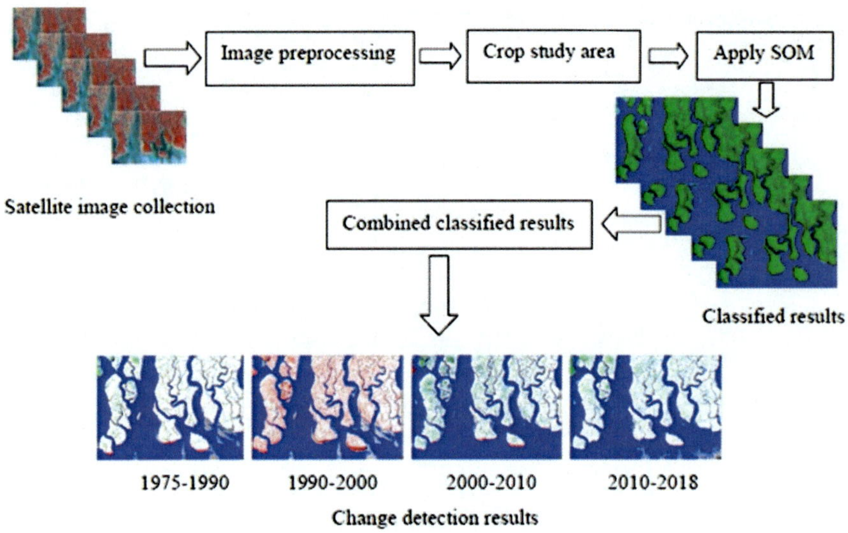

Fig. 3 Methodology used by the present study area

Table 2 Classification statistic (area in km^2) results for the year 1975, 1990, 2000, 2010, and 2018

Class name	1975		1990		2000		2010		2018	
	Area	%	Area	%	Area	%	Area	%	Area	%
Healthy vegetation	890.27	39.75	864.34	38.63	588.57	26.3	531.55	23.74	744.11	33.23
Unhealthy vegetation	165.58	7.39	191.73	8.57	324.53	14.5	414.53	18.51	256.17	11.44
Water bodies	1106.09	49.38	1009.01	45.09	1180.1	52.74	1150.36	51.37	1108.23	49.49
Wet land	78.01	3.48	172.65	7.72	144.53	6.46	142.79	6.38	130.71	5.84

disaster and some manmade activities like illegal cutting of trees, etc. The forest areas were increased because of re-growth of the trees and some plantation program were initiated. Water body also gradually increased from 1975 to 2018 due to the rising of sea level. Wet land areas were gradually increased because the deforestation happed. Figure 4 depicts the distribution of forest cover area in percentage for different time periods. The distribution of net vegetation area of various time periods is displayed in Fig. 5.

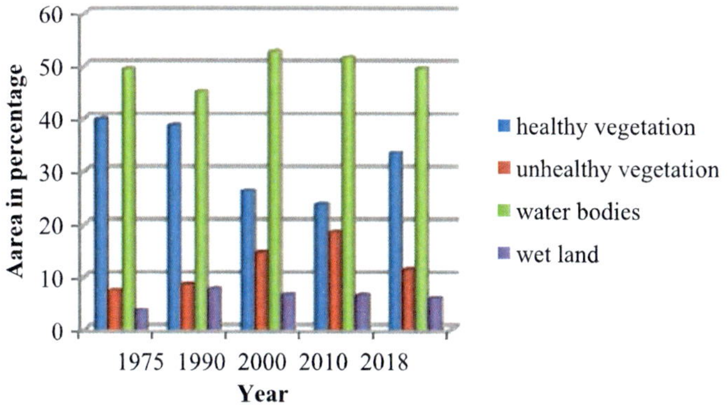

Fig. 4 Year-wise distributions of forest cover area in percentage

Fig. 5 Distribution of net vegetation area for different time period

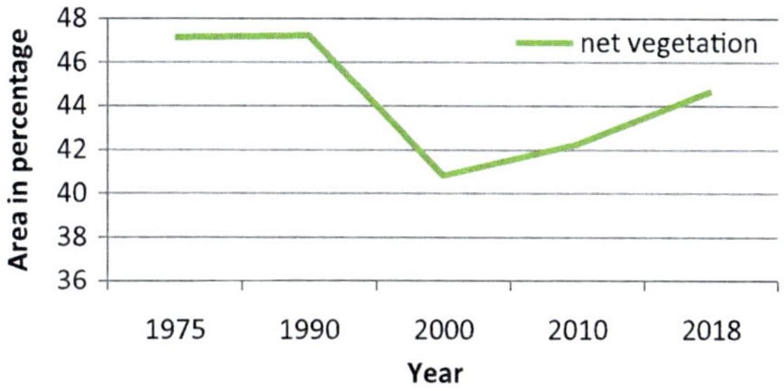

3.2 *Change Detection Analysis from 1975 to 2018*

Change detection results are illustrated in Table 3 for the time spans of 1975–1990, 1990–2000, 2000–2010, and 2010–2018. During the study period 454.67–777.51 km^2 healthy vegetation, 73.4–230.73 km^2 unhealthy vegetation, 970.77–1126.16 km^2 water bodies and 15.62–63.91 km^2 wet land was unchanged. The most significant changes were occurred during the period of 1975–1990, 126.35 km^2 water bodies were converted into the wet land. In the period of 1990–2000, 87.97 km^2 unhealthy vegetation areas were translated into the wet land areas and 153.27 km^2 wet land areas also converted into the water bodies. The 49.17 km^2 water bodies were converted into the wet land during the period of 2000–2010. The important changes were detected during the period of 2000–2018, 45.23 km^2 wet land were converted into the water bodies. Figure 6 shows that the change detection maps of different time periods. From this map, it is clearly seen that most of the deforestation were happened during the period of 1990–2000. It is also observed that major forest areas were degraded along the shoreline, sea surface, and the small channels on the southern region of the Sundarban.

Table 3 Change detection matrix (km^2) during the period of 1975–1990, 1990–2000, 2000–2010, and 2010–2018

Class name	Healthy vegetation	Unhealthy vegetation	Water bodies	Wet land
1975–1990				
Healthy vegetation	777.51	90.7	8.61	13.46
Unhealthy vegetation	71.2	73.4	3.34	13.23
Water bodies	4.25	4.52	970.77	126.35
Wet land	12.05	23.3	22.88	19.78
1990–2000				
Healthy vegetation	571.22	246.23	14.72	32.17
Unhealthy vegetation	15.99	73.96	13.83	87.95
Water bodies	0.51	1.44	998.27	8.79
Wet land	0.85	2.9	153.27	15.62
2000–2010				
Healthy vegetation	454.67	118.05	6.27	9.58
Unhealthy vegetation	67.67	230.73	3.01	20.04
Water bodies	1.34	2.71	1126.16	49.17
Wet land	7.49	62.77	10.36	63.91
2010–2018				
Healthy vegetation	522.19	6.52	1.36	1.49
Unhealthy vegetation	219.93	187.77	3.04	3.79
Water bodies	0	0.14	1058.49	91.72
Wet land	1.99	61.76	45.23	33.71

The change detection statistic shows that the decreased forest areas were converted to open land and water bodies. The forest areas were increased due to initiation of new plantation programs, re-growth of the plants, motivation of the local peoples, observation of the forest, etc. Sundarban ecosystem is threatened by various natural and human-induced pressures. The main causes of forest degradation are increased population and their livelihood depended on the forest resources. It also many factors to degradation of Sundarban ecosystem such as overharvesting of forest resources, morphological changes of coastal land use, unrestricted tourism, increased man-animal clashes, reduction of upstream water supply, salinity level increased, tree disease, fire in forest, change in climate, sea-level rise, natural calamity, pollution, oil spillage, inadequate knowledge to preservation of forest and lack of proper planning and management.

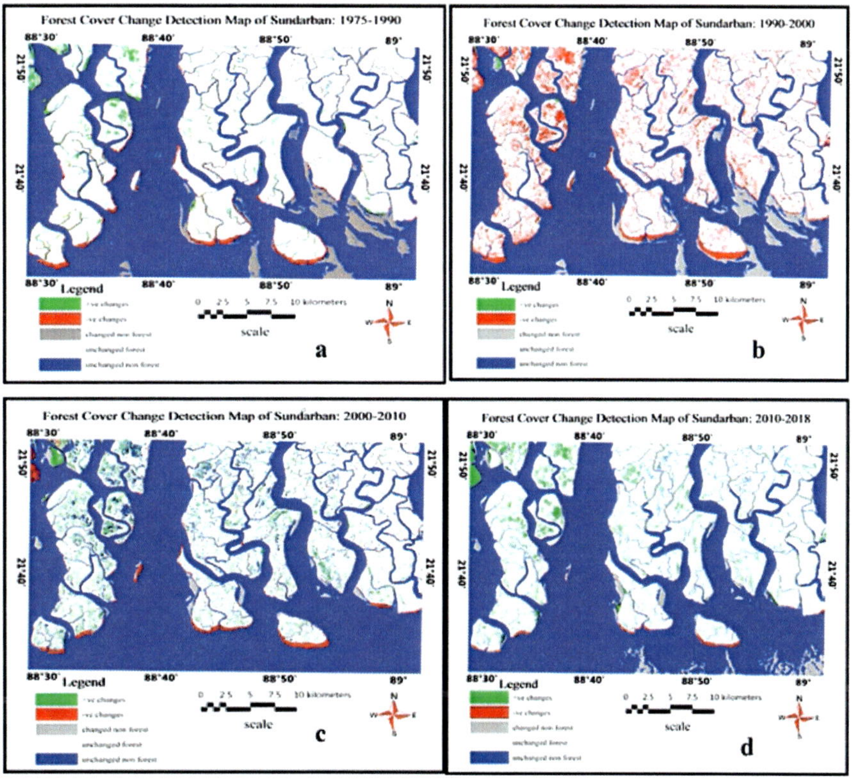

Fig. 6 Change detection map of **a** 1975–1990, **b** 1990–2000, **c** 2000–2010 and **d** 2010–2018

3.3 Classification Accuracy

The performance of the image classification was measured by using overall accuracy (OA) and Kappa statistics (KS). Post classifications of images, error matrix were estimated from the reference data and the classified data. Reference data were acquired from the visual interpretation of original image or Google Earth observation data where trace on the actual map. Overall accuracy was obtained from the ratio between the total number of pixels are correctly classified and the total number of pixels is used in this classification technique. Overall accuracy towards 100% indicates the good level of classification performance. Another non-parametric tool is used to measure the accuracy such as Kappa statistic. Generally, its value varies from 0 to 1. Kappa values towards 1 indicate that good level of agreement. The resulting overall accuracy were 88.91%, 89.32%, 90.32%, 89.93% and 92.17% and Kappa statistic were 0.85, 0.86, 0.87, 0.86 and 0.89 of the images 1975, 1990, 2000, 2010 and 2018 respectively.

3.4 Comparison with Other Classification Techniques

The classification accuracy of SOM method with the other existing methods is presented in Table 4 and Fig. 7 for different time periods. From this table, it is clearly examined that overall accuracy and Kappa statistic of SOM method is higher than the MLC (Maximum Likelihood classifier) then followed by FCM (Fuzzy *C*-means). The overall accuracy of SOM method is better because its learning rate is within the acceptable range for various sample sizes. Whereas much more learning rate is required for the FCM method whenever the size of the sample is too large and computational cost is huge. The MLC method is based on the supervised classification technique and needs a training sample to classify the image. For preparation of training sample is required lot of time. Therefore, SOM method is more suitable for classification of multi-spectral satellite images.

Table 4 Results of image classification accuracy for various methods

Year	SOM		FCM		MLC	
	OA (%)	KS	OA (%)	KS	OA (%)	KS
1975	88.91	0.85	81.73	0.74	80.91	0.73
1990	89.32	0.86	85.26	0.78	84.05	0.77
2000	90.32	0.87	80.51	0.73	82.79	0.76
2010	89.93	0.86	81.99	0.75	82.67	0.76
2018	92.17	0.89	82.59	0.76	84.12	0.78

OA overall accuracy, *KS* kappa statistic, *FCM* fuzzy *C*-means, *MLC* maximum likelihood classifier

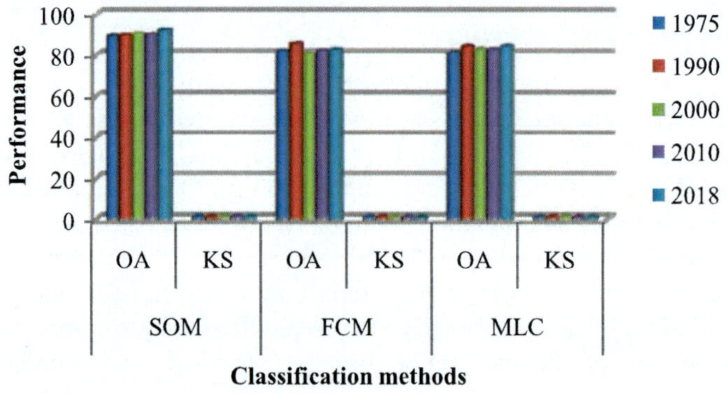

Fig. 7 Classification performance for different techniques

4 Conclusions

In the present work change detection of Sundarban reserve forest using self-organizing maps and remote sensing data is presented. The classification results (overall accuracy and Kappa statistic) shows that this method is more superior to other existing classification techniques due to the reasonable learning rate for different sample size. In this study, the overall accuracy was achieved between 88.91–92.17% that indicates the good level of classification accuracy. Kappa statistic results are showing that good level of agreement between the classified data and the reference data. Study revealed that net forest areas were gradually declined by about 2.47% from 1975 to 2018, while it was not uniform over the entire period. At the same time other geographical features also correspondingly changes. These changes may affect the ecosystem of the Sundarban forest. Most of the regeneration has been observed in the intertidal zone. The future scope of this paper is to prediction of forest using some mathematical model and fragmentation will be analyzed. This study may helpful for decision makers and planers to survive the natural forest ecosystem of Sundarban reserve forest.

References

1. Akhand, A., Mukhopadhyay, A., Chanda, A., et al.: Potential CO_2 emission due to loss of above ground biomass from the Indian Sundarban mangroves during the last four decades. J. Indian Soc. Remote Sens. **45**, 147–154 (2017)
2. Bansal, S., Srivastav, S.K., Roy, P.S., Krishnamurthy, Y.V.N.: An analysis of land use and land cover dynamics and causative drivers in a thickly populated Yamuna River Basin of India. Appl. Ecol. Environ. Res. **14**(3), 773–792 (2016)
3. Kundu, K., Halder, P., Mandal, J.K.: Forest cover change analysis in Sundarban delta using remote sensing data and GIS. In: Mandal, J., Sinha, D. (eds.) Intelligent Computing Paradigm: Recent Trends. Studies in Computational Intelligence, vol. 784, pp. 85–101. Springer (2020)
4. Ghosh, S., Nandy, S., Patra, S., et al.: Land cover classification using ICESat/GLAS full waveform data. J. Indian Soc. Remote Sens. **45**, 327–335 (2017)
5. Kundu, K., Halder, P., Mandal, J.K.: Forest covers classification of Sundarban on the basis of fuzzy *C*-means algorithm using satellite images. In: Mandal, J., Mukhopadhyay, S. (eds.) Proceedings of the Global AI Congress 2019. Advances in Intelligent Systems and Computing, vol. 1112, pp. 515–528. Springer (2020)
6. Ghosh, S., Nandy, S., Senthil Kumar, A.: Rapid assessment of recent flood episode in Kaziranga National Park, Assam using remotely sensed satellite data. Curr. Sci. **111**(9), 1450–1451 (2016)
7. Kundu, K., Halder, P., Mandal, J.K.: Estimation of changes of vegetation cover in Sundarban using multi-temporal satellite data. Adv. Modell. Analy. D (AMSE) **23**(1), 19–26 (2018)
8. Chakraborty, S.D., Yogesh, K., Mitra, D.: Assessment of land surface temperature and heat fluxes over Delhi using remote sensing data. J. Environ. Manage. **148**, 143–152 (2015)
9. Mitra, D., et al.: Forest cover change prediction using hybrid methodology of geoinformatics and Markov chain model: a case study on sub-Himalayan town Gangtok, India. J. Earth Syst. Sci. **123**(6), 1349–1360 (2014)
10. Jorma, L., Markus, K., Erkki, O.: PicSom-self-organizing image retrieval with MPEG-7 content descriptors. IEEE Trans. Neural Netw. **13**(4), 841–853 (2002)
11. Gwo-Fong, L., Lu-Hsien, C.: Identification of homogeneous regions for regional frequency analysis using the self-organizing map. J. Hydrol. **324**, 1–9 (2006)

12. Thomas, V., Erzsebet, M., Barbara, H.: Neural maps in remote sensing image analysis. Neural Netw. **16**(3–4), 389–403 (2003)
13. Theo, S., Alistair, F., Hamid, B.: Automatic registration of complex images using a self-organizing neural system. In: 1998 IEEE International Joint Conference on Neural Networks (IJCNN'98), pp. 1–6. Anchorage, Alaska, USA (1998)
14. Peter, D., Peggy, A., Anthony, S., Mohamad, M.: Self-organised clustering for road extraction in classified imagery. ISPRS J. Photogramm. Remote Sens. **55**(5–6), 347–358 (2001)

Reducing Bullwhip Effect in Distributed Supply Chain Management by Virtual Data Warehouse and Modified-Prophet

Partha Ghosh, Leena Jana Ghosh, Narayan C. Debnath, and Soumya Sen

1 Introduction

Immoderate forecasted demand that moving upstream can make supply chain [1, 2] inefficient. This distribution channel circumstance is known as bullwhip effect [1–3]. The effect of bullwhip distortion gradually increases with the increasing distance from the emerging position of demand. It generally caused due to sequential increasing waves that move upstream in inventory management [3] at every level of supply chain. Surplus demand at every stages of supply chain generates excess stock that results in waste of resources. Therefore, it can effect in the profitability for the organization. Nowadays, the effect of bullwhip is increasing due to the popularity of online retailing [4] where real-time multi-criteria [5] optimization is indispensable. Several new business plan like "extra discount on specific days," "flash sale" [1], etc. can suddenly increase sale beyond the expectation. Similarly, business plan like "free return policy" [4] can lead to waste of resources. Therefore, handling fluctuating business trend only through traditional data warehouse [6, 7] is quite impossible. It requires a real-time decision-making [8, 9] which is beyond the scope of traditional data warehousing. Hence, the present business scenario demands more up-to-date and convenient supply chain management that can cope the situation rapidly.

Maintaining warehouse virtually in a distributed manner is known as virtual data warehouse (VDW) [7]. In VDW, according to the business trend classification, small virtual warehouses [6] can be maintained over traditional data warehouse. Also, in order to distinguish between various abrupt business scenarios, fluctuating business

P. Ghosh (✉) · S. Sen
A. K. Choudhury School of IT, University of Calcutta, Kolkata, India

L. J. Ghosh
JC Edutech, Naihati, West Bengal, India

N. C. Debnath
Eastern International University, Thu Dau Mot, Vietnam

J. K. Mandal and J. K. Roy (eds.), *Proceedings of International Conference on Computational Intelligence and Computing*, Algorithms for Intelligent Systems,
https://doi.org/10.1007/978-981-16-3368-3_21

growth and outliers [6] can be stored separately in VDW [6, 7]. Therefore through VDW, abrupt business growth can be efficiently managed in real time [7]. Another reason of bullwhip effect is poor forecasting [10, 11] of sale. Currently, when inconsistent business growth are noticed due to several new business plans, proper forecasting is a big challenge. It requires present trend [12] analysis as well as handling seasonality [13] and holiday effects [12] too. Several forecasting models like ARIMA [14], SARIMA [13], Prophet [12], etc. are proposed over the time. Among these, ARIMA is an uni-variant model that produces erroneous results during long-range forecasting [10]. SARIMA is a modified version of ARIMA that is applicable only upon data set that has huge seasonal effect. A powerful forecasting model Prophet [12] is designed by Facebook [12] in order to forecast their social networking business. Prophet uses Bayesian curve fitting method to forecast the time series data. Therefore, unlike other models, regular data quantification is not required. In order to apply forecasting model to restrict the bullwhip effect of supply chain some advance architectural support and software modifications are required. For example, sale of a particular product depends not only upon the present business sale but also upon the present performance of that product. "customers' feedback" reflects the present performance of a product in the market. Therefore, incorporating "customers' feedback" in forecasting model may be one of the possible addition for mitigating the gap between actual and forecasted sale. Also, the forecasting model should react very quickly in case of sudden business growth or fall down. In this paper, a modified forecasting model based on Prophet [12] is incorporated at every stages of supply chain in order to crosscheck the incoming demands from its lower level. The abrupt business growth is handled through the concept of virtual data warehousing [7]. Experimental results revelation the advantage of proposed model.

The paper organization is as follows. Related work is at Sect. 2. A brief discussion upon supply chain management and bullwhip effect is at Sect. 3, followed by a brief discussion on Prophet forecasting model at Sect. 4. Proposed methodology is at Sect. 5. Case study is at Sect. 6, followed by a comparative analysis at Sect. 7. Finally at Sect. 8, we conclude.

2 Related Work

The present business trend [1] changes rapidly due to the incorporation of several new business plans. As a consequence, supply chain management [10, 12] faces several abashment by this varying business growth that enlarges bullwhip effect [1–3]. Presently, the effect of bullwhip in supply chain increases due to the opulence of online retailing [4]. A price-sensitive demand forecasting [1] is proposed that emphasize more upon discount in order to reduce bullwhip effect of supply chain. They proposed a dual-channel [1] based supply chain management in order to monitor the influence of price discount in bullwhip effect. This concept is enhanced by incorporating fuzzy control model [2] and several recycling channels. Researchers further show that lead time [2] between order and supply can also increase the effect of bullwhip. Several research works are performed to reduce the lead [3] time in supply chain. Further, several distributed supply chain management [4] techniques are proposed for online retailing. Relationship between uncertainties like lead time, demand and yield are replenished to alleviate the gap between retailer-managed and vendor-managed [4] inventory that in turn reduces the effect of bullwhip. These models [3] further used several forecasting [10] methodologies like ARIMA [14, 15], SARIMA [13], etc. in order to forecast sale at warehouse level or manufacturing unit level. They established their claim through standard error measuring techniques [11]. But, none of the methodologies considered the parameter customers' feedback [16] for sale forecasting that plays a vital role in future sale. Especially in case of irregular business sale, forecasting must emphasize more upon present customers' feedback [12] about that product. Another limitations of previous methodologies are that none of them have considered how to cope fluctuating business growth in real time. Managing fluctuating business growth in real time is beyond the capability of traditional data warehouse [6]. To generate real-time decision making, the concept of virtual data warehouse (VDW) [7, 16] comes in focus. It consists of small virtual warehouses over traditional data warehouse that is capable of handling abrupt business trend changes. Therefore, an updated model of supply chain management is required that can manage all types of forecasting issues as well as can react quickly in case of abnormal business trend change. These in turn shall reduce the effect of bullwhip in supply chain management.

3 Distributed Supply Chain Management and Effect of Bullwhip

Due to online retailing [4], centralized production and supply system become less advantageous for the large companies. Rather, in order to grab global market, companies are interested for decentralized distribution of their productions. As a result, supply chain management [1] in a distributed manner becomes indispensable. A logical diagram of distributed supply chain management is shown in Fig. 1.

Bullwhip effect [2] rises due to the sequential inflated inventory management that moves and grows upstream gradually in every level of supply chain. In distributed supply chain management [3], the summation of lower levels inventories is actually clubbed in the next higher level. For example, Bullwhip effect in Level 2 = $\sum_{i=1}^{k}$ (Bullwhip effect in Level 3_i). Due to this additive feature, the manifestation of bullwhip effect is more in distributed supply chain management. Hence, a proper up-to-date architecture is required in order to mitigate the effect of bullwhip in distributed supply chain management.

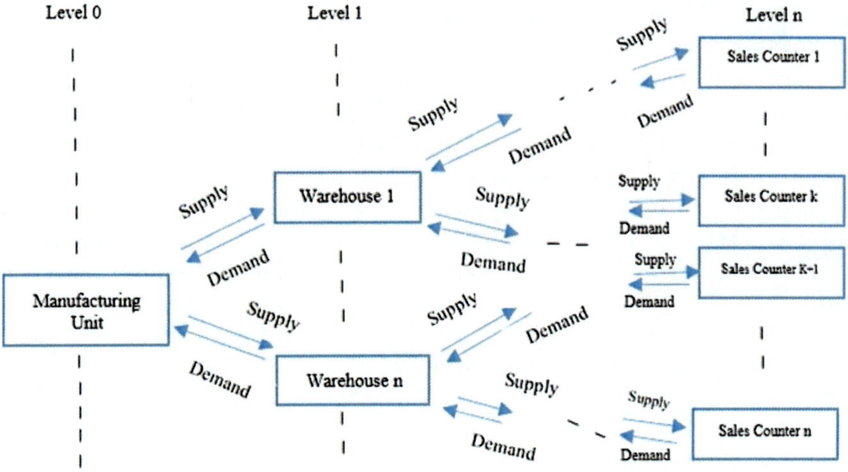

Fig. 1 Distributed supply chain management

4 Prophet Forecasting Model

Prophet [12] forecasting model is designed by Facebook [12] in order to forecast the progress of their social networking business. It is found in research that Prophet is a reliable model that uses the method of Bayesian curve fitting for forecasting time series data. The additive formula of Prophet [12] model is shown Eq. 1.

$$f(t) = g(t) + s(t) + h(t) + \varepsilon_t \tag{1}$$

Here, $g(t)$ = Trend of time series data, $s(t)$ = Seasonality, $h(t)$ = Holiday effects, and ε_t = Error terms. But, in case of sale forecasting, it misses another important parameter "customer feedback." Prophet can be used significantly for reducing bullwhip effect if it incorporates another parameter "customer feedback" for sale forecasting.

5 Proposed Methodology

In order to prevent the impact of bullwhip effect, business professionals should incorporate a compact business prediction model at every stage of supply chain. The model should handle the outliers caused by seasonal and holiday effects or by any special offer. Simultaneously, it should emphasize more upon customers' feedback in order to generate fine forecasting. In this paper, we have proposed a modified structure for supply chain management in order to minimize bullwhip effect as follows:

1. At every level, data outliers are stored in separate virtual data warehouses (VDWs) in order to handle seasonal and holiday effects.
2. A forecasting model is applied that emphasize more upon customer feedback and able to manage seasonality or special offers through VDWs.

These in turn generate demand prediction that should match the summation of incoming demands from lower level. If it does not match up to a permissible threshold limit, then the system automatically propagates a "correction demand" message to its lower levels. That is in level 0:

$$\text{Demand forecasting in Level 0} \approx \sum_{i=1}^{k} (\text{Incoming demand from Level } 1_i)$$

Similarly in level 1:

$$\text{Demand forecasting in Level 1} \approx \sum_{i=1}^{k} (\text{Incoming demand from Level } 2_i)$$

Here "k" is the number of lower level modules. This process continues up to the before last level. The logical diagram of proposed model is shown in Fig. 2.

The workflow diagram of proposed methodology is depicted in Fig. 3.

The proposed algorithm at every level is as follows:

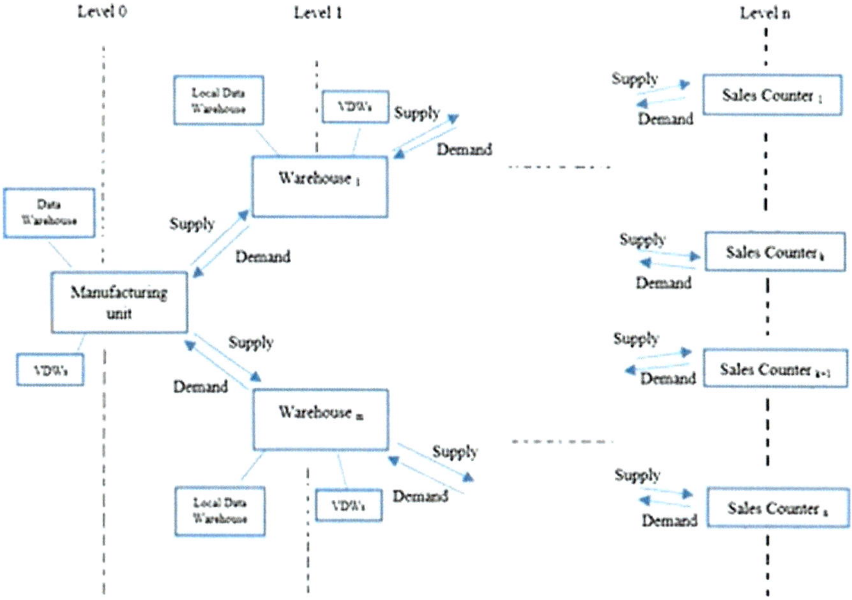

Fig. 2 Logical diagram of proposed methodology

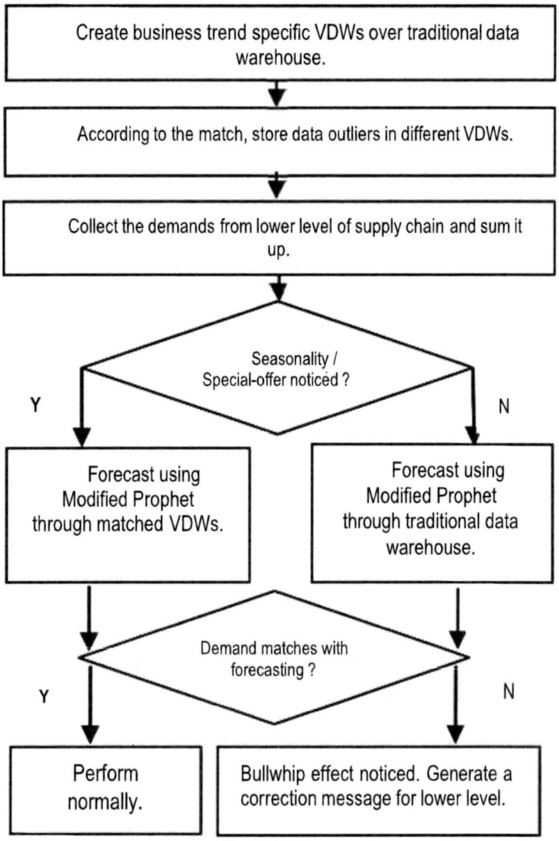

Fig. 3 Workflow diagram of proposed methodology

Step 1: Begin
Step 2: Identify outliers in historical data and store them in separate VDWs according to their calculated mean
Step 3: Loop for i=1 to number of nodes in lower level

Total_Demand = Total_Demand + Demand in ith node of lower level
end loop

Step 4: If any seasonality or special-offer noticed

For sale forecasting, apply Modified_Prophet( ) in matched VDWs
Else
For sale forecasting, apply Modified_Prophet( ) in Data-Warehouse
End if

Step 5: Take threshold limit as user input
Step 6. If | Forecasted sale − Total_Demand | <= threshold limit

Perform normally

Else
a) Bullwhip effect noticed.
b) Generate "correction demand" message to lower levels
End if

Step 7: End

Algorithm Modified_Prophet

Step 1: Begin
Step 2: n=numberOfReview
Step 3: w=0
Step 4: Loop for i=1 to n

r=take_ith Review
w=w+r
End Loop

Step 5: w=w/n
Step 6: $g'(t)=g(t) \times w$
Step 6: $f(t)=g'(t) + s(t) + h(t) + \varepsilon_t$
/*Where, $g(t)$, $s(t)$, $h(t)$ and ε_t *represent the same term as in prophet (Ref. equation-1)**/
Step 7: Generate forecasted result according to f(t)
Step 8: End

6 Case Study

We have applied our proposed model upon a standard real-life dataset "Amazon_Fashion" available at "https://nijianmo.github.io/amazon/index.html". Experiments are conducted on an Intel Core i5 processor and 16 GB of RAM. Another 16 GB SSD is also used as RAM. Different software are used Python 3.6, Matplotlib 3.1.2 used for data visualization, scikit-learn 0.21.3 used for machine learning, and pandas 0.25.3 used for large data manipulation.

In order to fit this data in our proposed model, we kept the whole data at level 0 (Manufacturing level, reference Fig. 2) and subdivided this time-series data as training data and the last one year data as test data. Output is shown in Fig. 4.

Next, at level 1 (warehouse level, reference Fig. 2) we subdivides the data into two parts. An alternating timestamp fashion is used in this case. For example, if time stamps of data are {1, 2, 3, 4}, we put {1, 3} in warehouse 1 and {2, 4} in warehouse 2, respectively. Next, at level 1, we subdivide each warehouse level data as training data and the last one-year data as test data. We have applied our proposed model both at level 0 (manufacturing level) and at level 1 (warehouse level). Finally, by the standard error detection model mean absolute percentage error (MAPE) [11] we measure the overall bullwhip effect at level 0. We have compared our proposed

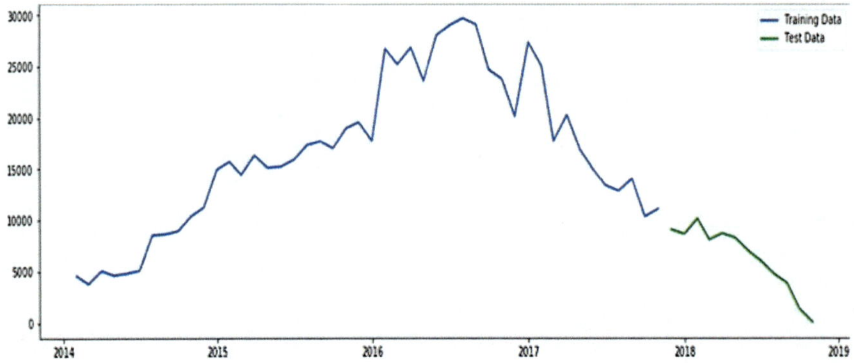

Fig. 4 Original time-series data divided as training and test data with respect to time stamp

Fig. 5 Mean absolute percentage error (MAPE) values

Model	MAPE
Prophet without VDW	52.00
ARIMA without VDW	28.32
SARIMA without VDW	32.35
Modified-Prophet with VDW	20.82

model (modified-Prophet with VDW) with other forecasting models and without virtual data warehouse (VDW). The result is shown in Fig. 5.

Hence, by the comparative result shown in Fig. 5, we can conclude that the proposed model can reduce the bullwhip effect in supply chain in a great extent in comparison with other models.

7 Result-Set Analysis

In this section, we analyze the result of case study to explain how much performance improvement is achieved by this proposed model.

We observed that the overall gap between forecasted demand and actual sale at level 0 is significantly less in standard error detection techniques MAPE [1, 11]. That in turn will reduce the overall bullwhip effect. A statistical comparative analysis is shown in Fig. 6.

From Fig. 6, we can conclude that the proposed framework provides around 60% improvement over the framework that uses Prophet model for forecasting and does not maintain VDWs. It provides around 14% improvement over the framework that uses ARIMA and does not maintain VDWs, and its improvement over SARIMA having no VDWs is around 22%.

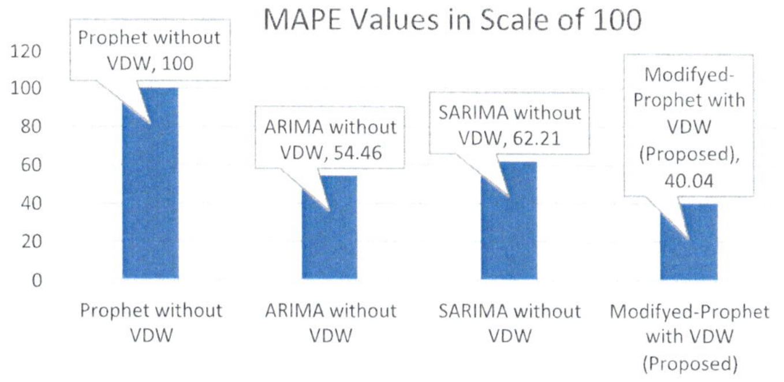

Fig. 6 Performance improvement comparison in the scale of 100

8 Conclusion and Future Works

Surplus inventory management can generate bullwhip effect in supply chain that hampers the profitability of the organization. This paper proposed a novel framework for reducing bullwhip effect by incorporating the concept of virtual data warehouse at every level of supply chain management. Further, the proposed model uses a modified-Prophet forecasting model of sale that emphasizes more upon customers' feedback which is an important parameter especially in the environment of online retailing. The experimental result reveals that in proposed model, bullwhip effect is significantly less compared to models that do not maintain virtual data warehouse.

In the future, this work can be extended in the form of customer relationship management. Customers' satisfaction level analysis through text processing may be another extension of this research work.

References

1. Gao, D., Wang, N., He, Z., Jia, T.: The bullwhip effect in an online retail supply chain: a perspective of price-sensitive demand based on the price discount in E-commerce. IEEE Trans. Eng. Manage. **64**(2), 134–148 (2017). https://doi.org/10.1109/TEM.2017.2666265
2. Zhang, S., Li, X., Zhang, C.: A fuzzy control model for restraint of bullwhip effect in uncertain closed-loop supply chain with hybrid recycling channels. IEEE Trans. Fuzzy Syst. **25**(2), 475–482 (2017). https://doi.org/10.1109/TFUZZ.2016.2574910
3. Lee, S., Kim, Y.J., Cheong, T., Yoo, S.H.: Effects of yield and lead-time uncertainty on retailer-managed and vendor-managed inventory management. IEEE Access **7**, 176051–176064 (2019). https://doi.org/10.1109/ACCESS.2019.2957595
4. Le, M.,Yingying, Z., Lu, Z., Ling, C.: Research on the impact of products exchange policy on bullwhip effect of remanufacturing closed-loop supply chain. In: 2017 Chinese Automation Congress (CAC), Jinan, pp. 3590–3596 (2017). http://doi.org/10.1109/CAC.2017.8243404
5. Ghosh, P., Sen, S., Cortesi, A.: Skyline computation over multiple points and dimensions. Innovations Syst. Softw. Eng. (2021). https://doi.org/10.1007/s11334-020-00376-1
6. Ghosh, P., Sadhu, D., Sen, S., Debnath, N.C.: Service modeling for virtual data warehouse. In: International Conference on Computer Applications in Industry and Engineering (2017)

7. Kholod, I.I., Efimova, M.S., Kulikov, S.Y.: Using ETL Tools for Developing a Virtual Data Warehouse (2016)
8. Ni, J., Li, J., Mc Auley, J.: Justifying recommendations using distantly-labeled reviews and fine-grained aspects. In: Empirical Methods in Natural Language Processing (EMNLP) (2019)
9. Ghosh, P., Goto, T., Sen, S.: Taxicab geometry based analysis on skyline for business intelligence. Int. J. Softw. Innovation (IJSI) (2018). https://doi.org/10.4018/IJSI.2018100107
10. Kalla, R., Murikinjeri, S., Abbaiah, R.: An improved demand forecasting with limited historical sales data. In: International Conference on Computer Communication and Informatics (ICCCI) (2020). http://doi.org/10.1109/ICCCI48352.2020.9104090
11. Khan, M.A., et al.: Effective demand forecasting model using business intelligence empowered with machine learning. IEEE Access (2020). https://doi.org/10.1109/ACCESS.2020.3003790
12. Taylor, S.J., Letham, B.: Forecasting at scale. PeerJ Prepr. **5**, e3190v2 (2017). http://doi.org/10.7287/peerj.preprints.3190v2
13. Vagropoulos, S.I., Chouliaras, G.I., Kardakos, E.G., Simoglou, C.K., Bakirtzis, A.G.: Comparison of SARIMAX, SARIMA, modified SARIMA and ANN-based models for short-term PV generation forecasting. In: IEEE International Energy Conference (ENERGYCON) (2016). http://doi.org/10.1109/ENERGYCON.2016.7514029
14. Fattah, J., Ezzine, L., Aman, Z., El Moussami, H., Lachhab, A.: Forecasting of demand using ARIMA model. Int. J. Eng. Bus. Manage. (2018). https://doi.org/10.1177/1847979018808673
15. Huyen, D.T.T., Binh, N.T., Tuan, T.M., Trung, T.Q., Nhu, N.G., Dey, N., Son, L.H.: Analyzing trends in hospital-cost payments of patients using ARIMA and GIS: case study at the Hanoi Medical University Hospital, Vietnam. J. Med. Imaging Health Informatics (2017)
16. Ghosh, P., Som, S., Sen, S.: Business intelligence development by analysing customer sentiment. In: International Conference on Reliability, Infocom Technologies and Optimization (Trends and Future Directions) (2018). http://doi.org/10.1109/ICRITO.2018.8748517

Development of an IoT-Enabled Aqueous Sulphur Sensor with a rGO/AgNp Composite

Brady Shearan, Fowzia Akhter, and S. C. Mukhopadhyay

1 Introduction

Our lives are currently being transformed by the advancement of wireless communication architectures. This transformation is due to the emergence of wireless sensor networks (WSNs) and the Internet-of-Things (IoT) concept, providing means for the rapid development of high-performance wireless sensing systems [1, 2].

The early detection of contaminants within a water supply is one of the most important parameters of environmental monitoring. Hydrogen Sulphide (H_2S) is a toxic gas most commonly found in wastewater systems. H_2S exerts an offensive smelling odour and can be harmful to vision, skin and respiratory system and also significantly reduces the lifespan of liquid waste service infrastructure it is carried through [3].

H_2S is traditionally detected utilizing semiconducting metal oxides or more recently through transition metal chalcogenides (TMDs) whilst in gas phase, but very few have the capability to detect H_2S in liquid phase [4]. A highly sensitive sensor for H_2S detection in aqueous sewerage samples using a microfluidic gas channel coupled with a metal-oxide semiconductor has been presented [5]. These types of sensing systems consist of an automated sampling system connected to often bulky vaporization chambers.

Recently, graphene-based chemiresistor have been leading research within H_2S detection, though little research has been presented around implementation in aqueous samples [6]. These systems aim to remove the requirement of a vaporization chamber, significantly reducing the cost, complexity and energy efficiency of a detection system.

B. Shearan (✉) · F. Akhter · S. C. Mukhopadhyay
School of Engineering, Macquarie University, Sydney, NSW 2109, Australia
e-mail: Brady.Shearan@students.mq.edu.au

J. K. Mandal and J. K. Roy (eds.), *Proceedings of International Conference on Computational Intelligence and Computing*, Algorithms for Intelligent Systems,
https://doi.org/10.1007/978-981-16-3368-3_22

Graphene has seen use within in numerous recent gas sensing applications, due to its well documented desirable range of physical properties [7]. Though graphene itself is expensive to produce, therefore reduced graphene oxide (rGO) as a low-cost alternative has seen extensive use within electrochemical sensing due to its similar properties [8]. A comprehensive review of recent rGO sensing applications has been presented [9]. A number of low-cost rGO based sensors for detection of H_2S in gas form have been developed [10–13], though there is little research surrounding aqueous applications. It is desirable to be able to detect H_2S if present in water supplies though currently there are few existing sensors which can achieve this detection without extensive power consumption and requiring numerous additional components.

The aim of this research is to develop, a graphene-based chemical sensor for detection of Sulphur ions in aqueous solutions for use alongside a portable wireless sensing system. The system should be developed for use in both immediate on-site wastewater quality testing and continuous, real-time online monitoring.

2 Materials and Methods

2.1 Synthesis of rGO-AgNp-PE

25 mg of silver acetate and 50 mg of rGO were mixed together with 15 mL of DMF inside a flask. The solution was sonicated for 20 min at approximately 90 °C. After cooling, the DMF was removed using a rotary evaporator, leaving behind a dark black powder. In order to remove any remaining unbound silver nanoparticles or any remaining silver acetate, the sample had to be washed with water and ethanol three times, respectively. The ethanol was dried by lightly spraying nitrogen across the surface of the solution. AgNp coated rGO was recovered as a dark black powder. Finally, 2 mg of rGO or rGO-AgNp powder was mixed with 1ml of DMF and 0.5 ml of a PE (30 mg/ml in Toluene) mixture. The combined mixture was then sonificated for 30 min.

2.2 Interdigital Sensor and Preparation of Sensing Surface

An interdigital sensor similar to prior research is used. The planar interdigital sensor utilized in this experiment consist of a 1-5-50 configuration (Fig. 1a). Before coating the sensing solution, the sensing surfaces of the interdigital sensor was washed with methanol to remove all unwanted particles. Then, 30 μL of the prepared suspensions were pipetted on the sensing surface and dried for 24 h at 80 °C. Figure 1b shows the coated sensor prior to coating.

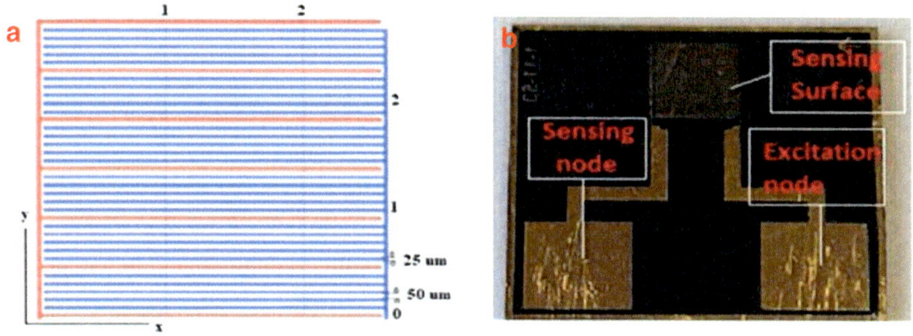

Fig. 1 **a** Interdigital sensor. **b** Sensor electrode layout

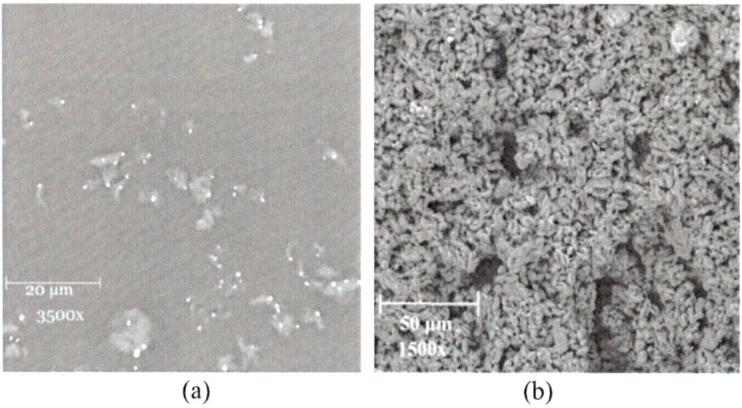

Fig. 2 **a** Reduced graphene oxide coated with silver nanoparticles. **b** Sensor electrodes coated with rGO-AgNP-PE

2.3 Characterization by Scanning Electron Microscope

It is important to analyse the surface morphology and particle size of the composite to ensure that both the silver nanoparticles have been absorbed onto the surface of the graphene as seen in Fig. 2b, and to confirm that the graphene has fit into the polymer matrix as exhibited in (c).

2.4 Sample Preparation and Measurement

Initially, a 1000 ppm (parts per million) Sulphur solution was prepared mixing Sodium Sulfite (Na_2SO_3) with deionized water. Serial dilution method has been used to prepare 0.5, 1, 2 and 3 ppm samples. The respective concentrations were calculated to maintain the same Sulphur ion concentrations as found in equivalent concentrations of H_2S. The sample water has been used to perform the experiments.

Electrochemical impedance spectroscopy (EIS) is the technique utilized to analyse the sensors characteristics. The sensor was connected to a 'Hioki IM3536' LCR Meter via a gold contact clamp connector [14]. During the measurement, the sensor was submerged in the sample for two minutes to allow for Sulphur absorption. The sensor was then washed with deionized water and allowed to dry before the next measurement. The measurements were taken utilizing a frequency sweep between 10 and 100,000 Hz at a 1 V peak to peak. Measurements were taken three times, respectively, and averaged.

2.5 Design and Development of the Portable Sensing System

A portable system was designed to measure Sulphur concentrations from samples at any desired location. The proposed system consists of a box in which all electronics are stored and a small cut-out for a liquid crystal display (LCD). The proposed system consists of an AD5933 impedance analyzer, Sulphur sensor, LCD, Arduino Uno Microcontroller, charging shield, LiPo battery and a LoRa shield for wirelessly transmitting data to the Thingspeak server. Figure 3a shows the connection diagram of the system and (b) displays the final system. The concentration of Sulphur measured by the sensor system is stored in the Thingspeak cloud. This will allow for an expert opinion if the water quality deteriorates.

Both the client and the server scripts are programmed within the Arduino IDE. After obtaining the and interpreting the sensor data, a Sulphur concentration is displayed on the LCD and then transmitted through the TTN (The Things Network) via a Dragino gateway, to the designated Thingspeak channel.

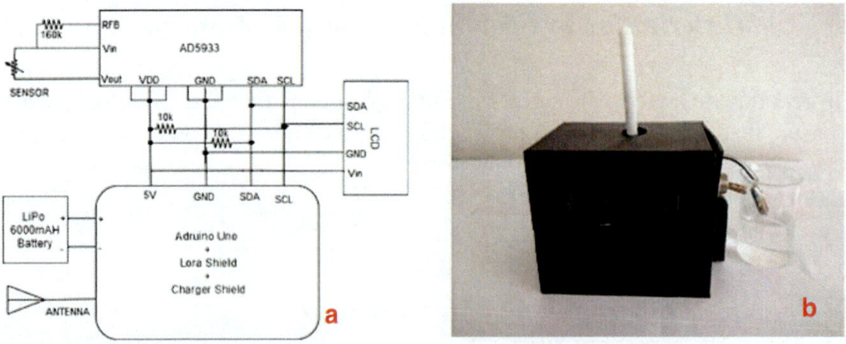

Fig. 3 **a** Connection diagram of the system, **b** Final portable system

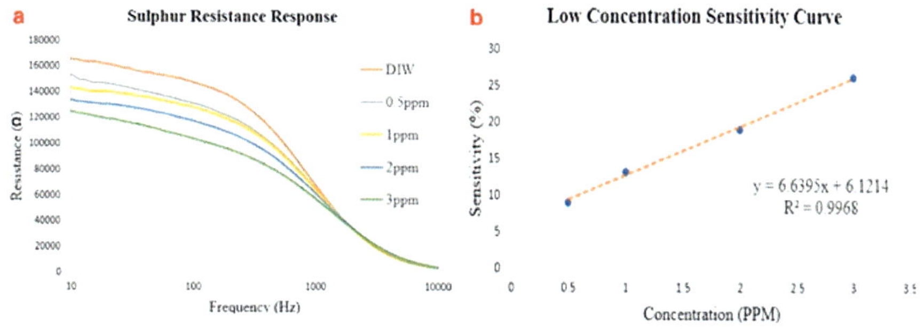

Fig. 4 **a** Variation in sensor's resistance for various sulphur concentrations. **b** Calibration curve for measuring sulphur concentrations from the sensor at 25 Hz

3 Results and Discussion

3.1 Sensor's Response Towards Sulphur Ions

The sensors surface was coated with rGO-AgNp/PE to observe the effect of the coating in achieving detection of differing Sulphur ions concentrations from 0.5 to 3 ppm. We can see in Fig. 4a that there is a distinguished difference in the measured resistance of the coated sensor when submerged in aqueous solutions with differing concentrations of Sulphur ions. The sensor shows considerable readings between 10 Hz and 1 kHz.

Therefore, calibration curves are developed at 25 Hz to determine Sulphur concentrations (0.5–3 ppm) with the sensor's resistance. The determination coefficient (R^2) from applying a linear fit is 0.9968 according to Fig. 4b, revealing a strong correlation between the Sulphur concentrations and the sensor's resistance. The following equation was utilized to calculate the sensitivity:

$$C(\text{ppm}) = \frac{R_{\text{sense}} - 6.1214}{6.6395} \tag{1}$$

3.2 Data Collection and Storage

The purpose of this research is to develop a portable Sulphur ions detection system in a real-life scenario. Therefore, the portable system is used to measure the Sulphur concentration in drinking water. The concentration of Sulphur measured by the sensor system is stored in the Thingspeak cloud. Figure 5, exhibits the measured concentration calculated by the final system and being uploaded wirelessly, with measurements being taken of the same 0.5 ppm sample. The measurements showed little fluctuation and stability over a 40-min period.

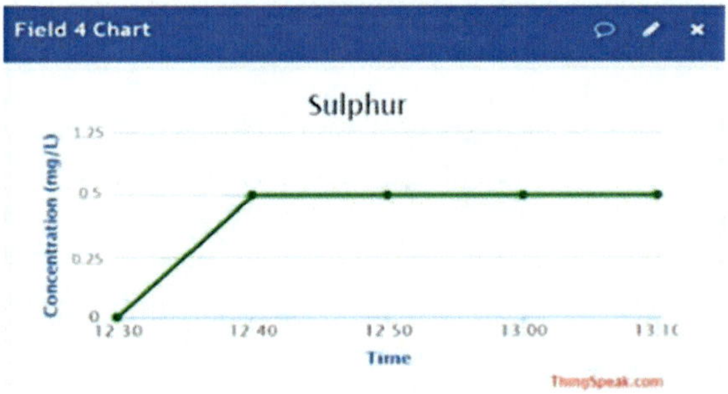

Fig. 5 Sulphur concentration stored on the Thingspeak server

4 Conclusion and Future Works

A low-cost sensitive IoT-enabled Sulphur sensor was developed and explained in this paper. The composite rGO-AgNp/PE was used as a sensitive coating on the sensors surface, and was able to detect differing ranges of Sulphur concentrations. EIS was utilized for the characterization of the developed coating, and a sensitivity curve was developed. The concentrations found in water can be stored wirelessly or analysed locally utilizing the developed system. The future work of this research is to enable the sensor to achieve accurate measurements in mediums of differing pH levels, temperature and with other analytes.

References

1. Akhter, F., Khadivizand, S., Siddiquei, H.R., Alahi, M.E.E., Mukhopadhyay, S.: IoT enabled intelligent sensor node for smart city: pedestrian counting and ambient monitoring. Sensors **19**(15), 1–19 (2019). http://doi.org/10.3390/s19153374
2. Akhter, F., Khadivizand, S., Lodyga, J., Siddiquei, H.R., Alahi, M.E.E., Mukhopadhyay, S.: Design and development of an IoT enabled pedestrian counting and environmental monitoring system for a smart city. In: Proceedings of the 13th International Conference on Sensing Technology (ICST), 2–4 Dec 2019, Sydney, Australia (2019). http://doi.org/10.1109/ICST46873.2019.9047695
3. Baratto, C., Rigoni, F., Faglia, G., Comini, E., Zappa, D., Sberveglieri, G.: ZnO and SnO_2 one-dimensional sensors for detection of hazardous gases, pp. 1–3 (2017)
4. Luo, Y., Zhang, D., Fan, X.: Hydrothermal fabrication of Ag-decorated $MoSe_2$/reduced graphene oxide ternary hybrid for H_2S gas sensing. IEEE Sens. J. **20**(22), 13262–13268 (2020)
5. Montazeri, M.M., De Vries, N., Afantchao, A.D., O'Brien, A., Kadota, P., Hoorfar, M.: Development of a sensing platform for nuisance sewer gas monitoring: hydrogen sulfide detection in aqueous versus gaseous samples. IEEE Sens. J. **18**(19), 7772–7778 (2018)
6. Yavari, A., Tahmooressi, H., Hoorfar, M., Mohaghegh Montazeri, M., Tasnim, N., Farahani, A., Kadota, P., Markin, P., Dalili, A., Taatizadeh, E.: A graphene-based chemical sensor for hydrogen sulfide measurement in water, pp. 1–4 (2019)
7. Zheng, H.: Application of graphene in electrochemical sensing. Curr. Opin. Colloid Interface Sci. **20**(5–6), 383–405 (2015). https://doi.org/10.1016/j.cocis.2015.10.011
8. Zheng, D., Hu, H., Liu, X., Hu, S.: Application of graphene in elctrochemical sensing. Curr. Opin. Colloid Interface Sci. **20**(5–6), 383–405 (2015)

9. Rowley-Neale, S., Randviir, E., Dena, A., Banks, C.: An overview of recent applications of reduced graphene oxide as a basis of electroanalytical sensing platforms. Appl. Mater. Today **10**, 218–226 (2018)
10. Peng, W.: Ultrasensitive ppb-level H_2S gas sensor at room temperature based on WO3/rGO hybrids. J. Mater. Sci. Mater. Electron. **31**(6), 5008–5016. http://doi.org/10.1007/s10854-020-03067-6
11. Some, X.: Highly sensitive and selective gas sensor using hydrophilic and hydrophobic graphenes. Sci. Rep. **3**(1), 1868–1868 (2013). https://doi.org/10.1038/srep01868
12. Zhang, Z.: Highly sensitive H_2S sensors based on metal-organic framework driven γ-Fe_2O_3 on reduced graphene oxide composites at room temperature. Sens. Actuators B Chem. **325**, 128804 (2020). http://doi.org/10.1016/j.snb.2020.128804
13. MalekAlaie, J.: Selective hydrogen sulfide (H_2S) sensors based on molybdenum trioxide (MoO3) nanoparticle decorated reduced graphene oxide. Mater. Sci. Semicond. Process. **38**, 93–100 (2015). https://doi.org/10.1016/j.mssp.2015.03.034
14. Akhter, F., Alahi, M.E.E., Siddiquei, H.R., Goonerate, C.P., Mukhopadhyay, S.C.: Graphene oxide (GO) coated impedimetric gas sensor for selective detection of carbon dioxide (CO_2) with temperature and humidity compensation. IEEE Sens. J. Early Access (2020). https://doi.org/10.1109/JSEN.2020.3035795

Extracting Operational Insights from Everyday IoT Data, Generated by IoT Sensors Over LoRaWAN

University Campus, Smart City Application

Ollencio D'Souza, Subhas Mukhopadhyay, Fowzia Akhter, Sam Khadivizand, and Erfan Memar

1 Introduction

A smart city [1] is a developed ecosystem with an innovative, environmentally friendly operational structure for living, working and conducting the daily social activity. It facilitates and automates relevant parts of the economy to improve the quality of life, health outcomes, security and safety. A systems approach, Fig. 1, is used to build smart city governance, responsible environmental practices and management of key factors required to meet the objectives of a smart city ecosystem, often delivered over converged communications channels, in the age of the internet.

The smart city service delivery operation is dependent on information systems to provide efficient and effective support, especially during event-driven emergency operations. However, communication infrastructures can be disrupted during emergencies [2] due to a congested or over-run wireless channel space. The adoption of LoRaWAN as a long-range low-power "always-on", non-packet switched network proved to be the right choice because of its reliable communications capability, reach and small payload transfer efficiency. **LoRa** [3] can demodulate signals which are -7.5 to -20 dB below the noise floor providing an economical and reliable non-line of sight solution to surmount current communication challenges. The design team

O. D'Souza · S. Mukhopadhyay (✉) · F. Akhter
School of Engineering, Macquarie University, Sydney, NSW 2109, Australia
e-mail: subhas.mukhopadhyay@mq.edu.au

O. D'Souza
e-mail: Ollencio.DSouza@students.mq.edu.au

S. Khadivizand
IEEE, Piscataway, USA

E. Memar
Marketik, Tehran, Iran

J. K. Mandal and J. K. Roy (eds.), *Proceedings of International Conference on Computational Intelligence and Computing*, Algorithms for Intelligent Systems,
https://doi.org/10.1007/978-981-16-3368-3_23

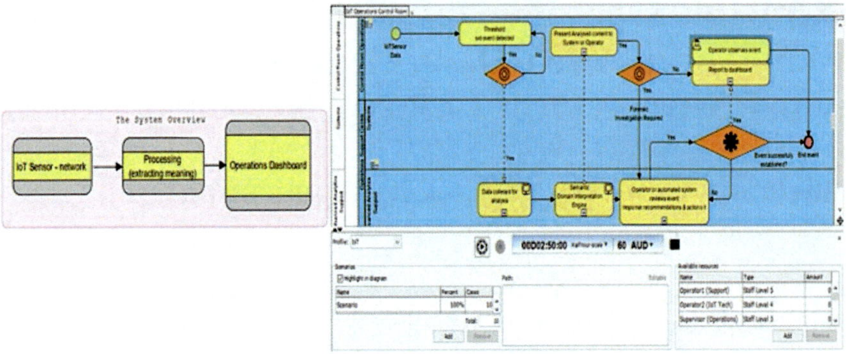

Fig. 1 Systems approach shows typical workflow and decision points with example resourcing and process automation

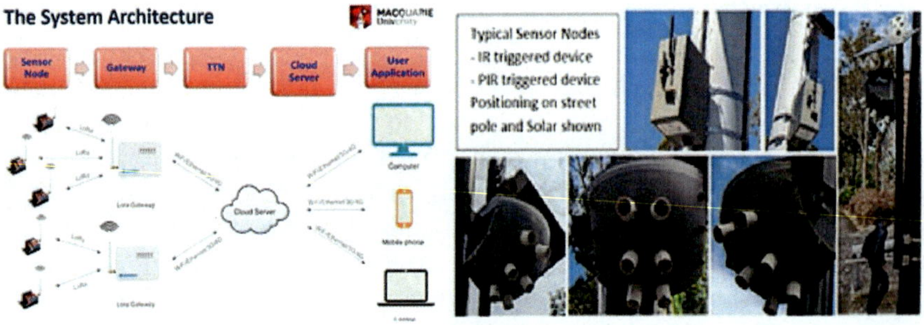

Fig. 2 IoT network with a customised sensor deployment example in pictures

took the strategic direction to adopt LoRaWAN, as the "network", Fig. 2, over which sensor data would be collected from carefully selected locations across the campus in precise time intervals. The custom-engineered sensors delivered traffic count, temperature and humidity data from which "air quality", "health index" ratings and other semantically recognisable, essential, operational performance measures are formulated for the site. The campus-wide installation of seventy-one LoRa "nodes", linked to the things network via two gateways, produces live data or access to stored history which could then be used to improve response to critical issues [4].

Operational integrity and campus safety [5] require timely data which, when analysed, provides valuable operational insights so operations staff can predict and respond promptly to events and issues at the site by mapping activity on campus [6], such as;

1. Establishing the "traffic flow".
2. Establishing the "air quality thresholds" (based on published guidelines).
3. Improving awareness of "gatherings" which could be breaking social distancing norms.
4. Setting threshold and responding to "congestion" breaches.

5. Establishing safe levels of "occupancy" at critical locations on campus.
6. Ensuring safety measures such as speed, parking and congestion based on counts (of people entering and leaving carparks).
7. Juxtaposing vehicular traffic with human traffic flows to reduce the risk of accidents (algorithmic calculation based on "other" data inputs).
8. Awareness of peak people traffic times of the day.
9. Crisis management knowledge, i.e. knowing the location of crowds.

Timely data from sensors providing status of conditions across the campus enables creation of an "early warning" system based on the thresholds set and response times required [1, 7].

2 Smart City Evolution

Digital transformation and the evolving semantic WEB are hastening the evolution of smart cities [8], and along with converged communications, these technologies create the multi-dimensional character of smart cities. In a well-designed managed environment, data is collected from "other" sources to access supporting information, such as maps or inputs from other installed site systems like CCTV and Access Control to improve decision making [9]. Such an ontology benefits citizens, policymakers, administrators and operations teams to meet objectives to deliver an effective operating environment [2]. The examples in Table 1 could be expanded to include many different direct sensor observations and post-analysis outcomes of algorithmically derived relationships between different groups of sensor-generated values. By tying two or more measured parameters together, the nexus between traffic count, temperature and humidity points to relevant "Indices" produced by regulatory bodies on health, safety and operational discipline. Meeting these parametric thresholds leads to "compliance" and another very important outcome, i.e. to maintain "operational integrity".

Table 1 demonstrates how semantically structured relationship between sensor inputs, activity profiles and "operational insights" are created and how they are derived from activity recorded by standard IoT sensors in the Macquarie University Smart City project.

3 The Operational Model

The operational model in Fig. 3 presents activity on campus, showing potential chaotic conditions to demonstrate the complexity of known activity and results of an important and serious period of activity on the site.

Table 1 Event, activity, and insights table

	Event filters	Activity profiles	Operational insights
1	Count	High count observed	Activity at this location has risen. Is it explainable?
2	Temperature	Higher than normal temperature observed	Are people "exposed", i.e. crowds in open spaces in danger of a heat stroke? Could "congestion" be an issue? Is health risk high?
3	Humidity	Higher than normal humidity	Has the health index been crossed (congestion, heat and humidity)? Are our medical response teams ready to respond?
4	Thresholds	Several key "alarm" threshold indices created from criteria based on observed criteria such as a numerical count per area PLUS humidity range PLUS temperature level crossed	When these indices are developed from known safe allowable environmental conditions and each step up could mean higher risk for the population from "fatigue" or health or personal risk (crowding) being prepared (risk anticipation, better awareness, outcome prediction, planned responses to match) are meaningful operational considerations
5	Planned scenarios	University timetables, curriculum activity, etc.	Operationally such activity should have a "predicted" range of criteria for each planned activity—an unscheduled departure from this could mean something unusual is happening and an investigation is warranted
6	Human/ vehicle	Detecting the "speed", from transitioning of an object from one point to another, count of larger objects and smaller objects	Operationally car parks are within the confines of the campus and give rise to the likelihood of "humans" and "vehicles" coming together—if these insights are available it could limit the risks, improve the response, and the after-effects of an incident occurring
7	Access restrictions	Time of day, entry and exit, no-go zones (Laboratories), etc. observations	Operational it would be unusual to see a lot of activity at six am in the morning at known sensitive locations on campus

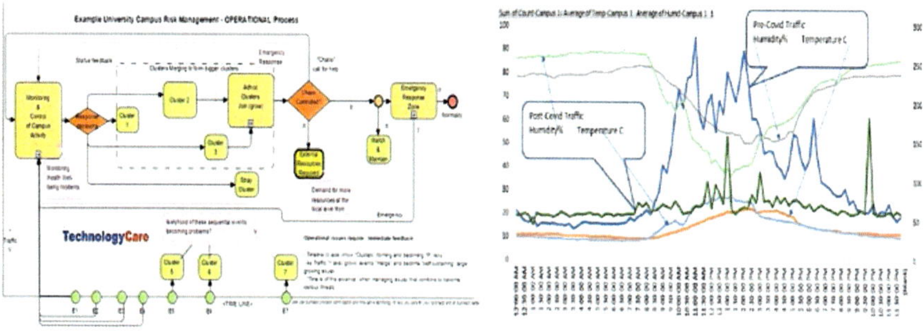

Fig. 3 Operational workflow and graphical COVID-19 inferences

4 The Health Nexus

The department of health and other researchers have established that the congruence of heat, humidity, wind and radiation affects human health adversely [10–14]. The "nexus" between temperature, humidity, "congestion" (crowds) beyond a prescribed threshold and it's relationship to human health is site-specific. Open spaces would allow airflow, whilst enclosed spaces might hold stagnant air in hot and humid conditions creating conditions for fatigue to set in and disease to spread [15].

5 Data from the Smart City System and the Semantic Model

The data provided in Fig. 4 (left) is a rich source of "observations" that have the potential to deliver unique insights using the semantic model analysis proposed on relevant

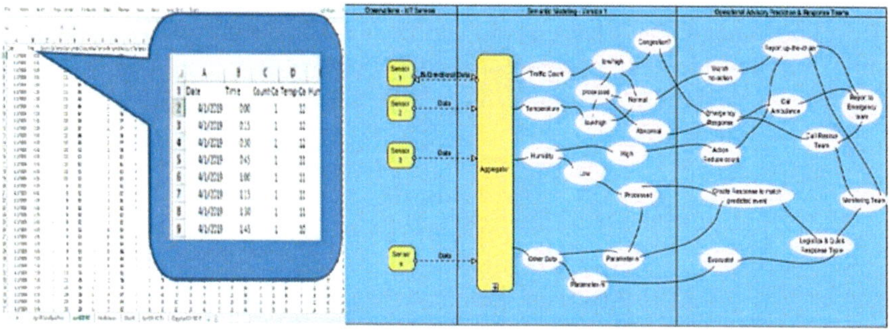

Fig. 4 Data gathering granularity and use in a semantically modelled process workflow

weather, air quality reference data and second-degree analysis of the known combination of parameters such as traffic movement over time. The semantic modelling process flow in Fig. 4 (right) presents a methodology to estimate direct and indirect (processed) data costs to manage system outputs from sensors and a finite set of manual resource. The process can be automated using an ontology-based resourcing methodology and is a template for the process workflow to be followed to provide tailored outcomes by automating data processing and verification before they are actioned.

5.1 Thermal Stress and Relevance to Campus Health and Safety

Human thermal comfort depends on environmental and personal factors. The four environmental factors are airflow (wind), air temperature, air humidity and radiation from the sun and nearby hot reflective surfaces.

Personal factors are the clothing being worn and the level of physical activity. The level of thermal sensation is also affected by acclimatisation/adaptation. Studies show that people who originate from hot climates are comfortable at higher temperatures than those coming from in cooler climates [16–18]. The mathematical model of heat balance in the human body should include the effects of temperature, humidity, wind speed and radiation [19].

5.2 Influence of Population Movement on Relative Humidity

The surface relative humidity is a key meteorological variable that has a profound impact on the environment and lives of city occupants, permanent or migratory. One study has shown that averaged over the study period, Beijing's urban relative humidity increased by 4.00% during the holiday festive week with 99% confidence, relative to the non-holiday background period. In relative terms during the festive period, the population movement resulted in a change of 76% of relative humidity, between urban and nearby non-urban areas [20].

6 The Dashboard

Data analysis generates unique insights, and these insights need to be visualised effectively. Not all insights are required at once, and hence, there needs to be a prioritised and hierarchical visualisation technique which provides an overall status of all areas on the site map. An event window presents threshold breaches and a

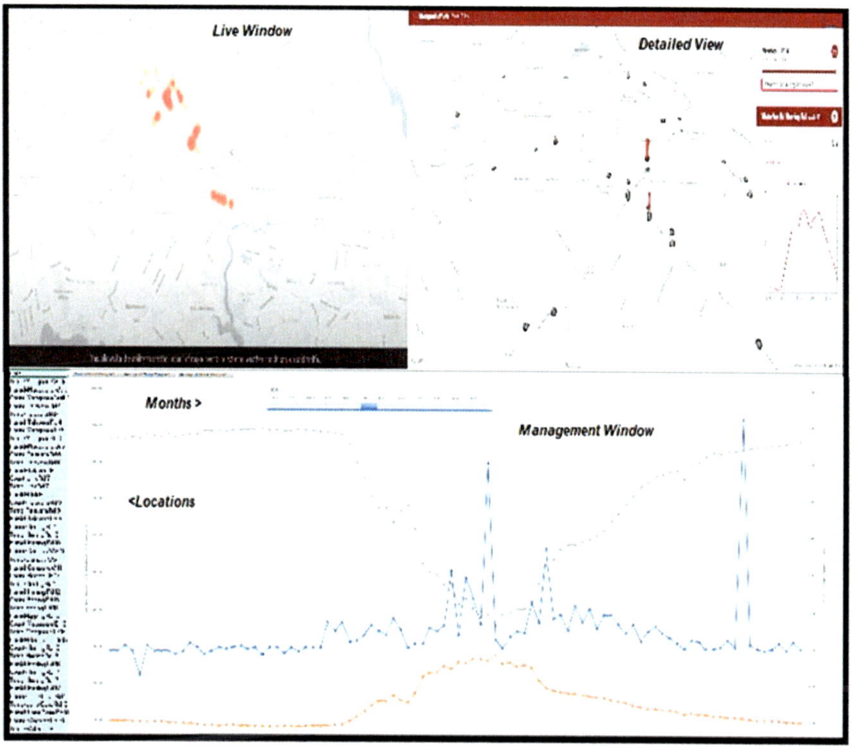

Fig. 5 Dashboard visualisation presents live "hotspots" (congestion), location accuracy (maps) and access to historical trends

live control window allowing monitoring and management of an event. The central operational control facility will determine what parameters are important to keep steady to maintain a stable operation. The trajectory of "congestion" traffic into an area via "hotspots" can be provided by analysing the rate of increase and comparing it with known trajectories from analysed history as in Fig. 5. Similarly, weather conditions could be based on predictions by the weather department linked to sense data at the site.

7 Conclusion

In a constantly changing operating environment, there are many stages of adoption of technology. At the first level receiving data consistently and regularly from sensors, reinforces operational confidence in the system. The reliability of data from LoRa sensors over the LoRaWAN network creates the resource for operational effectiveness. The second stage of acceptance is when initial insights become useful to manage critical operational issues like "overcrowding" and meeting mandated requirements, especially during difficult infectious times. At the third level, when data provides insights to correct and prevent potential future risks, the system becomes part of the operational fabric delivering significant value in terms of process optimisation possibilities, adding to the ontology of the Smart City environment.

Acknowledgements We would like to thank Macquarie University for funding the project. Our gratitude and respect for our Department Head and Supervisor Professor Dr. Subhas Mukhopadhyay. The efforts of the Coauthors and project team Mst Fawzia Akhter, Sam Khadivizand and Erfan Memar and their dedication to this project have made it possible to deliver positive outcomes.

References

1. Favale, T., Soro, F., Trevisan, M., Drago, I., Mellia, M.: Campus traffic and e-learning during COVID-19 pandemic. Comput. Netw. 107290 (2020)
2. Komninos, N., Bratsas, C., Kakderi, C., Tsarchopoulos, P.: Smart city ontologies: improving the effectiveness of smart city applications. J. Smart Cities **1**(1), 31–46 (2019)
3. "Predicting LoRaWAN capacity | DEVELOPER PORTAL." /library/tech-pa-pers-and-guides/predicting-lorawan-capacity/. Accessed 17 Nov 2020
4. Alvarez-Campana, M., López, G., Vázquez, E., Villagrá, V.A., Berrocal, J.: Smart CEI moncloa: an IoT-based platform for people flow and environmental monitoring on a smart university campus. Sensors **17**(12), 2856 (2017)
5. Charles-Edwards, E., Bell, M.: Estimating the service population of a large metropolitan university campus. Appl. Spat. Anal. Policy **6**(3), 209–228 (2013)
6. D'Souza, O.: A virtual presence/spl reg/, predictive analysis and pre-emptive intervention using wired and wirelessly networked, interactive, video, voice and control technology—a case study. In: Proceedings 39th Annual 2005 International Carnahan Conference on Security Technology, pp. 109–112 (2005)
7. Ling, J., Zhang, Y., Zhang, Z.: Human responses to high humidity in elevated temperatures for people in hot-humid climates. Build. Environ. **114** (2016). http://doi.org/10.1016/j.buildenv.2016.12.028
8. Abid, T., Zarzour, H., Laouar, M.R., Khadir, M.T.: Towards a smart city ontology. In 2016 IEEE/ACS 13th International Conference of Computer Systems and Applications (AICCSA), Nov 2016, pp. 1–6. http://doi.org/10.1109/AICCSA.2016.7945823
9. "What is humidity and why does it matter?," Blue Ox Heating & AIr, 06 Aug 2019. https://www.goblueox.com/blog/what-is-humidity-and-why-does-it-matter/. Accessed 22 Oct 2020
10. Fan, J., Stewart, K.: Modeling mobility and dynamics of scheduled space-time activities—an RDF approach. In: Human Dynamics Research in Smart and Connected Communities, pp. 81–105. Springer, Berlin (2018)
11. "Climate effects on human health." http://www.ciesin.org/docs/001-338/001-338.html. Accessed 31 Oct 2020
12. "Health index using temperature and humidity—GoogleSearch." https://www.google.com/search?q=health+index+using+temperature+and+humidity&rlz=1C1GCEB_enGBAU924AU924&oq=health+index+using+temperature+and+humidity&aqs=chrome..69i57.27837j0j7&sourceid=chrome&ie=UTF-8. Accessed 15 Nov 2020
13. "Temperature, sunshine, humidity and the coronavirus." https://www.wbur.org/weather/2020/03/26/temperature-sunshine-humidity-coronavirus. Accessed 22 Oct 2020
14. "Whatis the relationship between crowds, temperature and humidity—Google Search." https://www.google.com/search?q=What+is+the+relationship+be-tween+crowds%2C+tem

perature+and+humidity&rlz=1C1CHBF_enAU774AU779&oq=What+is+the+relationship+between+crowds%2C+temperature+and+humidity&aqs=chrome..69i57.61423j1j7&sourceid=chrome&ie=UTF-8. Accessed 22 Oct 2020
15. "Effects of humidity on the human body," UPMC HealthBeat, 23 June 2014. https://share.upmc.com/2014/06/effects-humidity-body/. Accessed 31 Oct 2020
16. Brooke, A.G., Bell Michelle, L., Peng Roger, D.: Methods to calculate the heat index as an exposure metric in environmental health research. Environ. Health Perspect. **121**(10), 1111–1119 (2013). http://doi.org/10.1289/ehp.1206273
17. Steadman, R.G.: Norms of apparent temperature in Australia. Aust. Met. Mag. **43**, 1–16 (1994)
18. "The steadman wind chill: an improvement over present scales | Weather and Forecasting | American Meteorological Society." https://journals.ametsoc.org/waf/article/13/4/1187/38306. Accessed 25 Nov 2020
19. "Australian Apparent Temperature (AT)." https://www.vcalc.com/wiki/rklarsen/Australian+Apparent+Temperature+%28AT%29. Accessed 25 Nov 2020
20. Zhang, J., Wu, L.: The influence of population movements on the urban relative humidity of Beijing during the Chinese spring festival holiday. J. Clean. Prod. **170**, 1508–1513 (2018). https://doi.org/10.1016/j.jclepro.2017.09.274

Handoff Analysis in Overlay Based GTP Tunneled 3G-WLAN Integrated Network

Asish Kumar Mukhopadhyay and Sajal Saha

1 Introduction

The current network communication has several properties in terms of bandwidth, mobility, frequency latency, and different communication channels. Handoff happens when switching from one physical network to another on a mobile device. Handoff, however, in heterogeneous networks environment increases handoff latency due to interface and protocol incompatibly. Handoff within heterogeneous networks is a critical issue when a user moves into a dissimilar network with dissimilar hardware and software standards with wide variety of communication devices, access networks and wide range of services. Under global perspective, integration of heterogeneous networks is a challenging task. Now-a-days, deployment of WLAN (as hotspot area) in public places and institutions has been very popular because of its high speed and bandwidth. Normally, the hotspot area also has cellular coverage. It has thus become interesting to integrate both the cellular and WLAN coverage for ubiquitous and seamless communication. Since 3G (UMTS/W-CDMA) is currently the market leader in cellular technology, 3G-WLAN integration providing seamless communication has specifically become the challenging task. The crux of the problem is the latency due to the handoff between heterogeneous networks (such as WLAN and 3G cellular) which is commonly known as vertical handoff (VHO). This is in contrast with the normal homogeneous inter-cell horizontal handoff (HHO).

We can reduce the vertical handoff (VHO) latency through the overlay network model. Overlay network model offering scalability, reliability, and congestion control

A. K. Mukhopadhyay (✉)
Department of E&CE, Bundelkhand University, Jhansi, India

S. Saha
Department of Computer Science and Engineering, The Assam Kaziranga University, Jorhat, India
e-mail: sajalsaha@kazirangauniversity.in

J. K. Mandal and J. K. Roy (eds.), *Proceedings of International Conference on Computational Intelligence and Computing*, Algorithms for Intelligent Systems,
https://doi.org/10.1007/978-981-16-3368-3_24

capability is poised to fulfill the essential requirements of QoS and seamless communication over heterogeneous networks.

2 Wireless Overlay Networks

Overlay network, also known as virtual network, is logical network constructed using the existing heterogeneous networks as substrate. A wireless overlay network (WON) [1] is overlaid on top of one or more existing networks where high bandwidth smaller coverage area network layered below low-bandwidth networks with wider coverage by strategically choosing network technologies according to the needs of any given location. Each user and network operator can share the benefit of improved coverage and high performance. Despite heterogeneity at the data link layer, end-to-end IP packet delivery is enabled in the overlay network. Overlay networks provide value-added network services by deployment of new functionalities and enabling few changes in the underlying network architecture. It can be used to modify the top layers without any changes of functionalities in the lower layers. The focus is to control and assign router resource and overlay link to different layers of the overlay network and avoid resource contention among the layers.

In overlay network, users' handoff both horizontally and vertically. Horizontal handoff is same as normal handoff in same kind of network. Vertical handoff (VHO) is a feature that denotes handoff between different layers of overlay network with different types of interfaces. A routing software is installed in the specific location of overlay network thorough direct link and encapsulation. In a wireless overlay network (WON), normally, high-bandwidth, smaller-coverage area network layered below the low-bandwidth networks with wider coverage areas with few deviations in the bandwidth. For instance, satellites are present in the top layer with high coverage area. User connected in the lower layer with high bandwidth lower coverage may transit to low-bandwidth high coverage dynamically when lower layer network coverage is not available. User can avail new services or platform to develop new applications through overlay network which are not present in the legacy network system. In addition, overlay networks uses IP Multicast support effective group communication, distribution, and avoid burdening the underlying network and utilize application layer information to provide better services. They can take the required resources available at overlay nodes to perform tasks which are not possible in conventional router, such as block chain operation, database lookups, file caching etc. However, overlaying may incur some management and communication overhead and redundant inter-layer functions due to additional layer interface between application layer and IP layer.

Multiple alternate overlay paths exist between a pair of source and destination, which is an essential aspect of this overlay architecture. As call requests a rive, the ingress overlay node decides the best overlay route to the destination based on the user's permission, objective, and network status.

3 Vertical Handoff in WON

Horizontal handoff is a standard feature in any wireless network and its management depends on that particular wireless network. Vertical handoff (VHO), on the other hand, is a special feature of overlay network with additional inter-layer challenges such as co-ordination and synchronization, while triggering VHO, incompatibilities in signal strengths and network conditions and metrics [2]. There are two way VHOs, one from high-bandwidth network (like WLAN) to low-bandwidth network (cellular) which is called upward vertical handoff and the other in the reverse direction which is called downward vertical handoff. One of the most difficult problems in the overlay network is to provide smooth vertical handoff, which requires the handoff procedure to be invisible to upper-layer applications. In order to ensure smooth handoff, a number of problems, such as handoff parameters, mobility management issues are needed to be taken care of [3]. VHO decision is an integral part of mobility management and depends on physical parameters such as signal strength, bit error rate, network coverage, handoff power requirements and also other parameters such as perceived QoS and the required QoS, call admission control scheme, inter-layer movement algorithms, cost and user preference [4]. VHO decision inputs are extracted from both the network and the user and therefore the mobile assisted approach is more solicited for handoff management. The mobility and layer management algorithms process this information and accordingly decide the most optimal network and the appropriate time to initiate the handoff. Schemes based on policy, fuzzy logic, and neural networks have been proposed [5]. The selected network admits the user if resources are available or may force a current user to handoff to another network to make way for the incoming connection request. The proposals are mainly concerned with handoff initiation and network selection.

The handoff management schemes pass through the three phases-handoff initiation, network selection and handoff execution. Handoff initiation is the process of monitoring the current network connection, recognizing the need for handoff and subsequently initiating it. The criteria used are signal strength, the coverage area, bit error rate and perceived QoS. The handoff initiation algorithm needs to process many parameters and decide whether a handoff to another system is required.

The aim of the network selection stage is to select a network connection that can satisfy the requirements of the network provider and also the user, such as low cost, good signal strength, optimal bandwidth, low-network latency and high reliability. The objective of the handoff execution stage is to change cell, code technology, or network conforming to the details resolved during the decision phase.

Many researchers have suggested several proposals in order to reduce vertical handover latency (VHO) latency. VHO performance of some significant WLAN-GPRS integration schemes are analyzed in papers [6–11]. Our suggested model is having better QoS support and reduced handoff latency [1, 12]. Nevertheless, the model only accept horizontal handoff intra-layer micro-mobility and vertical handoff inter-layer macro-mobility. However, when dissimilar networks coexist inside the same layer in a true heterogeneous scenario, both vertical and horizontal handoff can be done inside a layer, which has not been discussed in any proposal. Furthermore, apart from the network layer, the proposals are implemented in the transport, session, and application layers, while future all-IP networks would need network layer integration.

4 Handoff Analysis

Controlling inter-layer handoff is a priority for the Layer Manager. With the aim of preserving the optimal QoS, we consider two major handoff scenarios consistent with two major layers.

Handoff from cellular to WLAN: When a 2G user requires higher bandwidth, may link with the WiFi region with less cost. To ensure better QoS, the layer manager needs to sense both lower layers (PHY and MAC) of WLAN. The physical layer is used to detect reliable WLAN signal, while the MAC layer is used to detect WLAN network conditions-like available bandwidth and connection latency. Physical layer sensing is a general occurrence, but we only discuss MAC layer sensing in this paper. The LM listens and collects the network allocation vector (NAV) in the MAC layer from the control channels to estimate network parameters such as available bandwidth and access delay. It may be mentioned that NAV is the major mechanism in the MAC of the IEEE 802.11 WLAN to avoid collision. When a station listens to other stations transmission, it sets the NAV to busy state and keeps silent for time duration equal to the duration ID in the packet header. Thus, we find that the NAV busy state can efficiently indicate the system busy state or traffic load. Higher is the traffic, larger is the NAV busy occupation and vice versa. It depends on the traffic pattern and user number. Based on the NAV value in a certain time interval, it calculates the usable bandwidth and access latency with respect to a given packet size. We thus find that the MAC sensing scheme allows for a network-condition-aware handoff, and the QoS data allows for upper transport and application layer adaptation. Any one of the APs can be selected as the pilot access point (PA).

Handoff from WLAN to cellular: Unavailability of the WLAN is detected immediately as soon as the user moves out of the small coverage area of WLAN. It is then needed to switch the connection to the cellular network. Thus, the aim of the directional handoff is to switch to cellular before the WLAN link breaks, user prefers to stay under the aegis of WLAN as long as possible (due to higher bandwidth). The key issues in detecting the WLAN unavailability are accurate detection of the signal decay and weak signal. An adaptive configuration scheme can be used to set an effective RSSI threshold for poor signal detection, and an FFT-based solution can be used to detect signal decay.

Handoff from WLAN to GPRS/UMTS or vice versa: When a MN changes it is point of attachment from WLAN to 2.5G/3G in loosely coupled system packet is tunnelled through GPRS tunneling protocol (GTP) [13–17]. GTP (GPRS Tunneling

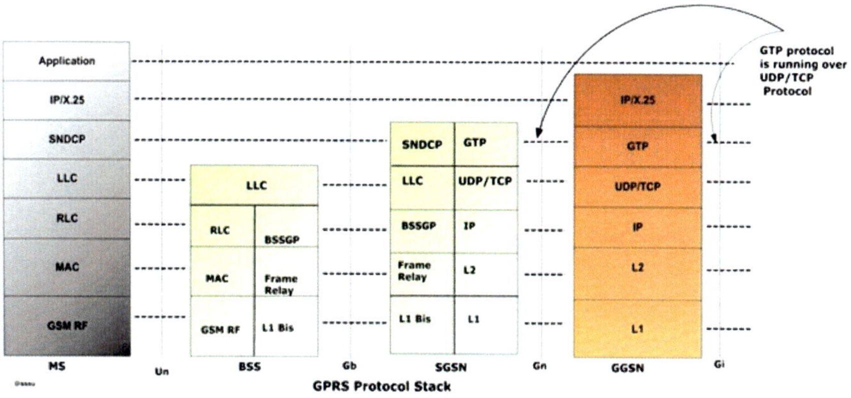

Fig. 1 GTP protocol attached to the application and transport layer of the protocol stack

Protocol) is the protocol between SGSN and GGSN in GPRS or UMTS backbone network. It includes GTP signaling, control (GTP-C) and user data transfer (GTP-U) procedures. It deals with different types of tunnels-one for signaling and control purpose and the other for user data. GTP protocol attached to the application layer and transport layer of the protocol stack as shown in Fig. 1. Data flow and control flow sequences are shown in Fig. 2.

GTP protocol includes following three steps:

1. GPRS attach through SIM card authentication
2. PDP context.
3. G-PDU transfer.

GTP Protocol: Algorithm

```
SIM Authentication
Step 1: SGSN receive a packet with the following input
Input: GTPType=8, Packetsize=500, start time=0
(GTP_App () takes the input from Ns Default.tcl)
Input: Ack time, RAND, IMSI
(MN takes RAND value from SGSN, Ack time Value from GTPbuf (), SGSN
takes RAND, SRES, Kc from AUC)
In this program we initialize
RAND=8
kc=2
IMSI=131;
time_=Scheduler: instance ().clock ()
Step 2:
At SGSN calculate
SRES=IMSI*kc+RAND;
Step 3:
SGSN send RAND to MN
At MN calculate
SRES'= IMSI*kc+RAND
Step 4:
```

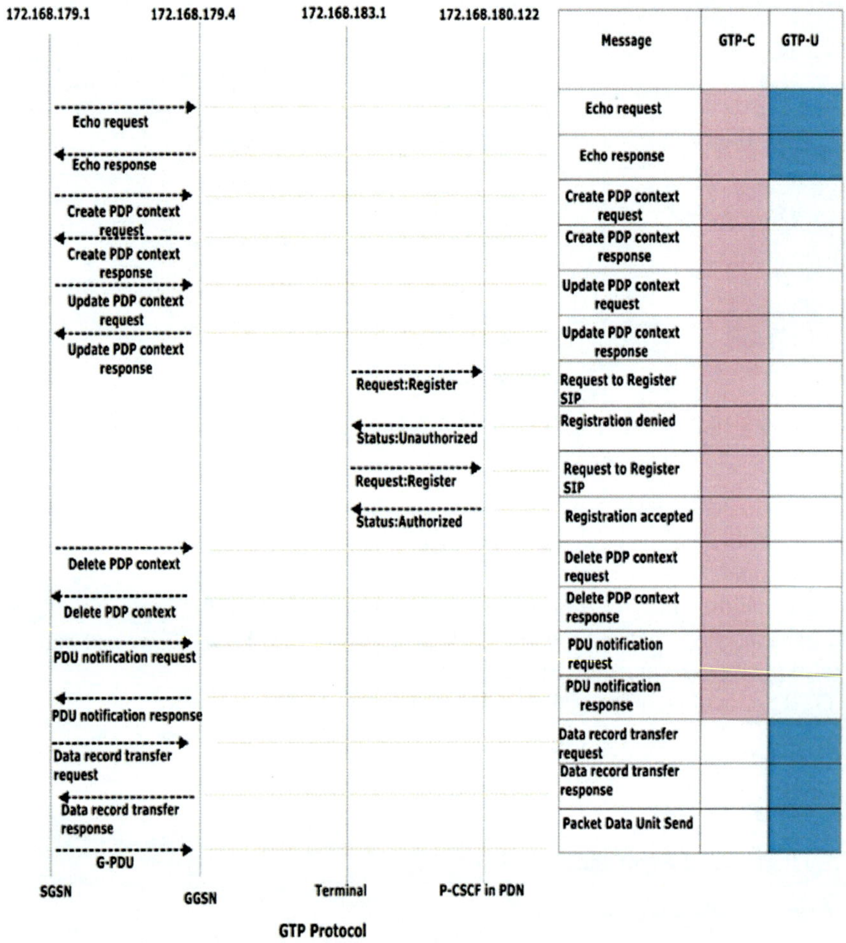

Fig. 2 Pdp context and G-PDU transfer timing diagram in GTP protocol

```
MN sends SRES' to SGSN
If SRES=SRES'
Then SIM is authenticated
```

PDP context

```
Step 1: Send echo request from SGSN to GGSN
Step 2: Send echo response from GGSN to SGSN
Step 3: Create PDP context request at GGSN
Step 4: Create PDP context response at SGSN
Step 5: Request to register SIP
    If Unauthorized
      Then registration denied
    Else
      SIP registered
Step 6: Delete PDP context request
```

```
Step 7: Delete PDP context response
G-PDU Transfer
Step 1: Send PDU notification request from SGSN to GGSN
   If tunnel is clear
    Then notification request accepted
   Else
    Notification request rejected
Step 2: Send PDU notification response from GGSN to SGSN
Step 3: Data record transfer request from SGSN to GGSN
Step 4: Data record transfer response from GGSN to SGSN
Step 6: Send G-PDU from SGSN to GGSN
```

5 Simulation Network Topology for GTP Tunneling and Handoff Delay Analysis in UMTS Network

Network Topology Configuration:

Step 1: Create 9 nodes name GGSN, SGSN, R3, RNC,2 Access Router,3 mobile Node in the following manner in NS2 environment.

GGSN(Domain=0, Cluster=0, Node=0)
GGR(Domain=0, Cluster=1, Node=0)
R3(Domain=0, Cluster=2, Node=0)
SGSN(Domain=1, Cluster=0, Node=0)

AR1(AccessRouter)(Domain=2, Cluster=0, Node=0)
AR2(AccessRouter)(Domain=3, Cluster=0, Node=0)
MN0(Mobile node)(Domain=1, Cluster=0, Node=0)
MN1(Mobile node)(Domain=2, Cluster=0, Node=0)
MN2(Mobile node)(Domain=3, Cluster=0, Node=0)
Step 2: Delivering packets from AR2 to GGSN
Attach-Agent UDP at transport layer to AR2
Attach-Application CBR at application layer to AR2
Attach Agent ping at IP layer to GGSN
GGR is treated as a simple GPRS node and HA (Home Agent) is attached with GGSN.
Connect AR2 and GGSN.
Attach-Agent GTP APP at application layer and GTP TCP at transport layer to create a GTP tunnel between GGSN to SGSN shown in the following Fig. 3.
Step 3: Delivering packets between SGSN to UMTS host MN0
Attach-Agent ping at IP layer to SGSN
Attach-Agent UDP at transport layer to MN0
Connect MN0 and SGSN.

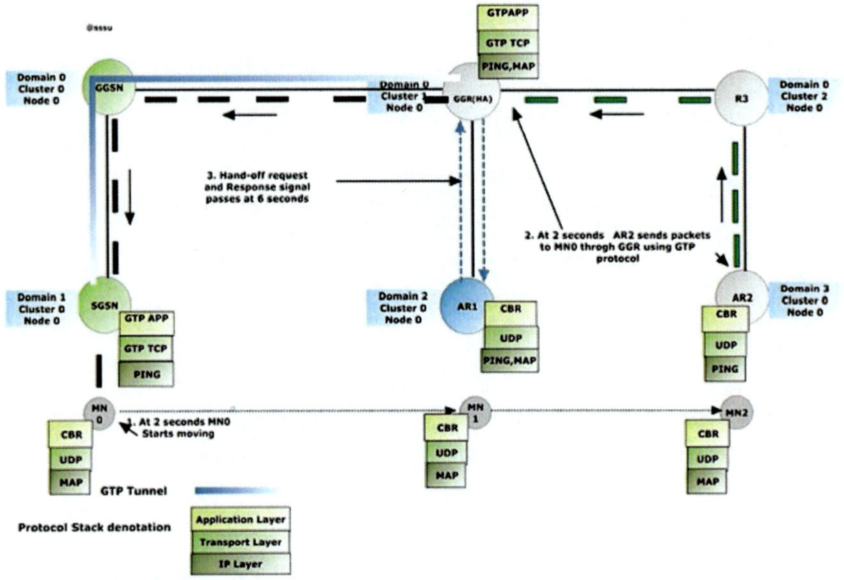

Fig. 3 Network topolgy for GTP tunnel in UMTS network

Step 5: Create tunnel between GGSN and AR1/FA
Attach-Agent ping at IP layer to GGSN
Attach-Agent Ping at IP layer to AR1
Connect GGSN and AR1.
Step 6: Packet delivery by AR1/FA to MN1
Attach-Agent Ping at IP layer to AR1
Attach-Agent UDP at transport layer to MN1
Connect MN1 and AR1.
Step 7: Configuration of MN1
Attach-Agent MAP at IP layer to MN1
Attach-Agent MAP at IP Layer to AR1
Connect MN1 and AR1 through MAP Agent
Step 8: Sending Handoff Request signal from MN0 to GGSN
Attach-Agent MAP at IP layer to mobilenode1
Attach-Agent MAP at IP layer to GGSN
Connect MN1 and GGSN through MAP Agent

6 Results and Discussion

For the simulation and analysis, we use software simulation tool network simulator (NS2) distribution. It is chosen as the simulation framework for this system due to the availability of the WLAN, UMTS, and GPRS patches. Initially, we created UMTS and WLAN network environment and perform the horizontal handoff of mobile with the varying packet size and traffic intensity. Default packet size is taken as 128 bytes. Figure 4 shows discontinuity in receipt of packets during the horizontal handoff in WLAN network. Figure 5 shows the horizontal handoff latency variation with change in packet size in the UMTS network environment. Then, we integrate the UMTS and WLAN network using layer manager with initial parameter taken as shown in Table 1 and the network topology configured as shown in Fig. 3. Figure 6 shows the vertical handoff latency from WLAN⟶UMTS network with varying packet size and Fig. 7 shows the Variation of handoff latency with varying traffic. We created a mathematical traffic model as shown below

$$f(x) = \frac{\sum_{k=1}^{n} \left\lfloor \prod_{i=1}^{k} \{x - ((n+1) - i)\} \right\rfloor \nabla^2 f(x_n)}{k!}$$

where

$\nabla^2 f(x_n)$	Newton's backward difference operator over $f(x_n)$.
x	No. of traffic.
k	Terminal simulation traffic no.

Figure 8 shows the comparison of handoff latency in the simulation environment and using the mathematical model above.

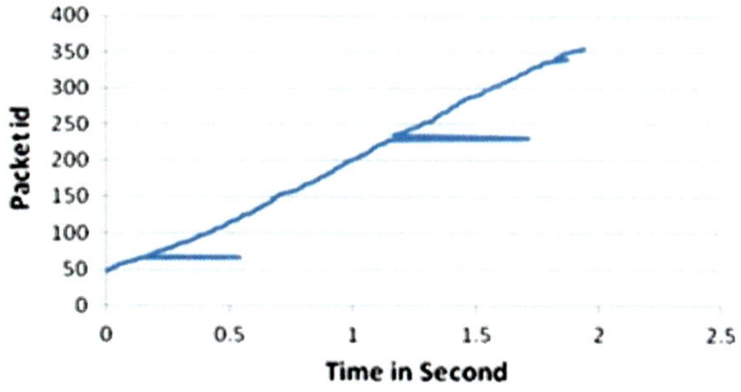

Fig. 4 Received packet id versus time in WLAN network

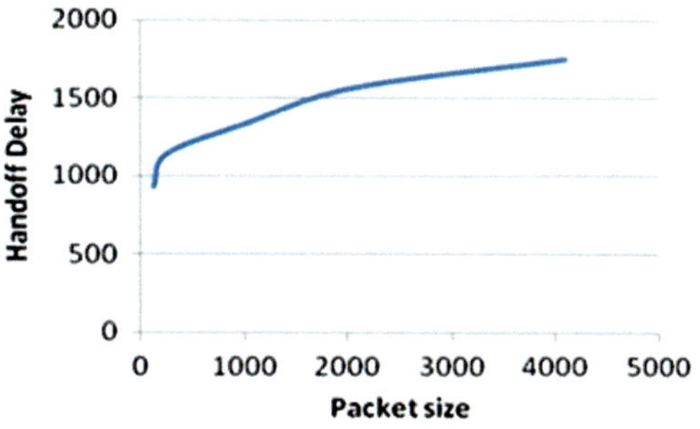

Fig. 5 Handoff delay in UMTS network with varying packet size

Table 1 Parameters used

Queue type	DropTail/PriQueue
MAC	MAC/802_11/UMTS Node B
Interface queue length	2000
TCP packet size in byte	128

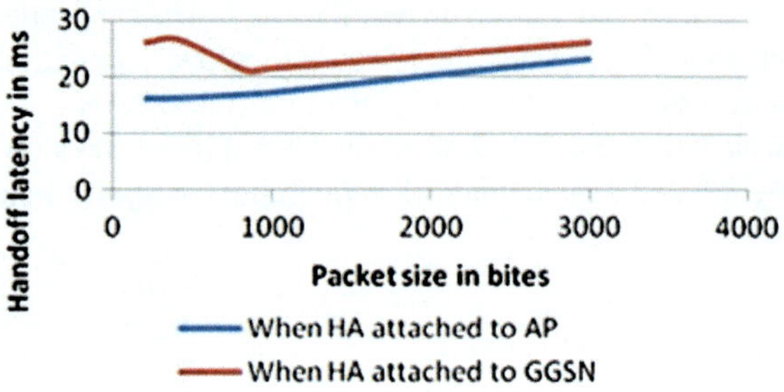

Fig. 6 Changes in uplink and downlink handoff latency with respect to packet size

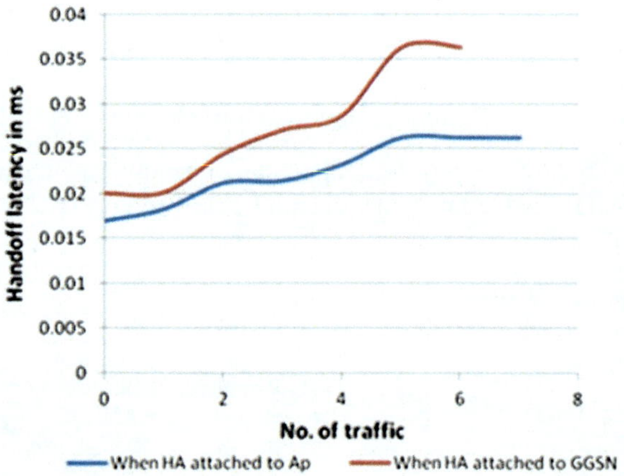

Fig. 7 Effect of handoff latency with increasing traffic

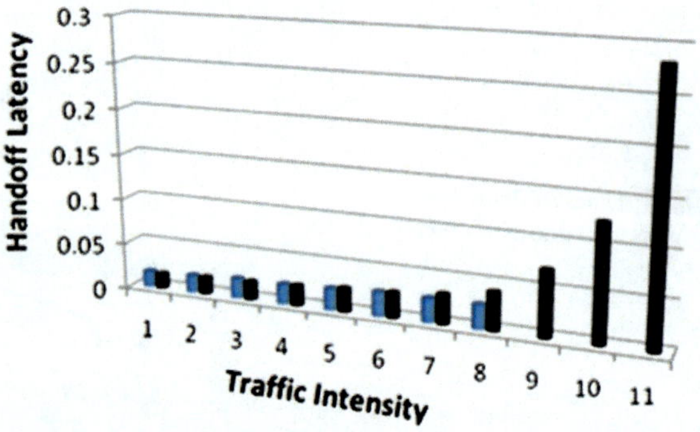

Fig. 8 Comparison of handoff latency with simulation environment (blue bar) and using mathematical model (black bar)

7 Performance Matrices

These metrics evaluate the delay to deliver multicast messages, the load on the network and the load imposed on end nodes and the routing complexity. Latency measures the end-to-end delay from the source to the receivers, as seen by the application. Bandwidth measures the application level throughput at the receiver.

Resource usage is a metric of the network resources consumed in the process of data delivery to all receivers. It is given by $\sum_{i=1}^{L} d_i * s_i$ where, L is the number of active links in data transmission, d_i is the delay in link i, and s_i is the link stress on link i. We assume that links with higher delay tend to be associated with higher cost.

Figure 4 shows the number of packets received by the MN during VHO from WLAN to UMTS network. Discontinuity in the curve shows the packet loss during handoff. It is seen that handoff delay time is 0.888073 s for packet size 128 which is much less than our threshold value as shown in Fig. 5. If we increase the packet size to 4096, we found the handoff delay is also less than 2 s so call drop will not occur.

8 Conclusion

Under true heterogeneity, call blocking and dropping probabilities, average vertical handoff latency are reduced, in the LM-based WON. In addition, the scheme is implemented at the network layer, opening the way for an all-IP network. This is a significant finding in comparison to previous WON analyzes [6–11]. Layer manager's logical functions can be implemented in a distributed manner in each layer's gateways. As a result, introducing the LM does not necessitate major hardware changes to current systems. The simulation results support the mathematical results.

References

1. Mukhopadhyay, A.K., Saha, D.: A general architecture of wireless overlay networks (WONs) for global mobility. In: Proceedings of First IEEE ANTS'07, Dec 14–15, 2007, Mumbai, India (2007)
2. Stemm, M., Katz, R.H.: Vertical handoffs in WONs. ACM MONET. Special issue: Mobile Networking in the Internet, pp. 335–350 (1998)
3. Zhu, F., McNair, J.: Multiservice vertical handoff decision algorithm. EURASIPJ. Wirel. Commun. Netw. **2006**(Article ID 25861), 1–13 (2006)
4. Wang, S.H., Kong, P.-Y.: A novel vertical handover scheme for integrated WLAN and cellular wireless networks. In: The 9th 1CCS 2004, Sept 2004, pp. 526–530 (2004)
5. Sharma, S., Baek, I., Dodia, Y., Chiueh, T.: Omnicon: a mobile IP-based vertical handoffsystem for wireless LAN and GPRS links. In: Parallel Processing Workshop ICPP 2004, pp. 330–337 (2004)
6. Zhang, Q., Guo, C., Guo, Z., Zhu, W.: Efficient mobility management for vertical handoff between WWAN and WLAN. IEEE Commun. Mag. 102–108 (2003)

7. Bernaschi, M., Cacace, F., Iannello, G.: Vertical handoff performance in heterogeneous networks. In: Proceedings of International Conference ICPPW'04, Aug 2004, pp. 100–107 (2004)
8. Wu, W., Banerjee, N., Basu, K., Das, S.K.: SIP-based vertical handoff between WWANs and WLANs. IEEE Wirel. Commun. 66–72 (2005)
9. Liu, C., Zhou, C.: An improved internetworking architecture for UMTS-WLAN tight coupling. In: IEEE Communication Society/WCNC 2005, pp. 1690–1695
10. Liu, C., Zhou, C.: HCRAS: a novel hybrid internetworking architecture between WLAN and UMTS cellular networks. In: IEEE Consumer Communications and Networking Congerence, Las vegas, Jan 2005
11. Choudhury, S., Saha, S., Mukhopadhayay, A.K.: A survey of vertical handover schemes in wireless networks. Published in McMillan Advance Research Series (ISBN 023-063-759-0), pp. 754–759. In: International Conference on Computer, Communication, Control and Information Technology (C3IT-2009), 06–07 Feb 2009
12. Mukhopadhyay, A.K., Saha, D.: Layer manager controlled mobility management on wireless overlay networks. In: Proceedings of 2nd IEEE ANTS'08, 15–17 Dec 2008, Mumbai, India
13. Buddhikot, M., Chandranmenon, G., Han, S., Lee, Y.W., Miller, S., Salgareli, L.: Integration of 802.11 and third-generation wireless data networks. In: Proceedings IEEE INFOCOM, pp. 503–512 (2003)
14. Rigney, C., et al.: Remote Authentication Dial in User Service (RADIUS). RFC 2865, IETF (2000)
15. Rigney, C.: RADIUS Accounting. RFC 2866, IETF (2000)
16. Calhoun, P., et al.: Diameter Base Protocol. Work in Progress—Internet Draft. IETF. draft-ietf-aaa-diameter-07.txt (2001)
17. Charging and Billing: 3G Call and Event Data for the Packet Switched (PS) Domain. TS 32.015, ETSI (2001)

Regulation of Blood Glucose Using Auto-Tuned PID Controller in Healthcare Systems

S. G. Rahul, R. Chitra, G. Srinivasa Sudharsan, A. Amruthavalli, and S. Sai Sudheer

1 Introduction

Non-invasive-based health monitoring systems are recently introduced to improve the quality of healthcare services by integrating the Internet of Things (IoT) technology. During the treatment, there is a need for real-time healthcare monitoring. In most conditions, patients discharged from hospitalization are firmly instructed to be on complete bed rest for a certain specified time. In such cases, such monitoring and regulating systems are highly beneficial. There is a tremendous growing demand to monitor and record various physiological parameters even though the patients are discharged for an extended observation [1]. Invasive treatment always requires a procedure, while non-invasive techniques do not result in direct contact with the patient's body such as body fluids, skin damage, or any physical inflammations. The essential parameters of the patients are being monitored through several non-invasive measures which also incorporates IoT technology for information transfer between doctors, hospitals, and patient care personnel to reduce complications and treatment delays [2]. A continuous contact is established between the patients and doctors by minimizing manpower efforts. Non-invasive monitoring techniques in health sectors help in its effective and efficient performance. The treatments' standards are highly enhanced in addition to improved health status. The non-invasive technique should always be given preference over an invasive method provided that it can record the parameter to be monitored at least as reliably and accurately as the latter. Monitoring

S. G. Rahul (✉) · R. Chitra · G. Srinivasa Sudharsan · A. Amruthavalli · S. Sai Sudheer
Vel Tech Rangarajan Dr. Sagunthala R&D Institute of Science and Technology, Chennai, India

A. Amruthavalli
e-mail: vtu14615@veltech.edu.in

S. Sai Sudheer
e-mail: vtu14649@veltech.edu.in

J. K. Mandal and J. K. Roy (eds.), *Proceedings of International Conference on Computational Intelligence and Computing*, Algorithms for Intelligent Systems,
https://doi.org/10.1007/978-981-16-3368-3_25

blood parameters easily aids in a faster disease diagnosis and the organ functionality enabling effective treatment [3].

2 Role of Insulin in Blood Glucose Regulation

The organ called pancreas is situated in the upper left abdomen. It aids to converts the food intake into energy [4]. The pancreas performs two significant functions: an exocrine function aids the digestion process and an endocrine function aids to regulate the sugar level in the blood [5]. The pancreas location is shown in Fig. 1.

The pancreas is hemmed in by several other organs of the digestive system. It is around 6–10 in. in length pear-like structure. The head part of the pancreas is situated toward the central abdomen [6]. In this region, stomach lets the food material into the intestine, and enzymes are released to the partially digested food [7]. Pancreas with good health condition secretes enzymes in a correct amount to digest the amount of food consumed. The chemical called insulin secreted by pancreas regulates the blood sugar as shown in Fig. 2. When the insulin is not secreted at right amounts, the blood glucose needs to be monitored at regularly, and insulin must be injected properly [8, 9]. The current medical treatment involves frequent measurements of glucose followed by insulin injections at equal intervals [10]. Such an invasive procedure is difficult and painful. The good amount of blood glucose for a healthy human body is 64–104 mg/dL or 3.6 and 5.8 mmol/L. When the value of blood glucose reaches beyond the limit, it makes the body unhealthy and results in hospitalization or death. So, an insulin pump can be used as a continuous injection into the human's body. And, to efficiently deliver the insulin, a suitable control system is required which is also attempted in this study.

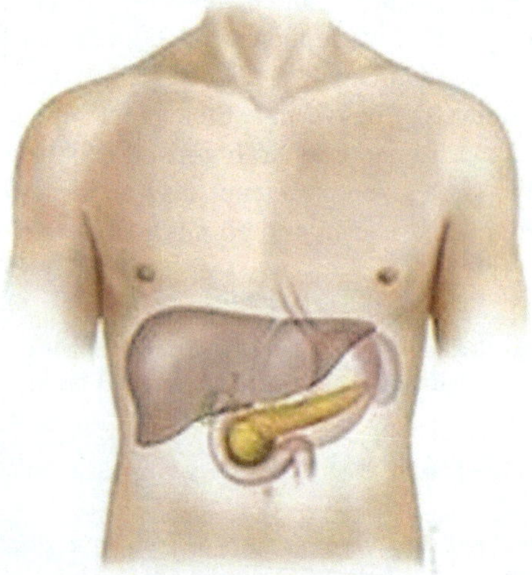

Fig. 1 Location of the pancreas in human body

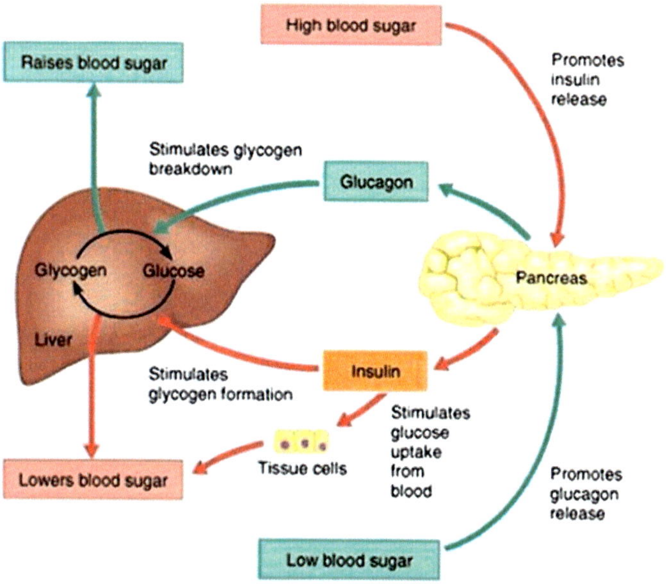

Fig. 2 Role of the pancreas in regulating blood glucose

3 Significance of Control System in Blood Glucose Regulation

The concept of the control system, derived from engineering, is incredibly helpful in the field of medical science and technology [11]. A control system results in the desired response by appropriately controlling the output parameter. Since one parameter is controlled by variation of another parameter, the control system has gained such a name. It can regulate the behavior of any system by appropriate design procedure [12]. Several research studies have been carried out in the filed control systems for the regulation of blood glucose which has paved a way for the artificial pancreas systems. Some of the literature studies are reported as follows. Various multiple model-based controllers together with the ubiquitous proportional-integral-derivative (PID) controllers were studied in past to specify insulin dosage in Type I diabetics and evaluated the performance and robustness of designed controllers using a detailed physiological model of a diabetic in consort with a meal disturbance model. K. Lunze, et al., reviewed on the essentiality to improve the control strategies by expounding the glucose-based insulin model to facilitate insulin treatment for patients with diabetic conditions [13]. Faiz-Ul-Hassan et al., proposed a closed-loop control system which can efficiently track the anticipated glucose levels with the lowest error and helps to normalize the high blood glucose level for diabetes patients [14]. Mounir Djouima et al. utilized Bergman minimal model to design the control laws and verified the performance of the super-twisting algorithm to act as a robust controller which can completely resist the internal and external disturbances in glucose regulation control systems [15]. Henrique Mohallem Paiva et al. projected a novel blood-glucose regulation methodology commissioning a fractional-order proportional-integral-derivative (FOPID) controller which resulted in a reduced output error when compared with the standard PID controller [16]. Likewise, various research activities are growing in the field of artificial pancreas which combines the technology of biomedicine and control system engineering. This motivation paves as a path to further research on an auto-tuned control system to regulate the glucose level in blood by appropriate insulin injection. The work carried is explained in the following sections.

4 Auto-Tuned PID Control Scheme for Blood Glucose Regulation

Section of insulin regulates the blood glucose level of a human body. When the pancreas does not produce the insulin at the required amount, the blood glucose level needs to be monitored frequently by invasive treatment. It is possible to develop a control system to automatically regulate the blood glucose level by changing the insulin injection rate. The set point for blood glucose level must remain within a specified parametric range for a healthy human body. The healthy range of blood glucose for a human being is 64–104 mg/dL or 3.6 and 5.8 mmol/L. When blood glucose values go out of limit, it is undesirable to health. The control system must restrict the flow of insulin as per the set point as a part of safety functioning.

4.1 Auto-Tuned PID Control Scheme for Blood Regulation

PID controller has been widely opted in most of the closed-loop systems as shown in Fig. 3. It is a feedback mechanism preferred in control system design. A PID controller measures an error value $e(t)$ and takes corrective action based on the

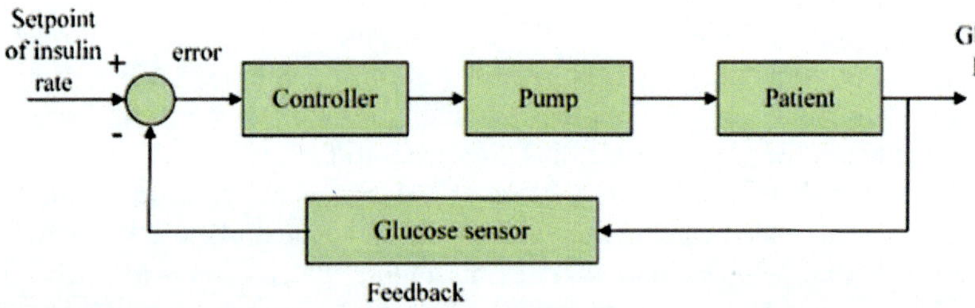

Fig. 3 Schematic representation of a closed-loop control system to regulate insulin

proportional kp, integral ki, or derivative kd terms. The design and selection of appropriate values of the three controller gains are called tuning. The most significant tuning method is Ziegler–Nichols tuning. Using the tuning algorithm developed by Abhijeet Kishorsingh Sukede and Jasmineet Arora in their study [17], an auto-tuned controller is designed which behaves like a more feasible choice of tuning for the systems with first-order lag characteristic such as our physiological model. The tuned PID parameters of PID controller are; kp = −1; ki = 0.05, and kd = 0 as shown in Fig. 4. The tuned response is illustrated in Fig. 5. The closed-loop control system consists of a glucose sensor to measure the level of blood glucose. The measured feedback signal is passed to a control system to compute the error which is essential to generate the necessary corrective action for insulin delivery. This is highly essential to maintain the blood glucose level in a stable range. Then, an electro-medical device called as final control element (FCE) such as a pump can inject the desired amount of insulin to the patient. The closed-loop control scheme is more trustworthy to maintain blood glucose and serves as an artificial pancreas.

Figure 6 illustrates the Simulink of a closed-loop control which regulates blood glucose. The base glucose of the body is set as 1200 for simulation, and meal glucose is fed as a pulse signal with lay since mean is consumed three times a day. Also, it is common for humans to consume snacks with some delays. These three signals are

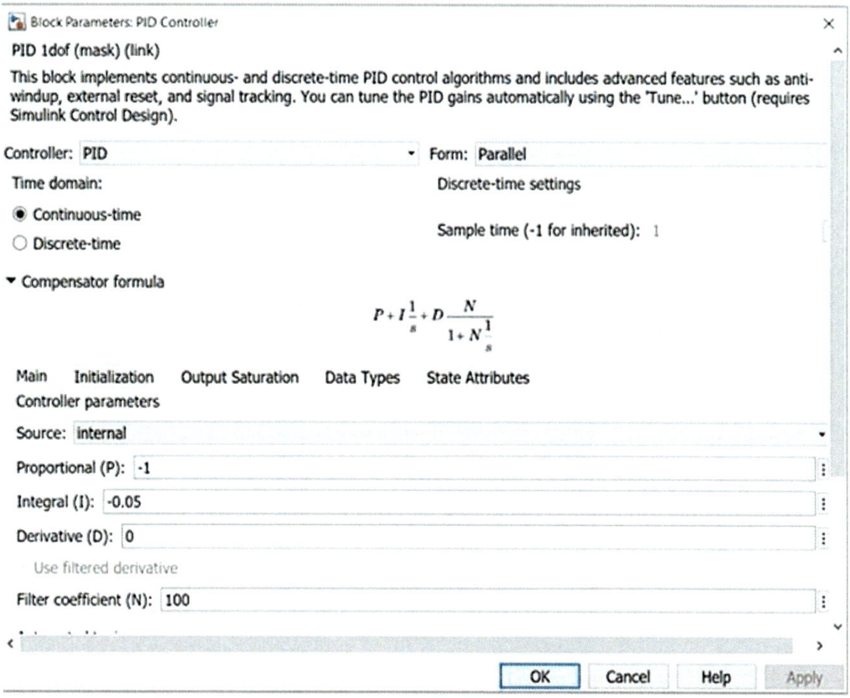

Fig. 4 Tuned PID parameters of closed-loop control

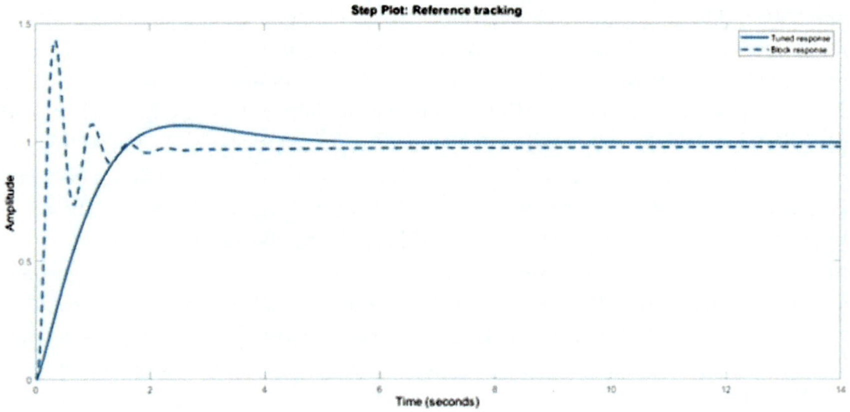

Fig. 5 Tuned response of PID controller

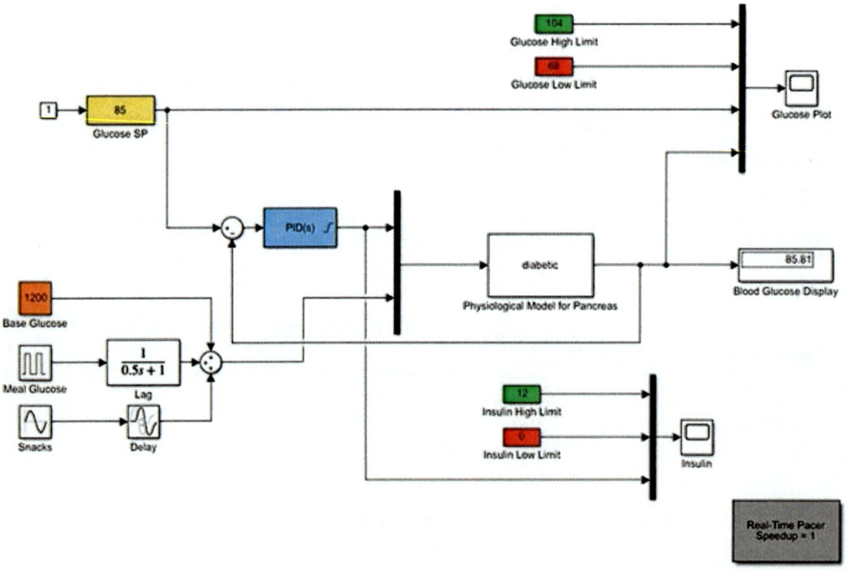

Fig. 6 Simulink model of the auto-tuned closed-loop control system

merged and fed to a summer. The desired glucose set point (SP) chosen as 85 mg/dL is then fed to the summing block. Alongside, the feedback signal is also fed to the summer. The error signal is computed at summer and fed to the PID controller. The controller then generates an appropriate control signal and passes it to the insulin pump and the diabetic patient model. The output signal from the patient displays the regulated blood glucose level due to controlled insulin injection. The amount of blood glucose is observed from the glucose plot of the Simulink.

The overall response of blood glucose after the successful implementation of the PID controller is shown in Figs. 7 and 8. It is observed that glucose amount in the blood increases after meal consumption as indicated by a brown marker. So, whenever the meal glucose rises, the sugar level also increases so an appropriate amount of insulin is required to absorb the blood sugar. Insulin is highly essential for energy conversion and utilization by body tissues. The high amount of sugar content is regulated only using insulin, and it is released when there is a drop in level of the blood sugar or when there is a required while consuming food. Hence, it is noticeably clear that insulin aids in balancing the blood sugar to keep the body healthy. The controller then triggers the control signal to inject insulin indicated

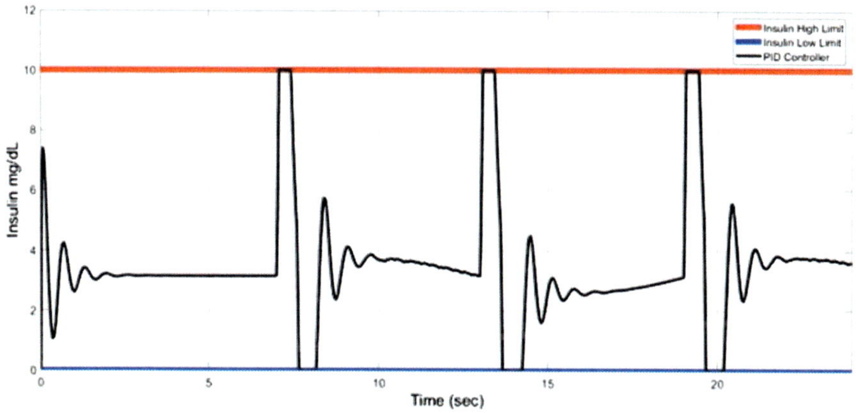

Fig. 7 Performance of the closed-loop control

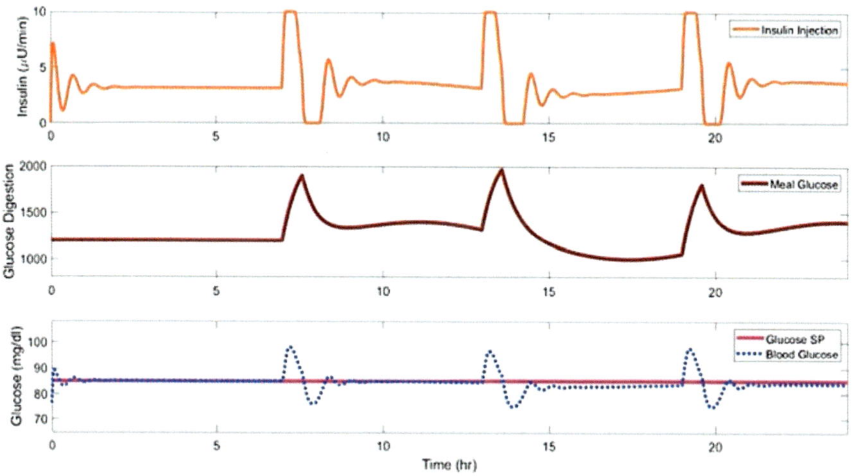

Fig. 8 Auto-tuned PID controller depicting blood glucose response got the studied physiological model

by the orange marker at the desired quantity to maintain the blood glucose level at 85 mg/dL. Whenever there is meal intake, insulin is secreted to absorb the sugar.

5 Conclusions

The non-invasive blood parameter monitoring systems are advantageous than invasive techniques. In this study, the authors worked in designing a control scheme of non-invasive blood glucose monitoring and regulation. The study is carried out as a software phase, and the results of the study pave a way to further enhancement for hardware development. The control system to regulate blood glucose by insulin injection is developed using MATLAB Simulink. Pancreases play a vital role in digestion and during conditions for improper insulin secretion. Using a pancreas mathematical model, a closed-loop auto-tuned PID control scheme is designed. The developed closed-loop control scheme regulates the blood glucose-insulin at the desired set point. The studied system replicates an artificial pancreas system and offers low complexity, and results are analyzed based on medical parameters. Implementing the control system in hardware is quite expensive which can be carried out as a scope of future work. Thus, the designed system could be fabricated as a hardware device in the future scope of studies to provide low complexity, low power consumption, and high portability for non-invasive patient monitoring.

References

1. Chen, S., Mulgrew, B., Grant, P.M.: A clustering technique for digital communications channel equalization using radial basis function networks. IEEE Trans. Neural Netw. **4**(4), 570–590 (1993)
2. Lamonaca, F., Polimeni, G., Barbé, K., Grimaldi, D.: Health parameters monitoring by smartphone for quality-of-life improvement. Measurement **73**, 82–94 (2015)
3. Ovadia-Blechman, Z., Meilin, A., Rabin, N., Eldar, M., Castel, D.: Noninvasive monitoring of peripheral microcirculatory hemodynamics under varying degrees of hypoxia. Respir. Physiol. Neurobiol. **216**, 23–27 (2015)
4. Messori, M., Incremona, G.P., Cobelli, C., Magni, L.: Individualized model predictive control for the artificial pancreas: in silico evaluation of closed-loop glucose control. IEEE Control. Syst. **38**(1), 86–104 (2018)
5. Wang, H.C., Lee, A.R.: Recent developments in blood glucose sensors. J. Food Drug Anal. **23**(2), 191–200 (2015)
6. Choi, H., Naylon, J., Luzio, S., Beutler, J., Birchall, J., Martin, C., Porch, A.: Design and in vitro interference test of microwave noninvasive blood glucose monitoring sensor. IEEE Trans. Microw. Theory Tech. **63**(10), 3016–3025 (2015)
7. Liao, W.C., Tu, Y.K., Wu, M.S., Lin, J.T., Wang, H.P., Chien, K.L.: Blood glucose concentration and risk of pancreatic cancer: systematic review and dose-response meta-analysis. Bmj **349**, g7371 (2015)
8. Roder, P.V., Wu, B., Liu, Y., Han, W.: Pancreatic regulation of glucose homeostasis. Exp. Mol. Med. **48**, 1–19 (2016)

9. Riddell, M.C., Zaharieva, D.P., Yavelberg, L., Cinar, A., Jamnik, V.K.: Exercise and the development of the artificial pancreas: one of the more difficult series of hurdles. J. Diabetes Sci. Technol. **9**(6), 1217–1226 (2015)
10. Schulz, T.C.: Concise review: manufacturing of pancreatic endoderm cells for clinical trials in type 1 diabetes. Stem Cells Transl. Med. **4**(8), 927–931 (2015)
11. Wang, Y., Zhang, J., Zeng, F., Wang, N., Chen, X., Zhang, B., Cobelli, C.: "Learning" can improve the blood glucose control performance for type 1 diabetes mellitus. Diab. Technol. Ther. **19**(1), 41–48 (2017)
12. Blauw, H., Van Bon, A.C., Koops, R., DeVries, J.H., PCDIAB consortium: performance and safety of an integrated bihormonal artificial pancreas for fully automated glucose control at home. Diabetes, Obes. Metab. **18**(7), 671–677 (2016)
13. Lunze, K., Singh, T., Walter, M., Brendel, M.D., Leonhardt, S.: Blood glucose control algorithms for type 1 diabetic patients: a methodological review. Biomed. Signal Process. Control **8**(2), 107–119 (2013)
14. Faiz-Ul-Hassan, M.A., Khaqan, A., Shuja, S., Tiwana, M.I., Qadeer-ul-Hassan, S.M., Riaz, R.A.: Closed loop blood glucose control in diabetics **28**(16) (2017)
15. Djouima, M., Azar, A.T., Drid, S., Mehdi, D.: Higher order sliding mode control for blood glucose regulation of type 1 diabetic patients. Int. J. Syst. Dyn. Appl. (IJSDA) **7**(1), 65–84 (2018)
16. Paiva, H.M., Keller, W.S., da Cunha, L.G.R.: Blood-glucose regulation using fractional-order PID control. J. Control Autom. Electr. Syst. **31**(1), 1–9 (2020)
17. Sukede, A.K., Arora, J.: Auto tuning of PID controller. In: International Conference on Industrial Instrumentation and Control, pp. 1459–1462. IEEE, India (2015)

Microwaves in Health Care for Breast Cancer Detection

R. Chitra, G. Srinivasa Sudharsan, S. G. Rahul, Seeram Sai Sudheer, and Archakam Amruthavalli

1 Introduction

Breast cancer is type of cancer among women in which breast tissue change and divide in an uncontrollable manner [1]. According to the cancer facts and figures 2021 given by American Cancer Society, there is an estimation of 281,550 new cases of invasive breast cancer diagnosed in women and 2650 cases diagnosed in men and, they estimated that 44,130 breast cancer deaths (43,600 women, 530 men) will occur in 2021. To reduce the death rate of this disease, it is to be detected early when the cancer is relatively small and has not spread to other parts of the body [2]. Nowadays breast cancer is detected through various diagnostic modalities such as X-Ray Mammography, Ultrasound, and Magnetic Resonance Imaging (MRI). But these techniques have some drawbacks as follows.

1.1 Limitations of Existing Breast Cancer Diagnostic Techniques

X-ray mammography is currently the most recommended method utilized by doctors for breast cancer examination. Diagnosis of breast cancer using X-ray mammography involves X-ray which causes the patient to be subjected to ionization which further increases the risk of cancer development. The breast is subjected to compression

R. Chitra · G. S. Sudharsan · S. G. Rahul (✉) · S. S. Sudheer · A. Amruthavalli
Vel Tech Rangarajan Dr. Sagunthala R&D Institute of Science and Technology, Chennai, India

S. S. Sudheer
e-mail: vtu14649@veltech.edu.in

A. Amruthavalli
e-mail: vtu14615@veltech.edu.in

J. K. Mandal and J. K. Roy (eds.), *Proceedings of International Conference on Computational Intelligence and Computing*, Algorithms for Intelligent Systems,
https://doi.org/10.1007/978-981-16-3368-3_26

while screening which causes pain to the patient who undergoes mammography. Also, it faces difficulty in locating tumor in dense breast [3]. It does not have enough sensitivity for patients having dense (thick) breast tissue and it cannot detect tumors whose size is less than a 1–2 cm. Ultrasound is a is safe screening tool because there is no harmful radiation, but it lacks enough spatial resolution. It could not be used to detect solid masses and the deeper lesions. Results produced by ultrasound are not always reliable since it is equipment and operator dependent [4]. MRI is not popular like other imaging techniques for breast cancer diagnosis since it is expensive. MRI is a time-consuming screening process for diagnosis of breast tumor in comparison with other diagnostic techniques [5]. The machine produces loud pounding and buzzing noises during the operation and calcification cannot be detected.

1.2 Microwaves in Cancer Detection

The requirement for a non-ionizing diagnostic technique which is non-invasive, low cost, less time-consuming, and accurate result producing has been increased. Microwaves can be used for the detection of the breast tumor. Basically, microwave imaging is termed as "seeing" the internal structure of an object by means of electromagnetic fields at microwave frequencies (300 MHz–300 GHz) [6]. Also, microwave has the potential to detect small tumors [7]. The applications of microwave technology were increasing in the field of biomedical engineering for diagnostic purposes. This technique uses the dielectric property contrast between normal and cancerous cells. Microwave interaction with the human body tissues causes the electromagnetic wave's amplitude to change as it propagates through it or disperses throughout the body tissues. Tissues electrical properties like relative permittivity (εr) and conductivity (σ) are used to design a phantom in CST environment. It is understood from various literature studies that the tissues permittivity is highly dependent on the water content. Low water content tissues (such as fat) have lower permittivity than high water content tissues like muscle, skin, heart, and cancer tissues while permittivity gives us the information of the tissues ability to store microwave energy, conductivity is related to the loss (dissipation) or attenuation (absorption) the signal suffers as it propagates through the tissue [8]. Therefore, as the microwave signal travels through different tissues on the breast it suffers both attenuation (amplitude loss) and reflection as it encounters discontinuity.

In this technique, an antenna is used to illuminate the breast phantom with microwaves, which travel through the breast and the microwave signal may be detected at receivers located on the opposite side of the breast [9]. Alternatively, reflections may be recorded at the transmitting antenna. With a tumor present, waves traveling through the breast encounter a change in electrical properties, causing the incident wave to scatter. This scattering changes the amounts of energy detected at the receivers and the transmitter. The objective of the proposed system is to identify the presence and location of tumor in breast phantom with the help of reflected signal by using a single antenna for both transmission and reception. This antenna transmits UWB pulses, which propagates into the breast, where it is reflected off due to the variation in electrical properties and is received by the same antenna. An ultra-wide band antenna is proposed to have different depth of penetration along with better spatial resolution. In comparison to the reflected signal with the transmitted signal, the presence of tumor is identified. Some of the literature reports on breast cancer detection are discussed as follows. Çalışkan et al., modeled a 3D breast structure with different permittivity and conductivity in HFSS by using Finite Element Method (FEM). The authors initially designed a microstrip patch antenna and then modified the patch and ground plane to increase the image quality. Parameters like electric field, magnetic field distribution, and current density on the antenna were evaluated [10]. Rafat Raihan et al., designed a wearable pentagon-shaped microstrip patch antenna which resonates from 2.4 to 2.4835 GHz to detect brain cancer. They embedded a 5 mm tumor in the phantom and compared the radiation pattern of both normal tissue and cancerous tissue [11]. Afyf et al., proposed a compact ultra-wideband (UWB) Planar antenna for Microwave thermography system. The authors performed simulation using two commercial software, CST Microwave Studio, and HFSS simulators. The simulated antenna exhibits the UWB characteristics from 1.3 to 6 GHz in HFSS software and from 2.01 GHz to 4.4 GHz in CST [12]. Ouerghi et al., presented a comparative study of five microstrip patch antennas for microwave imaging at 2.45 GHz frequency to select the suitable antenna for tumor detection. The choice was made by putting each antenna on the breast phantom to study the electric field, magnetic fields, and current density in Ansoft High Frequency Simulation Software (HFSS) [13]. From the literature review, it is understood that the techniques for tumor detection in skin, head, and breast tissues using patch antenna with design equations were mostly available. In this work, a circular patch antenna with wideband characteristics is investigated to identify the presence of tumor based on various parametric analysis.

2 Phantom Design

The breast phantom is designed in such a way that it must replicate the natural breast both in shape and the dielectric properties. First the phantom is designed in homogeneous manner without any abnormality and then a tumor is embedded in the same designed phantom. The tumor varies with the normal tissue by the dielectric values like permittivity, electrical conductance, thermal conductivity, etc. [14]. Properties of the materials used for skin, fat, and tumor are depicted in Table 1.

The breast phantom model with and without tumor created in CST environment are shown in Figs. 1 and 2 respectively. Initially, the phantom is designed with radius of 50 mm with 5 mm tumor centered at $x = 0$, $y = 0$ and $z = 20$ mm.

Table 1 Dielectric values of materials used in Phantom

Phantom layers	Permittivity (F/m)	Conductivity (S/m)
Skin	36.7	2.34
Fatty tissue	4.6023	0.5852
Tumor	67	49

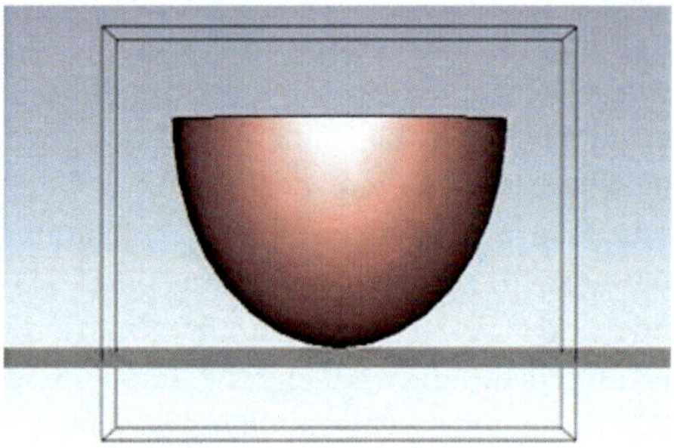

Fig. 1 Breast Phantom without tumor

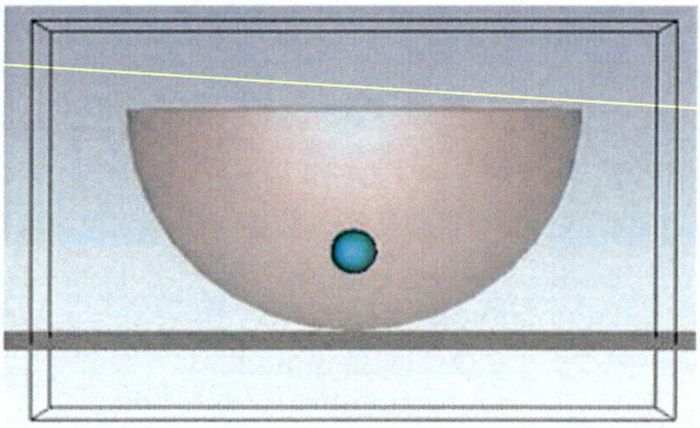

Fig. 2 Breast Phantom with tumor

3 Antenna Design

The most important part of the microwave-based tumor detection system is the antenna that is used [15]. Generally, the radar-based microwave detection is of two types—Multistatic and Monostatic. Multistatic is the one which uses two different antennas for transmission and reception purpose whereas the Monostatic type uses a single antenna for both purposes. To have less complexity, a monostatic antenna is designed. This antenna should act as a transceiver, i.e., it should transmit as well as receive the microwave signal. The micro-strip antennas are popular in a wide variety of applications due to their light weight, simple structure, low cost, and ease of fabrication It consists of a conducting patch and a ground plane on both side of the substrate. The radiating patch can be square, rectangular, circular, or any other shape. Antenna design is carried out using Computer Simulation Technology (CST) Microwave Studio software. Different types of antennas are designed to check whether it can fulfil the basic initial requirements like directivity, gain, impedance matching, reflection coefficient, and voltage standing wave ratio (VSWR) so that it can be used for further simulation with breast phantom.

3.1 Narrowband Antenna

Initially, a rectangular antenna is designed to resonate at a frequency of 2.5 GHz in which the length and width of the antenna are calculated using the below formulas. These two parameters will decide the characteristics of the designed antenna. Patch is the radiating element, where its feed is designed to have 50 ohms impedance for efficient energy transfer between source and antenna. Table 2 shows the design parameter of the narrowband antenna which are calculated using the above design equations. The perspective view of the narrowband antenna designed in CST environment is shown in Fig. 3.

Table 2 Design parameters of narrowband antenna

Parameters	Notation	Values(mm)
Patch width	Wp	36
Patch length	Lp	28
Ground width	Wg	2 * W
Ground length	Lg	2 * L
Substrate thickness	ts	0.035
Substrate height	hs	1.6
Feed width	Wf	3.137
Inset feed length in patch	Fi	8.85
Gap between feed and patch	Gfp	1

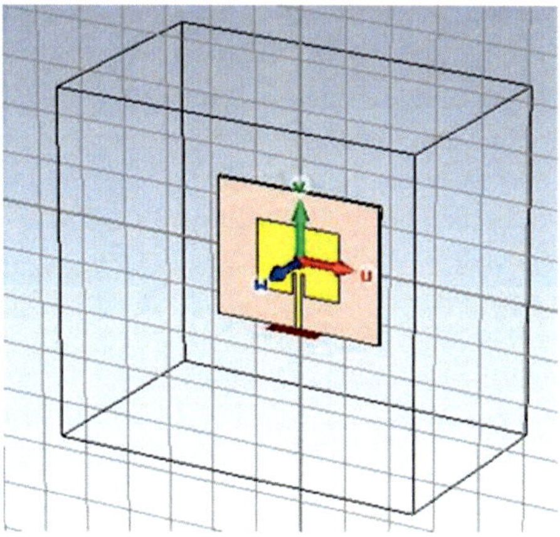

Fig. 3 Perspective view of narrowband antenna designed in CST

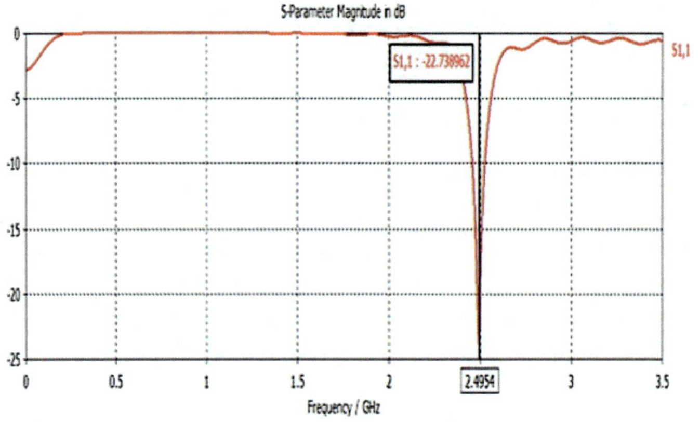

Fig. 4 *S* parameter of narrowband antenna

The scattering parameter of designed narrowband antenna is shown in Fig. 4. It is understood that the antenna is resonating at 2.5 GHz with a $S11$ value of -22.73 dB. Farfield results like directivity of 6.9 dB and gain of 4.5 dB are also observed in the antenna. But there are some limitations to the narrowband antenna. When an antenna is designed to resonate at a higher frequency (eg.,7 GHz), it could not be able to find the deeply seated tumor since the frequency is inversely proportional to the depth of penetration.

And, similarly in the case of lower frequencies, the penetration will be more with low spatial resolution which makes the antenna less efficient for tumor detection.

3.2 Ultra-Wideband Antenna

An ultra-wideband antenna is designed to achieve a better trade-off between spatial resolution and depth of penetration with the resonating frequency of 1.5–7.1 GHz. In order to have the sufficient bandwidth, circular patch is used in this section. Length of the ground is also changed to get the desired range of frequencies. The design and selection of suitable ultra-wide band antenna involve some optimization of antenna parameters which affects the bandwidth and radiation pattern. Figure 5 shows the effect of return loss for the patch radius which varies from 14 to 17 mm. It is understood that the radiation pattern is not having wideband characteristics for the patch radius 14 mm and 15 mm respectively. Similarly, the radius 17 mm is not used because of the higher edge frequency which will give side effects to the human body. So, the radius 16 mm is chosen for the design of wideband antenna.

The ground structure is modified for various length as shown in Fig. 6. It is clear that the antenna exhibits UWB characteristics as the ground length decreases. The ground length of value of 28.36 mm shows the wideband frequency range in the

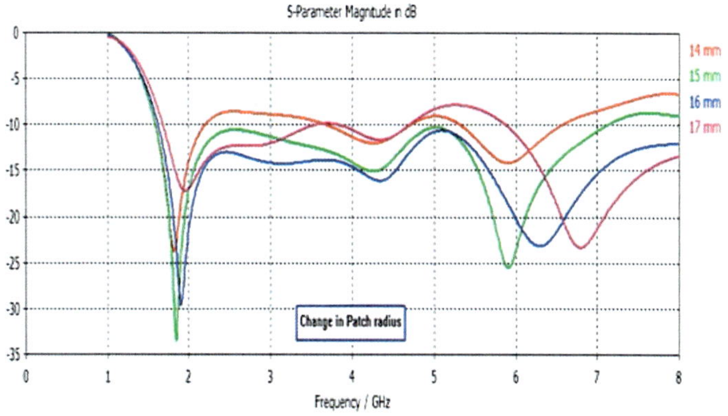

Fig. 5 Return loss for varying patch radius

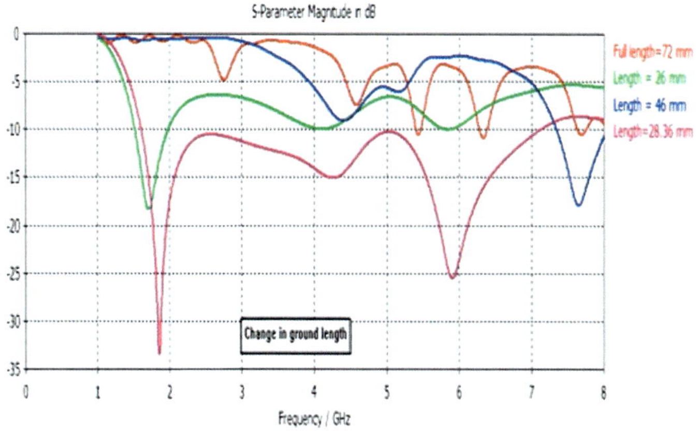

Fig. 6 S parameter for different ground length

graph which is suitable to find the deeply seated tumors and to achieve better spatial resolution. Table 3 shows the design parameters of the optimized wideband antenna.

Figure 7 shows the perspective view of the proposed circular antenna designed in CST Microwave Studio. To check the suitability of the antenna, both $S11$ and VSWR are plotted as shown in Figs. 8 and 9 respectively. From the various parametric analysis of antenna, it is understood that the proposed antenna is suitable for further simulation with the breast phantom.

Table 3 Design parameters of the optimized wideband antenna

Parameters	Notation	Unit (mm)
Patch radius	R	18
Ground length	Lg	28.36
Ground width	Wg	58.5
Substrate length	Ls	72
Substrate width	Ws	63
Substrate thickness	ht	0.035
Substrate height	hs	1.6

Fig. 7 Perspective view of circular wideband antenna

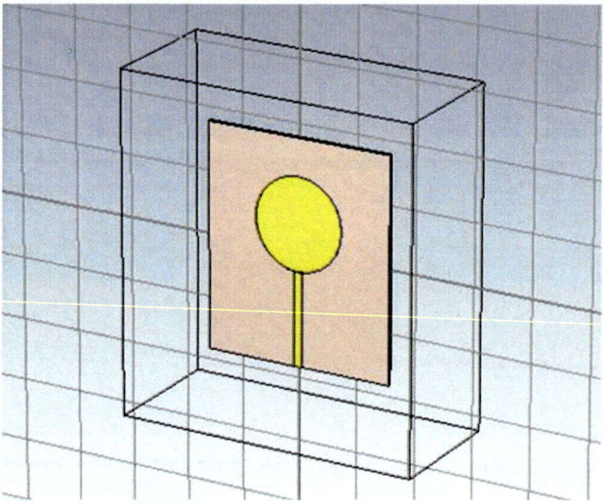

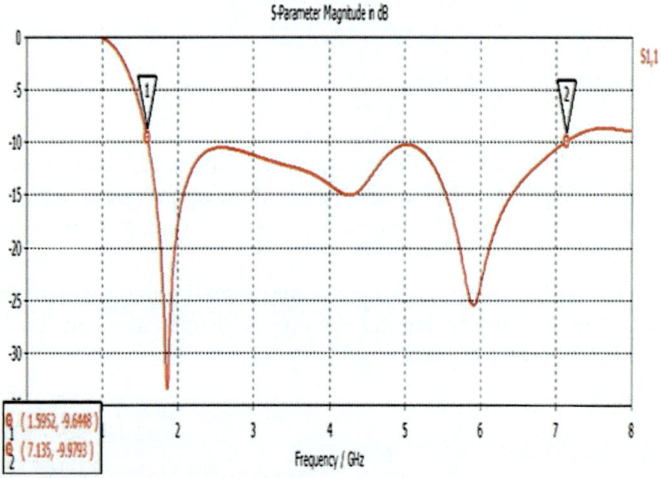

Fig. 8 S parameter of circular wideband antenna

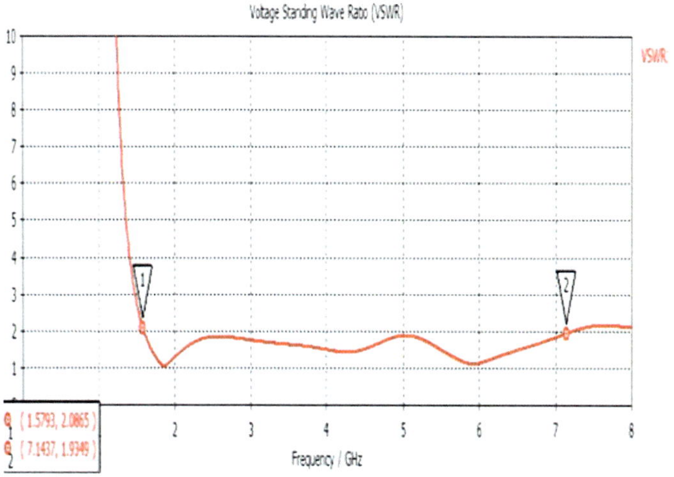

Fig. 9 VSWR of circular wideband antenna

4 Parametric Analysis of Antenna with Phantom

Once the antenna and phantom are designed separately, the simulation of antenna with phantom is carried out in CST environment. The input signal used in the antenna for simulation is a Gaussian pulse with the time in the range of nanosecond. When the antenna is simulated with the phantom having tumor, some variation is observed in the output port signal which is depicted in Fig. 10. It is understood that tissue

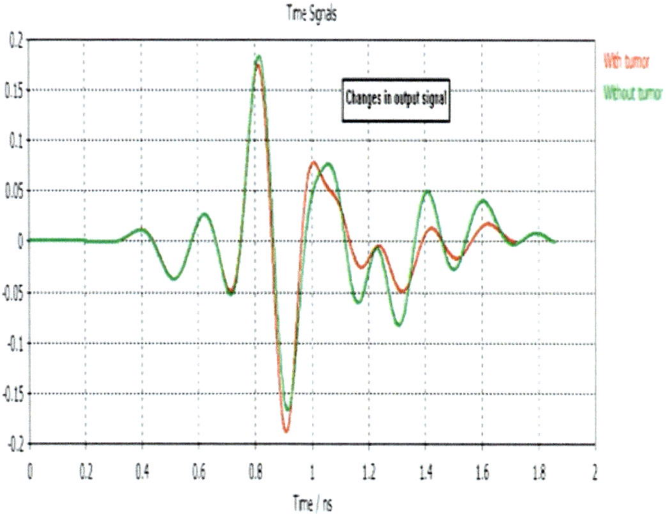

Fig. 10 Output signal for phantom without and with tumor

with different dielectric value is present in the normal tissue. Due to this change in dielectric values, the variation in output signal occurred and the presence of tumor is confirmed.

The presence of tumor in a normal tissue is identified by the absorption and reflection of the excited signal. When there is a healthy tissue without any malignant tumor, the absorption and reflection of the signal is less. If a tumor is present in the normal tissue, the absorbed and reflected power is more due to the change in dielectric values. Similarly, the comparison for various tumor size in terms of reflection is carried out in this work. From Fig. 11, it is understood that, as the tumor size increases, the reflection is more.

The SAR values are calculated at the frequency of 6 GHz using CST SAR calculation tool. Parameters like total SAR in W/kg, maximum SAR in W/kg, coordinate of the maximum SAR (X, Y, Z), absorbed power in W are used for the analysis. Table 4 summarizes the value of absorption rate for phantom without tumor and for phantom with tumor of radius of 5 mm embedded in 50 mm breast phantom. From the table, it is observed that the absorption rate is high for the phantom with tumor.

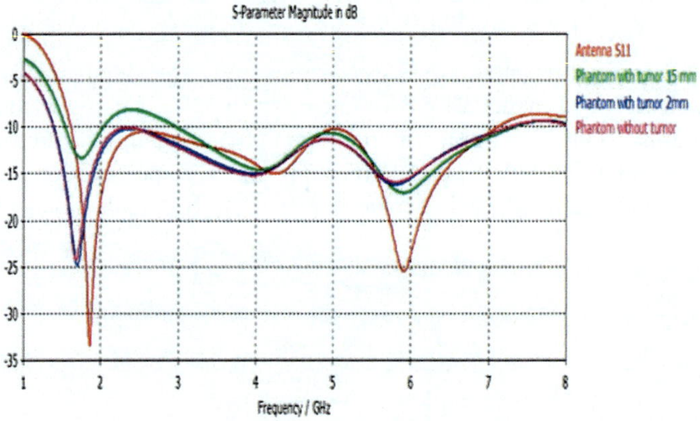

Fig. 11 S parameter for different tumor size

Table 4 Absorption rate for phantom with and without tumor for UWB range

UWB frequency range (GHz)	Absorption rate (W/kg)	
	Phantom without tumor	Phantom with tumor
2	5.21	5.33
3	5.90	6.00
4	6.01	6.31
5	4.05	4.07
6	4.76	4.80
7	5.02	4.80

Table 5 Absorption rate for different tumor size in UWB range

Tumor size (mm)	Absorption rate (W/Kg) at various frequency (GHz)			
	3	4	5	6
2	3.59	3.90	3.87	4.04
3	3.81	3.92	3.94	4.08
4	3.97	4.12	4.31	4.12
5	4.41	4.97	4.33	4.21
8	4.58	4.91	4.71	4.68
10	4.67	5.89	5.52	4.90

Table 5 shows the power absorbed by the tumor of different radius. It is observed that when the mass of the breast increases, the absorbed power also increases for different frequencies.

5 Conclusion

In this study, microwave is used with an objective of detecting a tumor at its early stage, so a tumor of diameter 5 mm in a 50 mm phantom is used during the simulation. The antenna is simulated with breast phantom for measuring the various antenna parameters. The narrowband antenna is not suitable to find the tumors which are located at different depth since frequency is inversely proportional to the depth of penetration. So, an ultrawide band is used with the resonant frequency which varies from 1.5 to 7.1 GHz and the Voltage Standing Wave Ratio (VSWR), a measure for the purpose of compatibility of any antenna is measured to be below 2. The return loss is found different for a tumor-affected breast phantom than a normal breast phantom. The shift of resonant frequency during the simulation of antenna with phantom indicates the existence of tumor inside phantom. Other parameters such as gain, directivity Specific Absorption Rate (SAR) are also taken into consideration for the tumor detection. The simulation is carried out for various tumor size and different depth of tumor to identify the presence and location of the tumor. In future, the simulation can be carried out to find the exact location of tumors using the absorbed power for multiple tumor cases.

References

1. Mohebian, M.R., Marateb, H.R., Mansourian, M., Mañanas, M.A., Mokarian, F.: A hybrid computer-aided-diagnosis system for prediction of breast cancer recurrence (HPBCR) using optimized ensemble learning. Comput. Struct. Biotechnol. J. **15**, 75–85 (2017)
2. Curry, S.J., Byers, T., Hewitt, M.: Potential of screening to reduce the burden ofcancer. In: Fulfilling the Potential of Cancer Prevention and Early Detection. National Academies Press (US) (2003)

3. Hellquist, B.N., Czene, K., Hjälm, A., Nyström, L., Jonsson, H.: Effectiveness of population-based service screening with mammography for women ages 40 to 49 years with a high or low risk of breast cancer: socioeconomic status, parity, and age at birth of first child. Cancer **121**(2), 251–258 (2015)
4. Ozmen, N., Dapp, R., Zapf, M., Gemmeke, H., Ruiter, N.V., van Dongen, K.W.: Comparing different ultrasound imaging methods for breast cancer detection. IEEE Trans. Ultrason. Ferroelectr. Freq. Control **62**(4), 637–646 (2015)
5. Roganovic, D., Djilas, D., Vujnovic, S., Pavic, D., Stojanov, D.: Breast MRI, digital mammography and breast tomosynthesis: comparison of three methods for early detection of breast cancer. Bosn. J. Basic Med. Sci. **15**(4), 64 (2015)
6. Alvarez-Berdugo, D., Jiménez, M., Clavé, P., Rofes, L.: Sci. World J. (Article ID 184526) (2014)
7. Hassan, A.M., El-Shenawee, M.: Review of electromagnetic techniques for breast cancer detection. IEEE Rev. Biomed. Eng. **4**, 103–118 (2011)
8. Chandra, R., Zhou, H., Balasingham, I., Narayanan, R.M., de Azevedo-Marques, P.M., Mencattini, A., Rangayyan, R.M.: Medical microwave imaging and analysis. In: Medical Image Analysis and Informatics: Computer-Aided Diagnosis and Therapy, pp. 451–466. CRC Press (2018)
9. Sill, J.M., Fear, E.C.: Tissue sensing adaptive radar for breast cancer detection—experimental investigation of simple tumor models. IEEE Trans. Microw. Theory Tech. **53**(11), 3312–3319 (2005)
10. Çalışkan, R., Gültekin, S.S., Uzer, D., Dündar, Ö.: A microstrip patch antenna design for breast cancer detection. Procedia Soc. Behav. Sci. **195**, 2905–2911 (2015)
11. Raihan, R., Bhuiyan, M.S.A., Hasan, R.R., Chowdhury, T., Farhin, R.: A wearable microstrip patch antenna for detecting brain cancer. In: IEEE 2nd International Conference on Signal and Image Processing (ICSIP), pp. 432–436. IEEE (2017)
12. Afyf, A., Bellarbi, L., Achour, A., Riouch, F., Errachid, A.: A novel low cost UWB antenna for early breast cancer detection. Am. J. Electromagnet. Appl. **3**(5), 31–37 (2015)
13. Ouerghi, K., Fadlallah, N., Smida, A., Ghayoula, R., Fattahi, J., Boulejfen, N.: Circular antenna array design for breast cancer detection. In: Sensors Networks Smart and Emerging Technologies (SENSET), pp. 1–4. IEEE (2017)
14. Karli, R., Ammor, H., Shubair, R.M., AlHajri, M.I., Alkurd, R., Hakam, A.: Miniature planar ultra-wide-band microstrip antenna for breast cancer detection. In: 16th Mediterranean Microwave Symposium (MMS), pp. 1–4. IEEE (2016)
15. Gupta, H., Maheshwari, V., Thakery, V.V.: Brain tumor detection by microwave imaging using planner antenna. Int. J. Bio-sci Bio-Technol **8**(5), 201–210 (2016)

Autonomous Ground Vehicle for Off-the-Road Applications Based on Neural Network

Alice James, Avishkar Seth, and S. C. Mukhopadhyay

1 Introduction

Autonomous ground vehicles are a growing research area due to their vast applications [1] in mining, military, farming, and many other off-the-road terrains. These vehicles or robots are equipped with smart sensing and mapping capabilities useful in data gathering and inspection [2, 3] One of the first approaches to the 'autonomy' in land vehicles was initiated by Dean A. Pomerleau [4], 'ALVINN', showcasing a neural network model using images for autonomous navigation. This approach was further enhanced by many researchers as the camera image quality and computation power grew over the years [5]. Consequently, the concepts of AI, convolutional neural network (CNN), and end-to-end deep learning were applied to the image data [6, 7]. Along with the camera as an input, a range of other sensors got added such as LiDAR, GPS, and ultrasonic sensors with obstacle avoidance, 360° environment mapping, and safety functionalities [8]. Further, for testing and prototyping these theories, miniaturized models were built which allowed learning, teaching, and developing of an AGV [9–12].

A. James (✉) · A. Seth · S. C. Mukhopadhyay
School of Engineering, Macquarie University, Sydney, NSW 2109, Australia
e-mail: alice.james@students.mq.edu.au

A. Seth
e-mail: avishkar.seth@students.mq.edu.au

J. K. Mandal and J. K. Roy (eds.), *Proceedings of International Conference on Computational Intelligence and Computing*, Algorithms for Intelligent Systems,
https://doi.org/10.1007/978-981-16-3368-3_27

2 Approaches to Steering an Autonomous Vehicle

The steering and throttle commands in an autonomous vehicle are controlled by different approaches of AI. In this paper, the three major methods of steering commands are discussed to understand the best-suited approach.

Traditionally, the steering controls were performed using PID [13] which uses a computer vision algorithm defined in a non-AI method using control theory. It uses an error function in a continuous loop based on constant feedback signal [14]. This approach proved to be limited in many application specific conditions due to its tuning speed and non-real-time capabilities. To overcome the limitations, the AI approach was introduced in the system. The basic AI method involves an 'agent' which is trained with a supervised learning algorithm and cautiously analyzed datasets. The aim is to recognize road patterns, obstacles in an image dataset to output steering control angles with FFN, and computer vision technology [9, 15].

The third approach allows a combination of standard robotics and modern end-to-end deep learning techniques to achieve driving controls [16–18]. The third approach was used in this paper along with dataset augmentation techniques [5, 19], to allow the 'off-road' capabilities. This technique allows to reduce training and computation times and enables real-time image analysis in complex environments [20] which can be done with unsupervised learning. CNN is a deep learning algorithm that uses feature extraction on image datasets, allocates distinctive weights, biases to the feature, and performs prediction. In this paper, the image datasets are trained with different CNN parameters such as filter dimensions, stride values, fully connected layers to achieve the lowest validation loss, and achieve better steering predictions.

3 System Summary

The following section summarizes the AGV model that has been used to implement the neural network algorithms. The functional requirements that have been conducted to determine the components have been discussed broadly in [10]. The AGV is operated principally using the Raspberry Pi 4, as it is a cost-effective and fast processing system. The prime input to the Raspberry pi is the camera that has been fitted on the 'Tower case.' Figure 1 illustrates the final assembled model of the AGV. The base plate and tower case have been custom designed and developed using a 3D printer and laser cutting machine, respectively.

The AGV is first trained by controlling the Raspberry pi wirelessly using a joystick/Web-interface [10], and the data collected from the camera that has been stored as a '.json' file is trained using different neural network models. These models have been tested when the AGV is driven in autopilot mode, and the results have been discussed further.

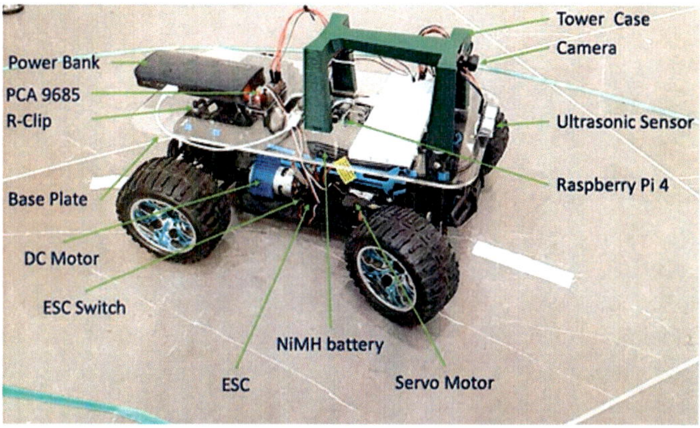

Fig. 1 Final assembled model of the AGV with custom 3D model design [10]

4 Implementation

This section elaborates on the implementation of the AGV, and it has been distributed into two main categories, i.e., the software setup and the tracks.

4.1 Software Setup

The software setup mainly discusses the designed and expected performance abilities of the AGV. The software message queue defined in [10] is an outline of the queue of the designated activities to be implemented by the AGV. As mentioned earlier, the vehicle is first trained by the user and images that correspond to the steering command are collected. This is then implemented in autopilot mode using the neural network algorithm training to the Raspberry pi.

The training provided by the neural network algorithms defines the steering and throttle commands to the AGV [10]. This implies, once the image is captured by the camera, the processor updates the servo and the DC motor defining the direction to the servo using pulse width modulation (PWM) signals and defining the PWM signal to the electronic speed controller (ESC) which commands the speed to the DC motor. These commands and images are stored in the SD as steering commands as datasets.

4.2 Tracks

Custom Track

This section describes the implementation and design of the tracks, it has been categorized as two types, i.e., the custom indoor track and the outdoor track. The AGV is primarily trained on the custom indoor track that has been illustrated in Fig. 2. This track has been used for training repetitively that collectively resulted in approximately 40 k datasets for training.

Outdoor Track

Figure 3 shows the outdoor track implemented to execute a systematic line follower using image analysis. The datasets collected were approximately 18k, which will be used further for training and augmentation.

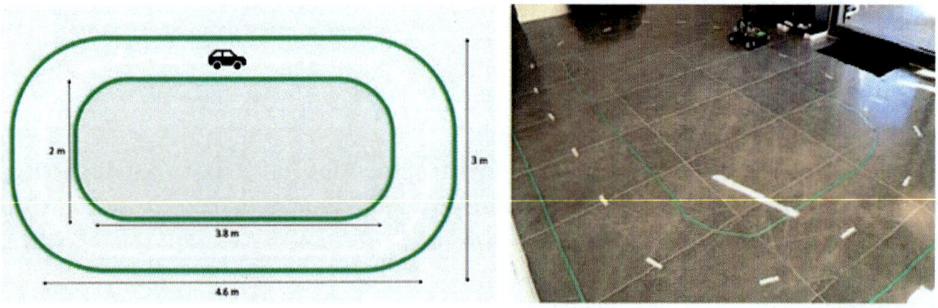

Fig. 2 Custom indoor track designed to accommodate the size of the AGV with left–right steering directions

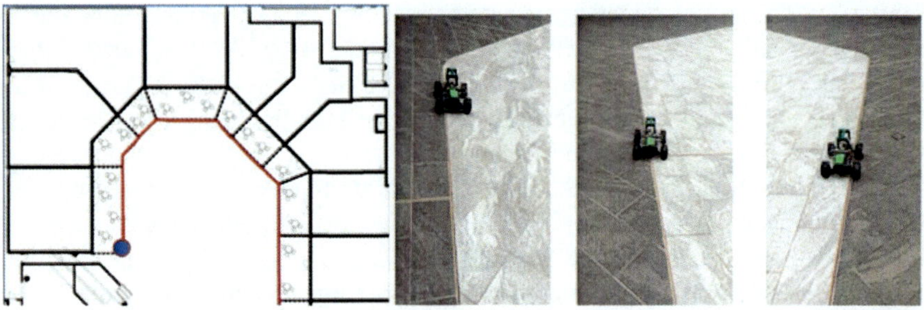

Fig. 3 Outdoor track implementing line follower, as observed in the leftmost map, the red line represents the center of the track

5 Results

5.1 Model Comparison

The following section summarizes the custom convolutional neural network (CNN) models that have been compared to determine the most optimum solution. Along with the custom CNN models, the AGV has also been trained using feed forward network (FFN) model and the NVIDIA PilotNet [5] model. Table 1 shows the different parameters set for the hidden layers of the CNN model and the corresponding validation loss. Thus, as observed in Fig. 4, the 'D' model is the most effective model, with the lowest validation loss, i.e., almost zero.

The best-selected models in custom CNN and FFN have been compared along with the PilotNet for the autonomy levels as discussed in [5]. The autonomy level

Table 1 CNN training model parameters comparison

Sr. no.	Convolution filters	Stride values	Fully connected layers	Parameters	Loss
A	10 × 3 × 3, 16 × 3 × 3, 40 × 3 × 3	2,2,2	400, 95, 20	810k	0.0152
B	10 × 3 × 3, 16 × 3 × 3, 40 × 3 × 3	2,2,1	90, 50, 10	710k	0.0223
C	10 × 3 × 3, 16 × 3 × 3, 20 × 3 × 3, 36 × 3 × 3	3,2,2,1	400, 100, 20	210k	0.0124
D	10 × 3 × 3, 16 × 3 × 3, 20 × 3 × 3, 32 × 3 × 3	3,2,2,1	900, 264, 32	1150k	**0.0107**
E	16 × 3 × 3, 20 × 3 × 3, 40 × 2 × 2, 60 × 2 × 2	3,2,2,1	400, 100, 20	310k	0.0136

Bold indicates the lowest validation loss amongst all models, that has been illustrated in Fig. 4

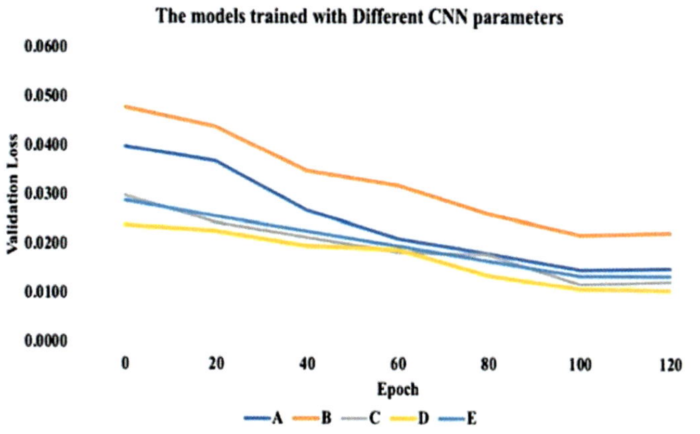

Fig. 4 Comparisons of the CNN models to determine the lowest validation loss

Table 2 Three neural network training model parameters comparison

Attribute	FFN-B-60	PilotNet	CNN_model_4
Duration (in seconds)	120	120	120
Interventions (Average)	16	3	1
Autonomy	20%	87%	94%

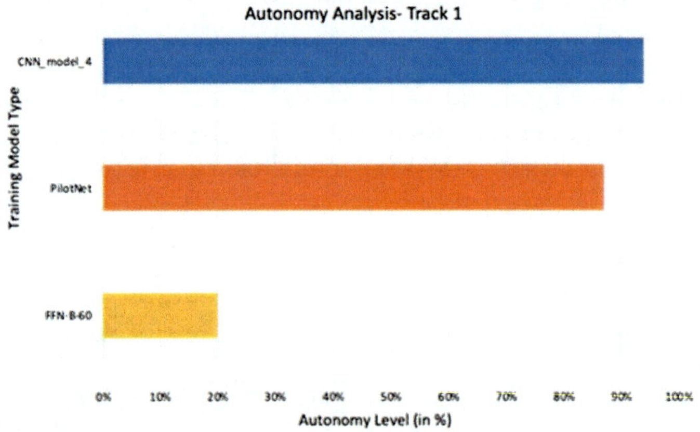

Fig. 5 Final comparisons of the best compared models of the CNN and FFN along with NVIDIA PilotNet to determine highest level of autonomy

indicates the ability of the AGV to traverse with minimum human intervention. The parameters assessed in this autonomy level is the duration traversed with a number of human interventions as given in Table 2. As observed in Fig. 5, the best model with the highest autonomy level is the custom CNN model selected earlier in this section.

The AGV traversed accurately for every 120 s with only one human intervention, thus offering 94% autonomy in custom CNN model-4, i.e., 'D'.

5.2 Data Augmentation

To compare and analyze the accurate amount of augmentation, the data was trained with three levels with increasing amounts of augmentation. Figure 6 shows the different augmentations applied such as fog, heavy rain, and gravel on road. In level 1, the dataset included augmentations like enhanced brightness and vertical flipping of the image. For Level 2, in addition to level 1 augmentations, darken images and noise were added as well. In level 3, additionally, speed images and heavy rain, and gravel were used to augment the data.

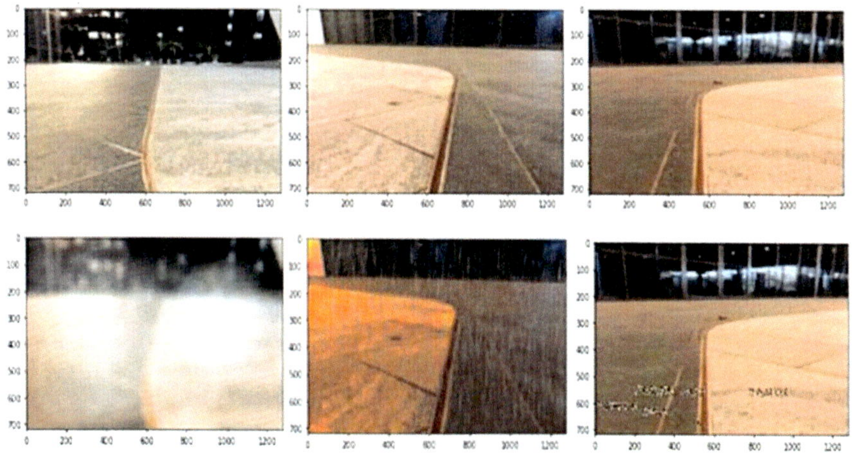

Fig. 6 Augmented images of the outdoor track to represent severe off-the-road conditions

The results showed that the best level of augmentation is Level 1, as it does not create redundant data which acts as noise in the system. And using application-specific augmentations was required depending on the environment.

5.3 *Steering Angle Analysis*

The steering angle analysis shows a trained and tested dataset of images before and after augmentation applied. The first graph plot is trained on a dataset of about 40k images and the second with about 18k images with and without augmentations. For ease of display plot, only 200 subsets of frames are taken. As seen in the first plot, applying augmentation level 1 significantly enhanced the prediction (orange and blue lines align). Similarly, the outdoor track performance was improved with augmentations, although with some minor errors still present due to the amount of driving data and user training provided as shown in Fig. 7.

6 Conclusion

This paper presents the autonomy results for the prototype AGV designed using neural network models to formulate a simple and appropriate platform. It has been observed that data augmentations level 1 was the most application suitable for the tracks tested. The outdoor track performance was still under development, even though augmentations were used, it did not significantly improve the prediction. The conclusion was that the indoor track-1 had superior performance after level 1

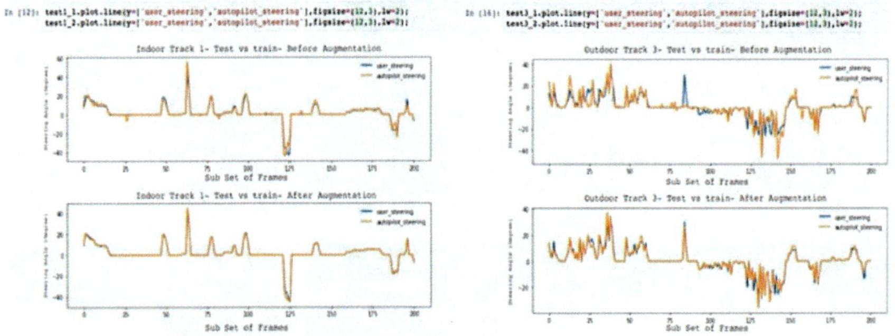

Fig. 7 (left to right) Final comparisons of the steering angle versus the number of frames for the indoor track and outdoor track where orange represents the autonomous steer and blue represents the user steering

augmentations as it was applied to a larger dataset. Thus, the steering angle analysis implies that different track or environments require different training datasets to increase the efficiency and reduce the validation loss. The results of the comparisons of training models and the accuracy of autonomy can aid the future work of the AGV, to traverse diverse paths.

Acknowledgements We are grateful for the facilities and laboratories provided by Macquarie University, Sydney. We acknowledge and sincerely show our gratitude to our Supervisor Dr. Subhas Chandra Mukhopadhyay for his guidance and profound perspective on this project. We would also like to acknowledge our parents Mr. James Joseph and Mrs. Gladys James and Late Mr. Nitin Seth and Mrs. Zenobia Seth for their constant motivation and inspiration.

References

1. Kelly, A., Stentz, A., Amidi, O., Bode, M., Bradley, D., Diaz-Calderon, A., Happold, M et al.: Toward reliable off road autonomous vehicles operating in challenging environments. Int. J. Rob. Res. 25. **5**(6), 449–483 (2006)
2. Behringer, R., Sundar S., Brian G., Richard E., Bob A., Wayne G., Robert D., David B.: The DARPA grand challenge-development of an autonomous vehicle. In IEEE Intelligent Vehicles Symposium, 2004. IEEE, pp 226–231 (2004)
3. Jurdak, R., Elfes, A., Kusy, B., Tews, A., Wen, Hu., Hernandez, E., Kottege, N., Sikka, P.: Autonomous surveillance for biosecurity. Trends. Biotechnol. **33**(4), 201–207 (2015)
4. Pomerleau, D.A.: Alvinn: An autonomous land vehicle in a neural network. In: Advances in neural information processing systems, pp 305–313 (1989)
5. Bojarski, M., Davide Del, T., Dworakowski, D., Firner, B., Flepp, B., Goyal, P., Lawrence, D., Jackel et al.: End to end learning for self-driving cars. arXiv preprint, arXiv:1604.07316 (2016)
6. Mishkin, D., Sergievskiy, N., Matas, J.: Systematic evaluation of convolution neural network advances on the imagenet. Comput. Vis. Image Underst. **161**, 11–19 (2017)
7. Glorot, X., Bordes, A., Bengio, Y.: Deep sparse rectifier neural networks. In: Proceedings of the fourteenth international conference on artificial intelligence and statistics, pp 315–323 (2011)

8. Himmelsbach, M., Mueller, A et al.: LIDAR-based 3D object perception. In: Proceedings of 1st international workshop on cognition for technical systems, vol. 1 (2008)
9. O'Kelly, M., Sukhil, V., Abbas, H., Harkins, J., Kao, C., Pant, Y.V., Mangharam, R et al.: F1/10: an open-source autonomous cyber-physical platform. arXiv preprint arXiv:1901.08567 (2019)
10. Seth, A., James, A., Mukhopadhyay, S.C.: 1/10th scale autonomous vehicle based on convolutional neural network. Int. J. Smart Sens. Intell. Syst. **13**(1), 1–17 (2020)
11. Roscoe, W.: Donkey Car: an opensource DIY self driving platform for small scale cars. (2019)
12. Bechtel, M.G., McEllhiney, E., Kim, M., Yun, H.: Deeppicar: A low-cost deep neural network-based autonomous car. In: 2018 IEEE 24th International conference on embedded and real-time computing systems and applications (RTCSA). IEEE, pp 11–21 (2018)
13. Sheppard, E.A., Mukhopadhyay, S.C., Sen Gupta, G.: Remote controlled solid state fireworks detonation system. In: Proceedings of the fourteenth electronics New Zealand conference ENZCon 2007, Wellington, New Zealand, ISBN 978-0-473-12951-4, pp. 95–99, 12–13 Nov 2007
14. Chandni, C.K., Sajith Variyar, VV., Guruvayurappan, K.: Vision based closed loop pid controller design and implementation for autonomous car. In: 2017 International conference on advances in computing, communications and informatics (ICACCI). IEEE, pp 1928–1933 (2017)
15. Dunbabin, M., Tews, A.: Toward robotic visual and acoustic stealth for outdoor dynamic target tracking. Proceeding of ARAA ACRA (2012)
16. Muller, U., Ben, J., Cosatto, E., Flepp, B., Cun, Y.L.: Off-road obstacle avoidance through end-to-end learning. In: Advances in neural information pro-cessing systems, pp 739–746 (2006)
17. Lake, B.J., Mukhopadhyay, S.C., Sen Gupta, G.: Electronics and magnetics for a smart battery charger for soccer robots. In: Proceedings of the fourteenth electronics New Zealand conference ENZCon 2007, Wellington, New Zealand, ISBN 978-0-473-12951-4, pp 47–52, 12–13 Nov 2007
18. Krizhevsky, A., Sutskever, I., Hinton, G.E.: ImageNet classification with deep convolutional neural networks mark. Commun. ACM **60**(6), 84–90 (2017)
19. Liu, D., Zhao, J., Xi, A., Wang, C., Huang, X., Lai, K., Liu, C.: Data augmentation technology driven by image style transfer in self-driving car based on end-to-end learning. Comput. Model. Eng. Sci. (2020)
20. Seelye, M., Gupta, G.S., Seelye, J., Mukhopadhyay, SC.: Camera-in-hand Robotic System for Remote Monitoring of Plant Growth in a Laboratory. In: Proceedings of 2010 IEEE I2MTC conference, Austin, Texas, USA, pp 809–814, 4–6May 2010

Monitoring of Automotive Response Based on Starter Motor Fault

Poulomi Ganguly, Surajit Chattopadhyay, and B. N. Biswas

1 Introduction

Fault in automotive systems can considerably lower vehicle performance and efficiency. They are the major contributors of vehicle breakdown that results in hefty expenditures in repair and maintenance. Also frequency of failure increases with age leading to loss of comfort, lowered operational efficiency with increased wear and tear of different automobile components. Thus, intelligent vehicle health-monitoring schemes are required for proper fault diagnosis in automotive systems [1].

A modern automobile is equipped with sophisticated electrical and electronic systems for control stability, safety gear adjustments and power supply operations. One of the main components of an automotive system along with the battery is its starter motor. The starter motor operation is very crucial for the operation of an automotive system. The starter motor helps to convert electrical energy into mechanical energy through a battery. Its main operation is to provide the cranking torque to the crankshaft of an IC engine (ICE) until the minimum starting speed is obtained. The torque generated by the starter motor is amplified through the use of a gear system. The starter motor generates high power for a very short duration of time. Each starting attempt by the starter motor can last about 1 s. During every ICE compression cycle, starter motor torque and hence force on the starter gear teeth are high. In the absence of any conventional over-current protection, short circuit may occur which may burn out the starter [2].

P. Ganguly (✉)
RVSCET, Jamshedpur, Jharkhand, India

S. Chattopadhyay
GKCIET, Malda, West Bengal, India

B. N. Biswas
SKFGI, Hooghly, West Bengal, India

J. K. Mandal and J. K. Roy (eds.), *Proceedings of International Conference on Computational Intelligence and Computing*, Algorithms for Intelligent Systems,
https://doi.org/10.1007/978-981-16-3368-3_28

The fault in automobile starter motor not only prevents the operation of the ICE but may cause other vehicular faults [3]. The service life of an average starter motor is roughly the same as that of its engine mounted on the automobile. Vehicles subjected to frequent starts should monitor their starting system regularly as with each start, the starter motor is subjected to resistance from compression, piston friction, earing friction, etc. However, the faults on an automobile starter motor occur not only due to motor failure but also due to other associated system failure as well. For example, motor fault can occur due to battery weakness, loosening of battery pole and broken cable connection. The most common faults that occur on starter motor and associated starter system are brush fault, open circuit fault, winding fault and short circuit fault [4].

The faults occurring on an automobile may either be electrical, mechanical or a combination of both [5]. Considering the time taken to normally detect faults in electric motors, condition monitoring of the motor along with its fault diagnosis can help to improve the security and reliability of the motor. Various fault diagnosis and detection methods have been researched and developed each with their own unique methods and technologies [6, 7]. With the increasing demand of online monitoring, fault diagnosis and its detection, conventional methods are gradually being replaced by artificial intelligence techniques as the conventional techniques fail to meet all demands that the diagnosis and monitoring system requires [8, 9]. Studies have been carried out on electric motor fault diagnosis using artificial neural network, integrated neural network as well as fuzzy logic system [10–12]. Fault studies have been successfully carried out using LVQ neural network as well as by using self-organizing map (SOM) method [13]. Compared to other ANN methods, LVQ has been found to be able to successfully classify faults and its diagnosis with greater accuracy [14].

Reduction of harmful emissions cannot be achieved without decreasing the operation time of the power unit and proper distribution of electrical energy. In this regard, cars with combined power system have been developed [15]. Cars are developed with an internal combustion engine connected to an electric machine that operates in both engine and generator mode. In the generator mode, the machine is responsible for the high-voltage on-board network and helps to charge the high-voltage battery [16]. The low-voltage network is supplied through DC-DC converters via the high-voltage network. The low-voltage network supplies electrical energy to on-board accessories while helping to charge the battery connected in parallel to the system.

However, very few works have been seen to deal with prediction of inductive nature of the winding using transients. This has motivated the author to deal with this challenge. This work presents an attempt to monitor inductive strength of stator winding using the transient response of the starter motor used in vehicles. This work has then focused on early and accurate diagnosis of starter motor fault in vehicle system. Commonly used automotive consisting of a battery, alternator, starter solenoid, a starter motor and additional on-board accessories system has been modeled, and study has been made. The fault diagnosis has been done by monitoring of current transients and measurement of maxima minima of transients. After system description, simulation, result, observation has been presented followed by conclusion.

2 System Description

A simple dynamic model of an automotive electrical system has been developed for this paper. The simulation has been modeled with automotive electrical, mechanical and thermal system. Figure 1 shows the block diagram layout of the automotive system. The system consists of a battery modeled as a constant voltage source in series with a resistance.

The starter motor used for the automotive system is a DC motor which has been parameterized based on stall torque and no-load speed. The starter motor specifications are shown in Table 1. It also consists of an alternator which supplies power to

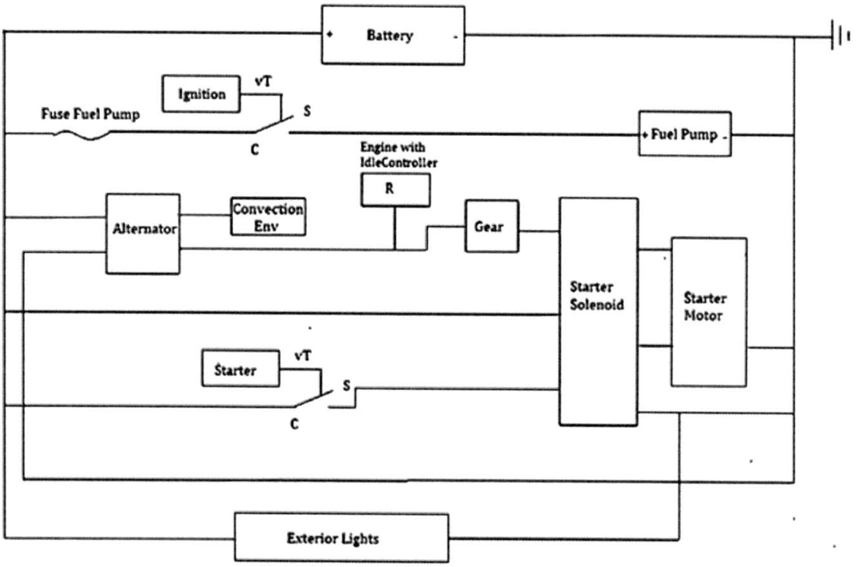

Fig. 1 Automotive electrical system block diagram representation

Table 1 Starter motor specification

Starter motor specification (DC motor)	
Armature inductance	1 mH
Stall torque	1 Nm
No-load speed	12,000 rpm
Rated DC supply voltages	12 V
Rotor inertia	0.001 kg m^2

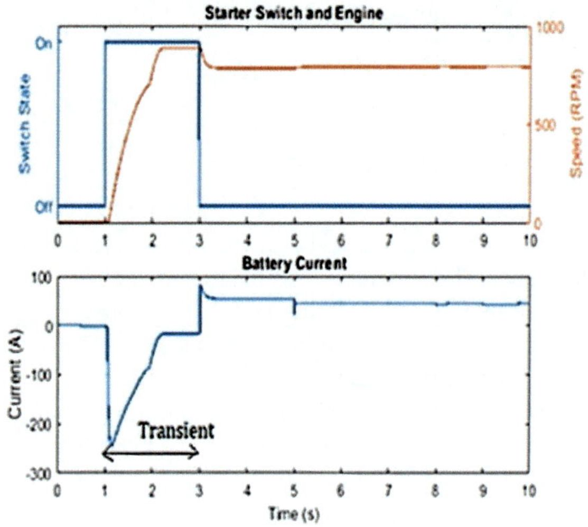

Fig. 2 Start-up sequence for the automotive electrical system

the starter solenoid via a gear, starter motor, starter and the exterior lights of the automobile. The starter motor load is the engine with varying load torque that depends on the automobile condition, environmental factors and multiple resonances [17].

During the occurrence of a fault on starter motor and the corresponding starter system, the peak current during cranking will increase. Here, a simulation of the automobile system has been performed considering starter armature coil fault which may occur due to insulation breakdown, degradation due to environmental condition or damage due to unprecedented incidents like an accident.

3 System Response Profile

The automotive electrical system has been modeled in MATLAB 2018 Simulink. The current response for starter cable fault has been recorded and plotted. A sample of the response characteristics along with the automotive start-up sequence has been shown in Fig. 2. Variation is mainly observed in the transient state (observed as shown in Fig. 2 from 1 to 3 s) of the response curve. Thus, main emphasis has been given to the data in the transient region of the response to study the effect of starter armature coil fault.

4 Variation of Response Profile

During a short circuit fault in the starter armature, the current drawn by the motors increases, and also this fault results in the starters failure to start the automobile engine. For every percentage reduction in starter armature inductance due to short circuit condition, different parameters of current response like maximum value, minimum value and slope of the curve (in the transient part of the response) have been monitored.

The transient response and shape of current signals have been observed. Significant difference has been observed in the transient part in time domain with healthy starter motor and faulty starter motor runs. This phenomenon has been considered as the basis for further mathematical computation to catch the variation and detect starter motor fault. The change in response profile has been recorded through the data saved in workspace every time in offline mode. From the data, the maximum and minimum response has noted from the saved workspace data file.

The slope of the curve is calculated by a drawing a tangent in the transient part of the current response curve. By definition, the slope (m) is calculated using the following formula:

$$m = \frac{\text{Change in } y \text{ axis data}}{\text{Change in } x \text{ axis data}} = \frac{\Delta y}{\Delta x} = \frac{y_2 - y_1}{x_2 - x_1}$$

Table 2 shows the response variation in terms of its maximum and minimum value and slope for stepwise degradation in starter motor armature coil (shown in terms of percentage compared to rated armature inductance of 1mH). The first data corresponds to the healthy starter armature coil while the last data in the table corresponds to complete short circuit condition of the armature coil.

Table 2 Response variation due to change in starter motor armature inductance

% Degradation of starter motor armature	Max (A)	Min (A)	Slope
0	92.142	−266.994	97.11884
10	92.1429	−268.569	97.40553
20	92.1439	−270.217	97.6912
30	92.1452	−271.197	97.97657
40	92.1467	−273.813	98.26102
50	92.1486	−275.791	98.52835
60	92.1509	−277.936	98.80994
70	92.1538	−280.3	99.09004
80	92.1575	−282.84	99.36934
90	92.1615	−286.359	99.64829
100	92.1652	−290.098	99.72882

5 Observation

From Table 2, it can be seen that there appreciable changes in the minimum value of the current drawn compared to the maximum current value. Also the slope of the response profile slope is seen to change from the normal operating condition to the short circuit condition. Variation of positive peak and negative peak of transient with respect to percentage degradation has been observed as shown in Figs. 3 and 4; they show gradual increase and decrease with respect to the reduction of inductive strength. From these two parameters, rate of change of response during the span of transients has been observed as presented in Fig. 5 which shows well variation suitable for scaling in the diagnosis of starter motor winding fault.

Inductive strength of stator motor winding varies with starter motor fault. Moreover, the observation presented here shows that change of inductive strength of the starter motor has changed the transient behavior of the current signal drawn by the starter motor. Thus, the characteristics of Fig. 5 directly reflect the reduction of the inductive strength of starter motor indicating the occurrence of fault in the starter motor.

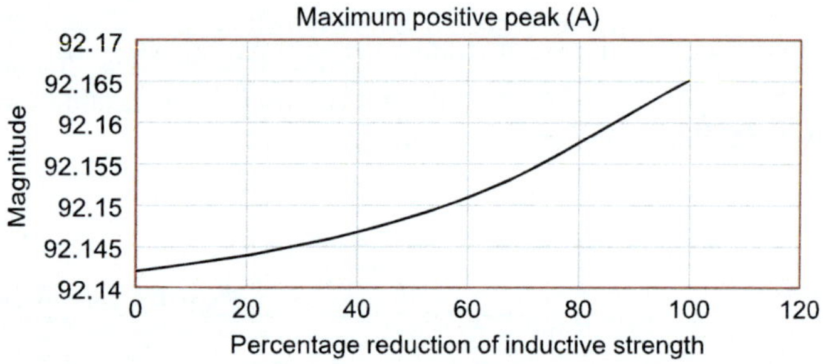

Fig. 3 Change in positive response peak with change in starter armature inductance

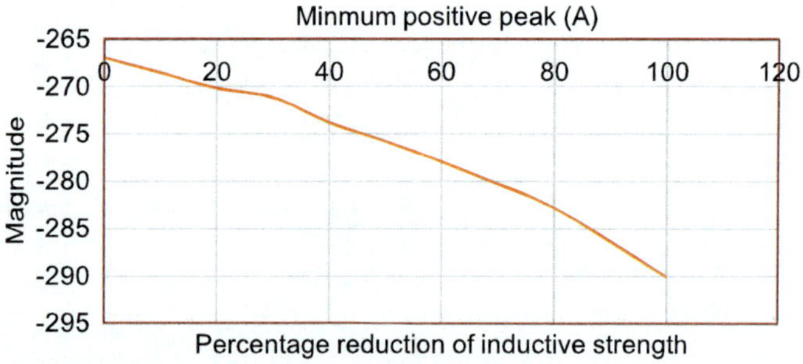

Fig. 4 Change in negative response peak with change in starter armature inductance

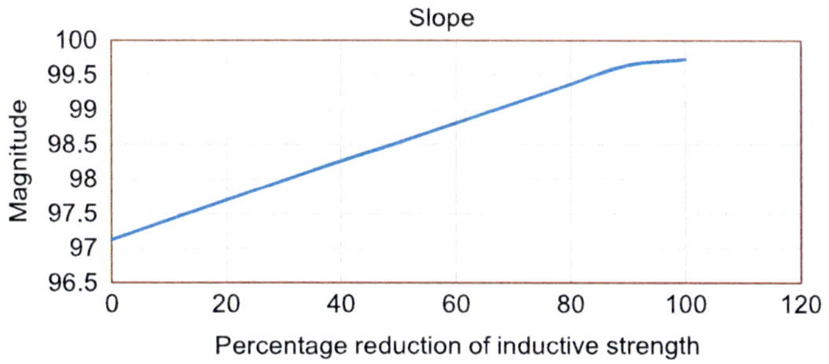

Fig. 5 Change in 'rate of change of response' (A/s) with change in starter armature inductance

6 Conclusion

This paper investigates the relation between inductive strength of starter motor winding with the transient response of the motor. From the observation, specific relation has been observed which is of a gradual change in nature. The relation shows linear variation up to 85% reduction of inductive strength of the starter motor. As faults in starter motor influence reduction of starter motor, the relation found is suitable for scaling and therefore may be used for prediction of winding fault in starter motors used in automobile applications. The analysis has been carried out in simulation, using real parameters in offline mode. It may be extended for online prediction.

References

1. Choi, K., Singh, A., Kodali, A., Pattipati, K.R., Sheppard, J.W., Namburu, S.M., Chigusa, S., Prokhorov, D.V., Qiao, L.: Novel classifier fusion approaches for fault diagnosis in automotive systems. IEEE Trans. Instrum. Meas. **58**, 602–611 (2016)
2. Murugesan, V.M., Chandramohan, G., Senthil Kumar, M., Rudramoorthy, R., Ashok Kumar, L., Suresh Kumar, R., Basha, D., Vishnu Murthy, K.: An overview of automobile starting system faults and fault diagnosis methods. ARPN J. Eng. Appl. Sci. **7**(7), 812–819 (2012)
3. QiMa Rajagopalan, S.S.V., Yurkovich S., Guezennec, Y.G.: A high fidelity starter model for engine start simulations. American control conference In: Proceedings of the 2005, American Control Conference (7): 4423–4427 (2005)
4. Bayir, R., Bay, O.F.: Serial wound starter motor faults diagnosis using artificial neural network. In: Proceedings of the IEEE international conference on mechatronics. IEEE, pp.194–199. Istanbul, Turkey (2004)
5. Bayir, R.: Condition monitoring and fault diagnosis of serial wound starter motor with learning vector quantization network. J. Appl. Sci. **8**(18), 31–48 (2008)
6. Vas, P.: Parameter estimation, condition monitoring and diagnosis of electrical machines, 1st edn. Clarendon Press, Oxford Oxford University Press, New York (1993)
7. Chow, M.Y.: Methodologies of using Neural Network and Fuzzy Logic Technologies for motor incipient fault detection, 1st edn. World Scientific Publishing Co, Singapore (1997)
8. Hajiaghajani, M., Toliyat, H.A., Panahi, I.M.S.: Advanced fault diagnosis of a DC motor. IEEE Trans. Energy Convers. **19**, 60–65 (2004)

9. Liu, X.Q., Zhang, H.Y., Jun, L., Yang, J.: Fault detection and diagnosis of permanent magnet DC motor based on parameter estimation and neural network. IEEE Trans. Ind Electron. **47**, 1021–1030 (2000)
10. Yang, B.S., Han, T., An, J.L.: ART-KOHONEN neural network for fault diagnosis of rotating machinery. Mech. Syst. Signal Process **18**, 645–657 (2004)
11. Marzi, H.: Real-time fault detection and isolation in industrial machines using learning vector quantization. Proc. Inst. Mech. Eng. Part B J. Eng. Manuf. (218), 949–959 (2004)
12. El Shafei, A., Hassen, T.A.F., Soliman, A.K., Zeyada, Y., Rieger, N.: Neural network and fuzzy logic diagnostics of 1x faults in rotating machinery. J. Eng. Gas Turbine Power **129**, 703–710 (2007)
13. Wang, C.C., Too, J.G.P.: Rotating machine fault detection based on HOS and artificial neural networks. J. Int. Manuf. **13**, 283–293 (2002)
14. AbuMahfouz, I.A.: A comparative study of three artificial neural networks for the detection and classification of gear faults. Int. J. General Syst. **34**, 261–277 (2005)
15. Kozlovsky, V.N., Aydarov, D.V., Vasilyev, M.M., Debelov, V.V.: Development of electric cars and cars with hybrid energy unit. Gruzovik **6**, 18–21 (2018)
16. Endachev, D.V., Skripko, L.A.: Calculation of the degree of charge of lithiumion batteries used in battery systems of electric vehicles and hybrid cars. Electron. Electr. Equip. Trans. **4**, 8–11 (2017)
17. Sajjad, S., Zaidi, H., Aviyente, S.: Prognosis of gear failures in DC starter motors using hidden markov models. IEEE Trans. Ind. Electron. **58**(5), 1695–1706 (2011)

Design, Fabrication, and Implementation of a Novel MWCNTs/PDMS Phosphate Sensor for Agricultural Applications

Fowzia Akhter, H. R. Siddiquei, Md Eshrat E. Alahi, and S. C. Mukhopadhyay

1 Introduction

Phosphate (PO_4^{3-}) ions are one of the essential minerals for plant growth. It contributes to cell growth and reproduction. Hence, a small amount of phosphorus loss may significantly impact the soil, such as reducing the growth of algae and freshwater weeds [1, 2]. Inadequate amount of phosphorus in lake water may also harm flora and fauna's life [3]. When phosphate concentration in lake water exceeds more than one ppm, the water's natural balance gets affected [4]. Therefore, it is vital to manage the phosphate level in water for effective and economical farming.

Numerous methods are available for detecting phosphate levels such as high-performance liquid chromatography (HPLC), UV-spectroscopy, and chromatography. These expensive techniques provide accurate results but complex, time-consuming and difficult to maintain [5, 6]. Some sensors have been developed in recent years [7–10] for phosphate detection, but there is still a lot to improve in terms of cost, response time, and detection range.

This research proposes a novel flexible MWCNTs/PDMS interdigital sensor for phosphate detection. This also discusses the design and development process of a portable smart sensing system with LoRa connectivity for water real-time water quality monitoring. The farmers can regularly check the phosphate level in farm water and get expert's feedback whenever necessary, irrespective of the farm's time and place.

F. Akhter (✉) · H. R. Siddiquei · S. C. Mukhopadhyay
Faculty of Science and Engineering, Macquarie University, Sydney, NSW 2109, Australia

M. E. E. Alahi
Shenzhen Institute of Advanced Technology, Chinese Academy of Science, Shenzhen, China

J. K. Mandal and J. K. Roy (eds.), *Proceedings of International Conference on Computational Intelligence and Computing*, Algorithms for Intelligent Systems,
https://doi.org/10.1007/978-981-16-3368-3_29

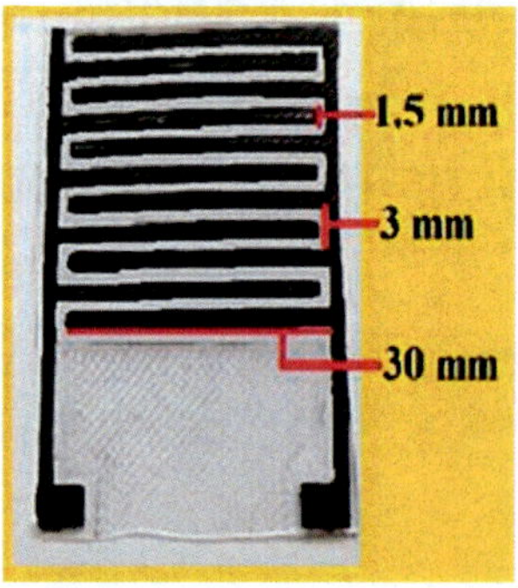

Fig. 1 Developed sensor patch

2 Material and Methods

This section represents the fabrication process and experimental procedures for phosphate ions detection using the developed sensor in water.

2.1 Sensor Fabrication Process

The base molds are designed and printed using the 3D printer. The developed molds are washed thoroughly by deionized water before starting the fabrication process. Conductive ink is prepared by mixing the MWCNTs powder, and PDMS are mixed at a 10:1 ratio. The ink is then cast on the mold to develop the electrodes. After that, the molds are cured at 50 °C for an hour to solidify the nanocomposites. Later, PDMS is poured on the mold for substrate formation. A casting knife is used to adjust the height of PDMS to 1000 μm. After that, the sample is desiccated until the trapped air bubbles are removed. Finally, the sample is heated at 50°C for five hours for solidification. The samples are then peeled off and used for phosphate detection experiment. Figure 1 shows the developed sensor which has six pairs of the interdigitated electrode.

2.2 Experimental Process

The phosphate detection experiments using the MWCNTs/PDMS sensor are carried out at ambient environment. Figure 2 shows the experimental setup, which includes HIOKI IM 3536 for EIS analysis, a laptop for data collection, phosphate sensor, and a stand for placing the sensor inside the solution. At the beginning of the experiment, different concentrations (0.1, 1, 10, 20, and 30 ppm) phosphate solutions are prepared.

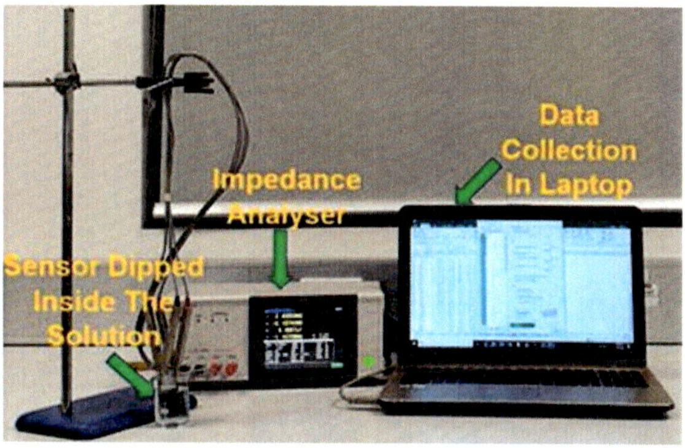

Fig. 2 Experimental setup for phosphate detection

2.3 *Portable System Development*

A portable smart phosphate detection system for water quality monitoring is designed and developed. It consists of an AD5933 impedance analyzer, an LCD for displaying sensor data, a phosphate sensor, an Arduino Uno microcontroller for processing data, a LoRa shield for data transmission, a power converter for charging the battery, a battery for energy supply, and an antenna. Once the system is powered, it initializes the AD5933 to operate the phosphate sensor. After reading the phosphate sensor resistance value toward the sample solution, the data is converted into appropriate con centration, applying a data processing algorithm. The phosphate concentration and water temperature data are shown on display and transmitted to the Thingspeak server immediately to store and process further. This Thingspeak server is accessible from any location. Therefore, the agronomist can check the dataset and provide necessary feedback to the farmers if any anomaly is found in the farm water. Figure 3 shows the system's workflow and the final prototype measuring phosphate concentration in sample water.

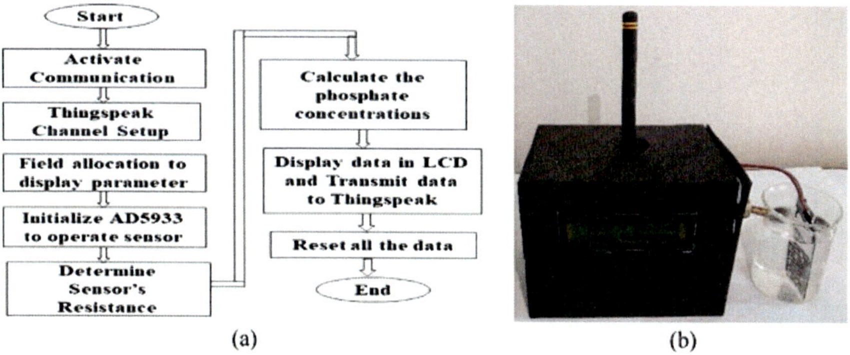

Fig. 3 **a** System's flowchart, and **b** portable system in its working condition

3 Results and Discussions

This section represents the sensor responses toward phosphate solutions and the method of developing calibration standard.

3.1 Response Toward Phosphate Solutions

The sensor is tested toward differently concentrated phosphate solutions. The LCR meter frequency is varied from 10 Hz to 10 kHz, and the sensor's response in terms of resistance, reactance, and impedance is recorded. After analyzing the data, it is found that the sensor resistance changes significantly compared to reactance. Therefore, frequency versus resistance curve for various frequency toward phosphate concentrations is included in this paper. Figure 4a shows the sensor's response to various concentrations of phosphate solutions. This change in response is because of the charge transfer resistance (R_{ct}), solution resistance (R_s), and double-layer capacitance (C_{dl}). The sensor shows distinguishable changes toward phosphate between 10 Hz to 10 kHz for a frequency sweep from 10 Hz ~ 100 kHz. While obtaining the calibration curve to measure unknown concentrations, a particular frequency

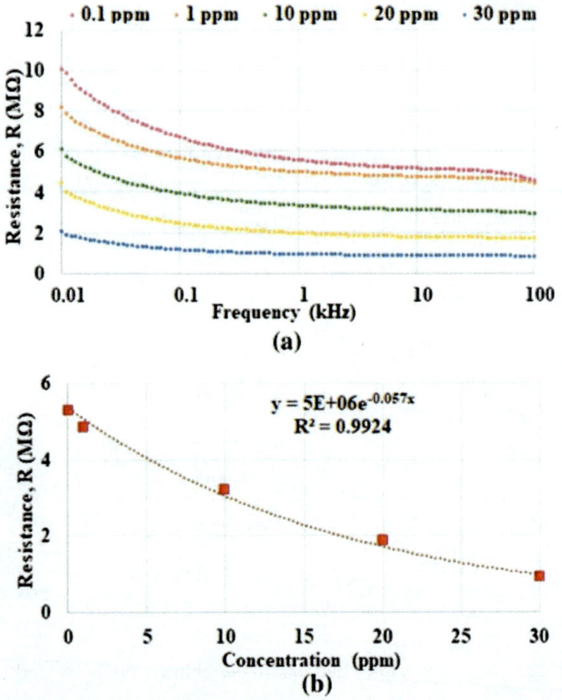

Fig. 4 **a** Change in sensor's resistance for differently concentrated phosphate solutions, and **b** the developed calibration curve at 1.1 kHz frequency

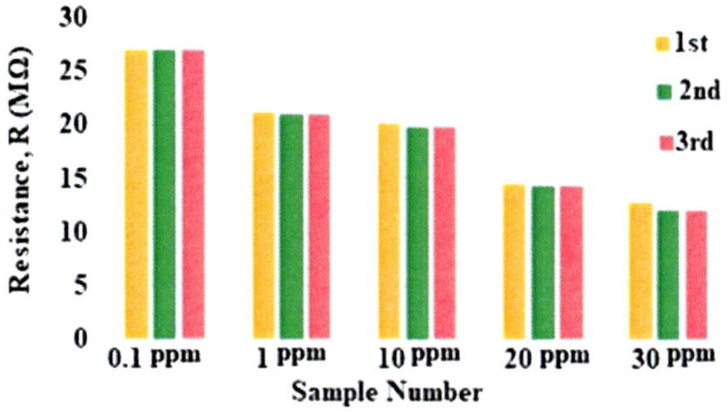

Fig. 5 Repeatability test of the sensor for various phosphate solutions (0.1–30 ppm)

of 1.1 kHz is chosen because it is in the region where all the concentrations are separable.

The calibration curve is derived by plotting curve taking all the measured concentrations versus the corresponding resistance. Figure 4b shows the calibration standards for measuring phosphate concentrations. The maximum determination coefficient (R^2) is found exponential fit. Although these regression analyses are complicated and time-consuming compared to the linear fit, these have been chosen for covering the entire concentration range of phosphate.

$$C_{\text{phosphate}} = \frac{15.42 - \ln(\mathrm{R_s})}{0.057} \tag{1}$$

In Equation 1, *C*phosphate (ppm) refers to the actual phosphate concentrations, and Rs (Ω) represents the sensor's resistance measured for a particular concentration. Equation (1) can be used utilized for determining phosphate concentrations in any unknown sample solution.

Repeatability test is essential to prove the reliability of the sensor. Therefore, the sensor response for a particular solution is recorded three times. Outcomes of this experiment are shown in Fig. 5. It is evident from the experimental results that resistance variation is not significant toward a sample solution. The relative standard deviation (RSD) of the dataset is found less than 2%. Thus, it can be claimed that the proposed sensor is highly stable and repeatable for phosphate solutions.

4 Conclusion

The fabrication and implementation of a novel flexible MWCNTs/PDMS interdigital sensor for phosphate detection are successfully presented in this paper. The sensor can detect a wide range of phosphate concentration from 0.1 to 30 ppm, signifying this sensor's application for various agricultural applications. The consistency in phosphate detection and fast response makes the sensor an ideal candidate for agrarian farms. Therefore, an IoT-enabled mobile sensor node with LoRa communication protocol is designed and developed. Anyone can use the system with minimal training. The farmers can regularly check the phosphate level in the farm water and share it with the agronomist on the Thingspeak server. The stored data will help the experts do any case study and provide necessary feedback to the farmers. Existence of this type of mobile sensor node in any agricultural farm will significantly impact the agricultural industry, detecting any abnormality in an early stage and taking necessary measures before the situation goes out of hand.

Acknowledgements The authors would like to thank Macquarie University for providing all the necessary facilities to carry out the research.

References

1. Environmental impact of nitrogen and phosphorus fertilisers in high rainfall areas, https://www.agric.wa.gov.au/high-rainfall-pastures/environmental-impact-nitrogen-and phosphorus-fertilisers-high-rainfall-areas, last accessed 13 Jan 2021
2. Nag, A., Alahi, M.E.E., Feng, S., Mukhopadhyay, S.C.: IoT-based sensing system for phosphate detection using Graphite/PDMS sensors. Sens. Actuators, A. **286**, 43–50 (2019). https://doi.org/10.1016/j.sna.2018.12.020
3. Sarwar, M., Leichner, J., Naja, G.M., Li, C.Z.: Smart-phone, paper-based fluorescent sensor for ultra-low inorganic phosphate detection in environmental samples. Microsyst. Nanoeng. **5**, 1–10 (2019). https://doi.org/10.1038/s41378-019-0096-8
4. Adelowo, F.E., Oladeji, S.O.: Spectrophotometric analysis of phosphate concentration in agricultural soil samples and water samples using molybdenum blue method. Braz. J. Biol. Sci. **3**, 407–412 (2016). https://doi.org/10.21472/bjbs.030616
5. Islam, S., Reza, M.N., Jeong, J.T., Lee, K.H.: Sensing technology for rapid detection of phosphorus in water: a review. J. Biosyst. Eng. **41**, 138–144 (2016). https://doi.org/10.5307/JBE.2016.41.2.138
6. Eshrat, Md., Alahi, E., Mukhopadhyay, S.C.: Detection methodologies for pathogen and toxins: a review. Sens. **1885**, 1–20 (2017). https://doi.org/10.3390/s17081885
7. Zhu, L., Zhou, X., Shi, H.: A potentiometric cobalt-based phosphate sensor based on screen-printing technology. Front. Environ. Sci. Eng. **8**, 945–951 (2014). https://doi.org/10.1007/s11783-013-0615-z
8. Kumar, P., Kim, D.M., Hyun, M.H., Shim, Y.B.: An all-solid-state monohydrogen phos-phate sensor based on a macrocyclic ionophore. Talanta **82**, 1107–1112 (2010). https://doi.org/10.1016/j.ta-lanta.2010.06.011
9. Akhter, F., Nag, A., Alahi, M.E.E., Liu, H., Mukhopadhyay, S.C.: Electrochemical detection of calcium and magnesium in water bodies. Sens. Actuators, A. 305, 1–10(2020). https://doi.org/10.1016/j.sna.2020.111949
10. Manováa, A., Beinrohrb, E.: Determination of phosphate in water by flow coulometry. Acta. Chimica. Slovaca. **13**, 102–107 (2020). https://doi.org/10.2478/acs-2020-0015

An IoT-Based Architecture for Edge Computing to Reduce the Latency and Bandwidth for Streaming Data Application

Chinmoy Bharadwaj, Sajal Saha, Kaushik Kamal Kalita, and Asish Kumar Mukhopadhyay

1 Introduction

IoT devices and its applications are increasing in various fields day by day to provide anything, anytime, and anywhere connectivity to the user. Realtime media streaming from a surveillance camera or cloud-based servers is one such application. However, bandwidth limitations are one of the key constraints to provide data from IoT sensors to users via cloud application servers (AWS cloud formation, AWS KINESIS, etc.) in a near real-time manner. Cloud application servers are being used to provide customized data to end-user as per requirement. However, bandwidth constraints and network traffic congestions limit near-real-time data transmission. Here remains a high demand for low latency, high bandwidth, high storage with high computation, high security in real-time IOT applications like video streaming, media streaming, live updates, health care, etc. So, proper fog and edge computing remain a significant part of research in knowledge for low latency among the data centers with a limited number of bandwidths for faster transmission of data. Owing to incredible growth in the accessibility of datasets through several data types, a real edge computing technique is categorically essential to aim at solving the fog, edge, cloud computing problems with low latency, and a limited number of bandwidths for better efficacy in real-time IoT applications. Even though several existing cloud computing techniques help in processing and storage of those data, but it is lagging in terms of less storage with less computation, high latency, low bandwidth, and less security for real-time applications. To check latency and bandwidth among the computing techniques, a

C. Bharadwaj (✉) · S. Saha · K. K. Kalita
Department of Computer Science and Engineering, The Assam Kaziranga University, Jorhat, India

S. Saha
e-mail: sajalsaha@kazirangauniversity.in

A. K. Mukhopadhyay
Department of E&CE, Bundelkhand University, Kolkata, India

J. K. Mandal and J. K. Roy (eds.), *Proceedings of International Conference on Computational Intelligence and Computing*, Algorithms for Intelligent Systems,
https://doi.org/10.1007/978-981-16-3368-3_30

better technique for solving the fog, edge, cloud computing must be developed with low latency and a limited number of bandwidths for better efficacy.

1.1 Motivation

The use of mobile devices has played a very important role in exercising the multi-media services. Media streaming analysis focused on the wireless LAN (WLAN) and mobile nodes that have been considered as an edge device.

The main motivation of this paper is to design an architecture for media streaming data using fog computing and edge computing. For this, we implement a riverbed academic modeler based on the Opnet network simulator to reduce the latency and bandwidth for fog, edge, cloud computing. By this, we are using the media services application based on the mobile node moving to the trajectory path between the devices for checking the latency.

1.2 Paper Organization

The paper is organized as follows: In Sect. 2, we introduce the related works of fog and edge computing, and the opportunities besides the challenges that need to be undertaken for the execution of fog and edge computing in multimedia services. The proposed research is discussed in detail in Sect. 3. There in Sect. 4, we have made a thorough analysis of the proposed structure and the experimental results obtained. Finally, Sect. 5 of the paper is the conclusion.

2 Related Work

Wilianto et al. [1] has discussed the development of information technology service management using a cloud computing approach to improve the performance of the administration system and online learning at STMIK IBBI Medan, Indonesia. The network topology is developed in cloud computing using Amazon AWS architecture and is designed, modeled, and simulated using Riverbed Modeler Academic Edition to obtained values of the parameters like delay, load, CPU utilization, and throughput for system administration and online learning. Tuli et al. [2] has established the efficacy of FogBus by creating a computing environment with it that integrates finger pulse oximeters as IoT devices with Smartphone-based gateway and Raspberry Pi-based Fog nodes for Sleep Apnea analysis. We also evaluate the characteristics of FogBus concerning other existing frameworks and the impact of various FogBus settings on system parameters through the deployment of a real-world IoT application. The experimental results show that FogBus is comparatively lightweight and responsive, and different FogBus settings can tune the computing environment as per the situation demands. Shih et al. [3] has explained the technical characteristics of Fog computing and its real-time applications. Fog Computing is an integration of cloud and IoT. It works for the interconnection of different edge devices and is used for processing data communications instead of telecommunications services in the network edge. On the other side, the edge computing platform worked by application processing to be performed by an edge server positioned between the cloud and the user. This allows for some of the work offloads from the cloud or the user by speeding up the application process. Chen et al. [4] has described the basic knowledge, common functions, and modeling method of the OPNET network packet switching routine simulation where the readers will be familiar with the operation of common editors and have a good command of the OPNET uses a three-layer modeling mechanism and simulation using basic OPNET debugging methods on an OPNET model of a large sensor network. Aamir et al. [5] has presented a VoIP traffic Quality of Service (QoS) evaluation in a multimedia network with other forms of network traffic, such as database, video, and internet is also deployed. In a heavily trafficked network, VoIP is equipped with various codecs and frame counts per packet to see how these changes affect VoIP QoS metrics such as jitter, queuing latency, packet end-to-end delay, and packet delay variance (end-to-end jitter) metrics. Core Network, Parvez et al. [6] presented a detailed survey on emerging technologies for achieving low latency communications, taking into account three distinct solution domains: Core Network, Radio Access Network, and Caching. Premsankar et al. [7] suggested a use case methods to evaluate the response delay in various implementation scenarios using a resource-intensive 3-D application as a paradigmatic example. The results of the experiments show that edge computing is needed to meet the latency challenge of augmented and virtual reality applications.

Panse et al. [8] has observed and summarized different use case scenarios of fog computing with contributions on security and privacy issues. Finally, this paper similarly highlights several open issues in fog computing thereby giving light on future research directions for implementing fog computing systems. Mohorko et al. [9] has presented simulation results concerning real-time simulations, which were completed in the framework of a military development nearby the modeling of Command-and-Control Information Schemes. A new OPNET Modeler module System-in-the-Loop was used, which allows for communication between real and simulated devices and networks, and represents a new simulation approach. The test network we constructed by means of these tools, wherever we connected computer equipment creating real-time video streams with a simulated mobile wireless 802.11 g network on a virtual ground. Aamir et al. [10] has emphasized QoS analysis in a wired IP network with more realistic enterprise modeling and presents simulation results of a few statistics not presented and discussed before. Four different applications are used, i.e., FTP, Database, Voice over IP (VoIP), and Video Conferencing (VC). Two major queuing disciplines are evaluated i.e., 'Priority Queuing' and 'Weighted Fair Queuing' for packet identification under Differentiated Services Code Point (DSCP). The simulation results show that WFQ has an edge over PQ in terms of queuing delays and jitters experienced by low-priority services. For high priority traffic, the dependency of 'Traffic Drop', 'Buffer Usage', and 'Packet Delay Variation' on selected buffer sizes is simulated and discussed to evaluate QoS deeper. In the end, it is also analyzed how the network's database service with applied Quality of Service may be affected in terms of throughput for internal network users when the server is also accessed by external users through a Virtual Private Network (VPN). Memon et al. [11] has evaluated the video streaming performance that has been performed by applying different simulation scenarios over the IP-based network. The core objective is to analyze network behavior and video quality. The network behavior includes packets loss, end-to-end delay, and throughput. The received video quality is compared and analyzed using application-level values, which include Peak Signal to Noise Ratio (PSNR) method and Mean Opinion Score (MOS) scale. The simulation results show different network behavior and quality of the received video depending on the network congestion and sent with dummy background traffic. Haratcherev et al. [12] has presented a communication architecture for video streaming over 802.11 that is capable of adapting to changes in the link quality and sharing of the wireless channel in various use scenarios. Seamless video streaming over wireless links imposes strong demands on video codecs and the underlying network. It is not sufficient that only the video codec or only the radio adapts to changes in the wireless link quality; efforts should be applied in both layers, and—if possible—synchronized. Also, the disturbing effect of possible background traffic over the same shared medium has to be taken into account. Experimental results show that substantial improvements in the quality of the video can be obtained by applying link adaptation and cross-layer signaling techniques. Musaddiq et al. [13] has used the OPNET IT Guru simulation tool for the performance of the IEEE 802.11g wireless local area network and IEEE 802.3i Ethernet network for a new campus building at Bahria University Islamabad campus has been analyzed. The newly constructed campus consists of two floors and accommodates around 300 users including students and faculty. The network on both floors has the same model and applications for traffic generation. The performance has been analyzed in terms of traffic, queuing delay, and throughput. The above-mentioned protocols are also evaluated based on the cost incurred for the purchase of equipment, installation, and maintenance for the campus. The simulation results and cost analysis show that the performance of network protocol IEEE 802.3i is more efficient and reliable but the implementation is complex and expensive, whereas the IEEE 802.11g protocol is easy to install and provides mobility but with low-grade QoS (Quality of Service). Sondi et al. [14] have evaluated the performance of a QoS extension of the OLSR protocol on voice communication using the OPNET simulator. The network model can represent various locations like railway stations, Campus Universities, or traffic jams. We describe the performance end-users involved in the voice communication session can expect, even in the presence of other traffics, like file transfer between other users in the same network. Khan et al. [15] has proposed an integration of Graph Query Language web framework based on the integration of an abstraction layer through constrained request or response structure which is targeted to overwhelmed this challenge in cloud-based IoT sensing applications. Inorder to achieve energy conservation and efficient data management in Internet of Things

(IoT) sensing applications, we have included some modules of efficient web framework with cloud services, in order to minimalize the number of rounded trips for data delivery and graph-based data representation. Further, the author implemented the proposed interfacing of abstraction layer in constrained applications, to develop a testbed using Z1 IoT motes, Contiki OS with Google cloud services. The experimental results show substantial reductions having minimum data delivery delay of 51.53% and minimum energy consumption of 52.88%, respectively in IoT applications involving sensor network. Al-Ghadanfary et al. [16] has simulated some of these protocols for multimedia applications. Various scenarios for studying the effect of network scalability were conducted. Furthermore, the performance of these protocols with fixed or moving nodes was also studied. The results showed that AODV protocol throughput performance is the best when the network nodes increased both for fixed or moving scenarios. OLSR, TORA, and GRP were the favorite protocols in terms of WLAN delay leaving AODV and DSR as the worst in this statistic. The results of this paper suggest the use of AODV routing protocol in large client mesh wireless networks for the video streaming application in which high throughput is required and WLAN delay is tolerated; while OLSR, TORA, and GRP are the preferred protocols for VoIP applications. Khan et al. [17] has presented an efficient regression model for predicting video quality non-intrusively prediction of H.264 encoded video quality over UMTS networks besides towards demonstrate their request to video quality monitoring and variation in mobile wireless streaming services. The model is predicted from a grouping of a set of objective parameters in the application and physical layer in relations to the Mean opinion Score (MOS). The application layer parameters measured are the content type, sender bitrate, and frame rate and the physical layer parameters are the block error rate demonstrated through two-state Markov model for a mean burst length of 1.75. The performance of the proposed metric is evaluated with unseen dataset with good prediction accuracy.

3 Proposed Research

A physical architecture has been created in Opnet and analyzed the quality of service parameters like throughput, latency, bandwidth, and jitter of the proposed model using the toolkit Riverbed academic modeler for connecting the media streaming data using WLAN and mobile nodes that have been considered as an edge device, while the mobile nodes are moving in a trajectory path between the devices for finding the latency and bandwidth in near-real-time IoT applications of Edge computing. Here the mobile nodes have been used for data streaming between the devices for setting the trajectory path while the mobile node is moving from one node to another node to measure the latency and bandwidth. The comparison between the various types of architectures indicates that there has been a variation in delay when the mobile node is moving along a trajectory path considering its near real-time movement. The data streaming has been possible between mobile devices for finding the delay and throughput (Figs. 1 and 2).

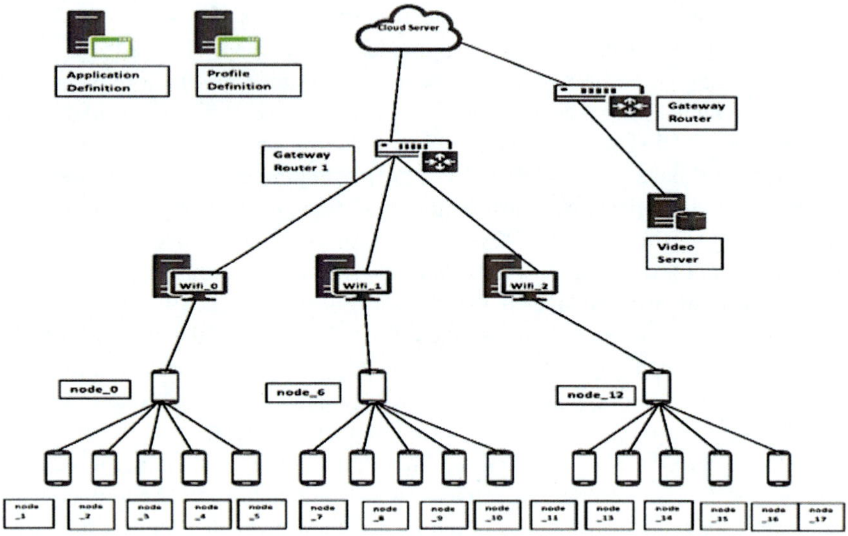

Fig. 1 The proposed architecture of a media streaming application

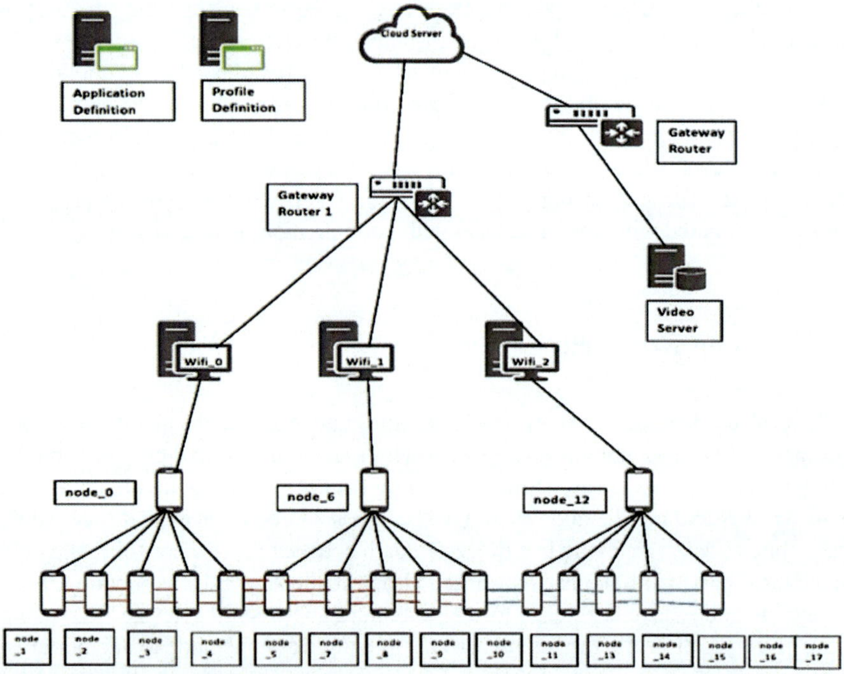

Fig. 2 The proposed architecture of a medium streaming application for setting the trajectory path between the mobile devices

4 Experimental Results

We have created a simulation environment in Riverbed academic modeler and analyze the data in MATLAB. In Opnet architecture media streaming devices are connected from the video server to the gateway router in a wired network and from the gateway router to the mobile nodes in a wireless network. Both WLAN and mobile nodes have been considered as an edge device. Here the mobile nodes have been used for data streaming between the devices for setting the trajectory path while the mobile node is moving from one node to another node to measure the latency and bandwidth. Network architecture and mobile node position and their movements are shown in Figs. 3, 4, 5 and 6 in the simulation environment. The simulation results in MATLAB have been done for finding the variation in delay transmission time as well as for the network bandwidth by using the datasets in the edge computing environment are shown in Figs .7, 8, 9 and 10.

Parameters considered for the simulation are as follows Table 1. The packet transmission end-to-end delay time for video conferencing, video streaming, and VoIP when data is streamed to mobile node from cloud server directly is shown in Fig. 7. The average throughput value of the network when data is streamed to mobile node from cloud server directly is shown in Fig. 8. The packet transmission end-to-end delay time for video conferencing, video streaming, and VoIP of the networks when data is streamed through the mobile edge devices is shown in Fig. 9. The average throughput value of the network when data is streamed through the mobile edge devices is shown in Fig. 10.

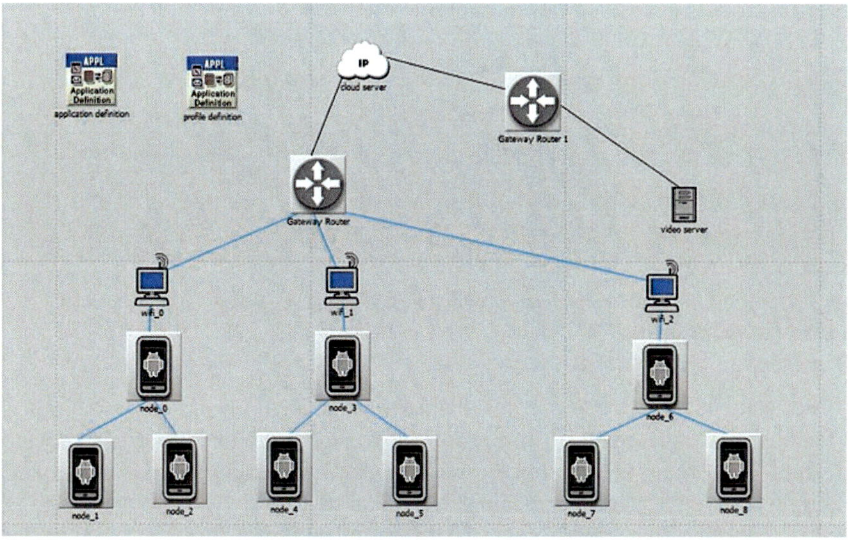

Fig. 3 Opnet IoT simulation architecture for a mobile node is considered as an edge device in Scenario 1

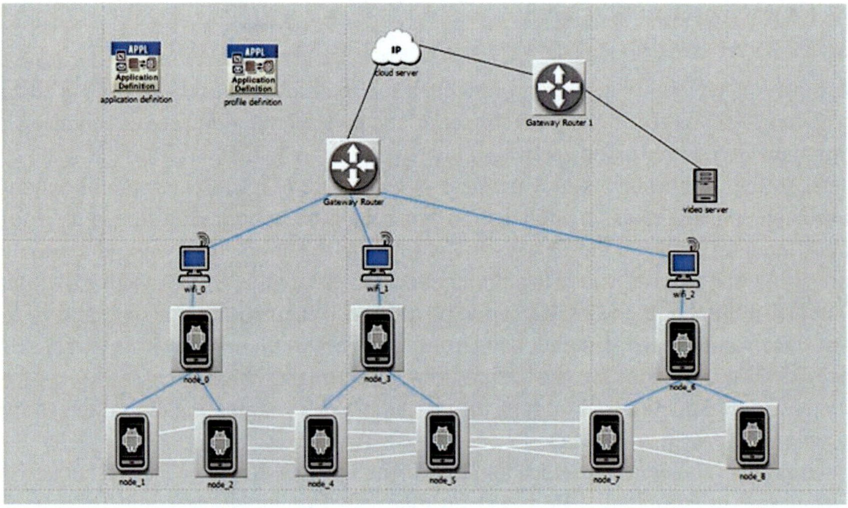

Fig. 4 Considering mobile node as an edge device for setting the trajectory path between the mobile devices in Scenario 2

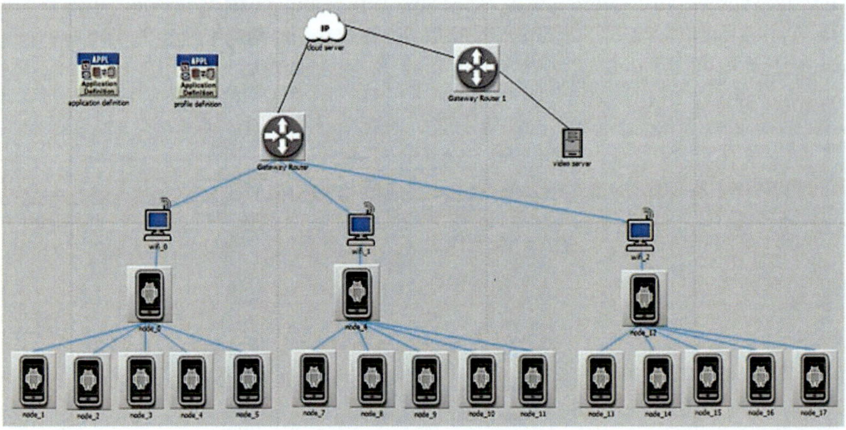

Fig. 5 Opnet IoT simulation architecture for mobile node considered as an edge device moving in Scenario 2 with increased traffic

It is evident from the above result that throughput remains unaffected when network architecture changed from direct cloud streaming to mobile edge device streaming. But there is a significant improvement in delay when data streaming changed from direct cloud streaming to mobile edge device streaming. Delay is further reduced when mobile node is moving. Therefore, there is a significant improvement in QoS when the mobile node is moving in a near real-time manner.

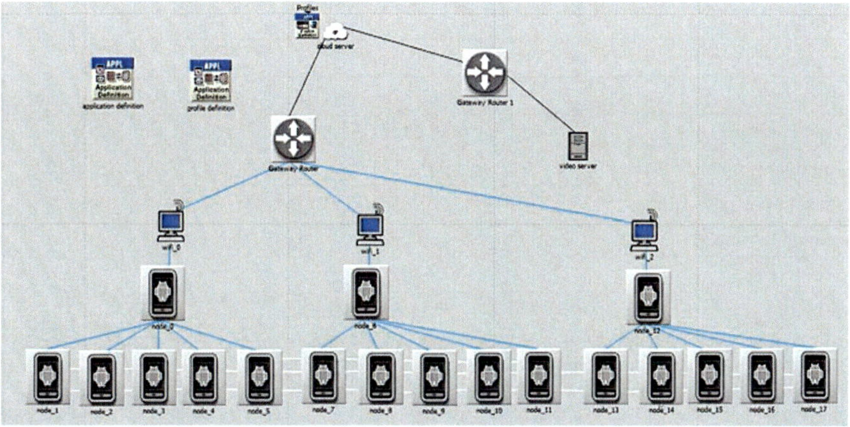

Fig. 6 Considering mobile node as an edge device moving when the nodes number has been increased for setting the trajectory path between the mobile devices

Table 1 Parameters considered for the simulation

Parameters	Values
Encoding techniques	PCM
Encoder scheme	G.723.1 5.3 K
Buffer size (bits)	256,000
Frame size information (bytes)	128 × 240 pixels
Type of service	Best Effort (0)
RSVP	Null
Video conferencing (high-resolution video)	128 × 240 pixels
Video streaming (type of service)	28
Voice over internet protocol (PCM quality speech)	Interactive Voice (6)
Quality of service (QoS)	Enabled (1) and Disabled (0)
CPU background utilization	Null
Server advanced server configuration	Sun Ultra 10 333 MHz: 1 CPU, 1 Core(s) Per CPU, 333 MHz, Solaris, System
WLAN physical characteristics	Extended Rate PHY (802.11 g)

(continued)

Table 1 (continued)

Parameters	Values
Data rate (bps)	54 Mbps
Channel settings	5 GHz Ch 153
Minimum frequency (MHz)	5.755
Transmit power (W)	0.005 W/6.99 dBm
Simulation time	30 min
No. of nodes	8 and 17
Mobility	Fixed + mobiles
Network area	100 m × 100 m
Packet size (bytes)	Exponential (1024)
Cache size	10
OPT transmit interval (s)	300 s
IP packet discard timeout (s)	10 s
Cellular mobility in a building	5
Mobility across geographical area	10

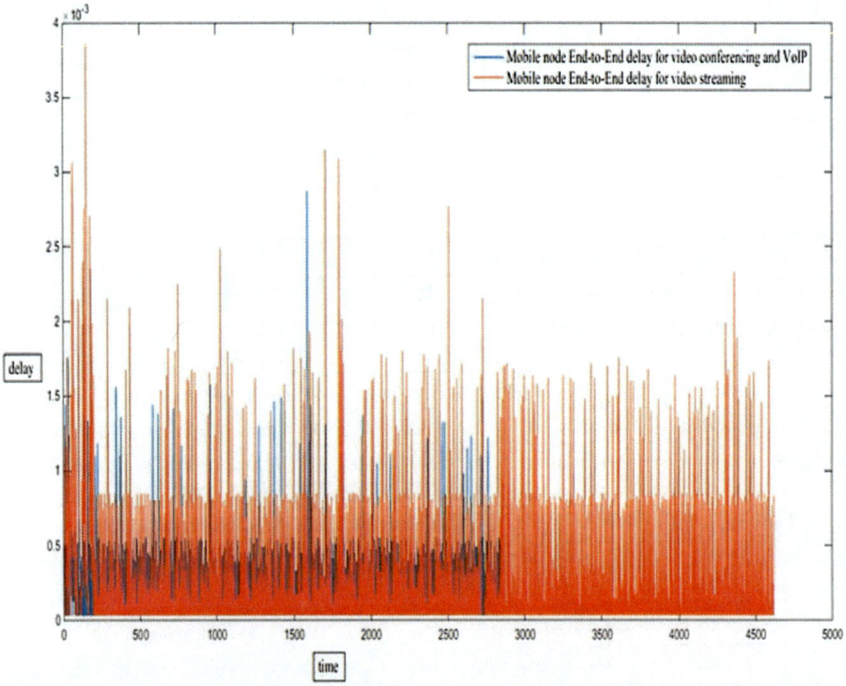

Fig. 7 The packet transmission end-to-end delay time for video conferencing, video streaming, and VoIP when data is streamed to Mobile node from cloud server directly

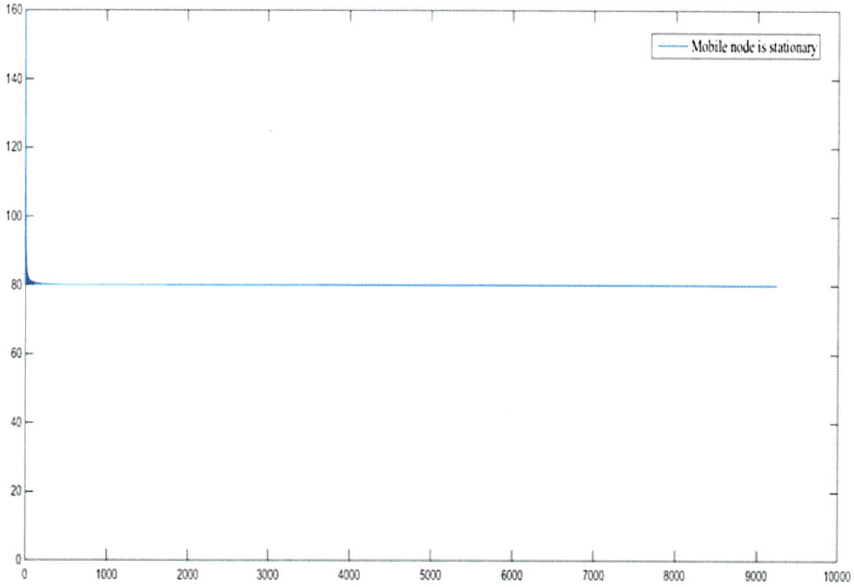

Fig. 8 The average throughput value of the network when data is streamed to Mobile node from cloud server directly

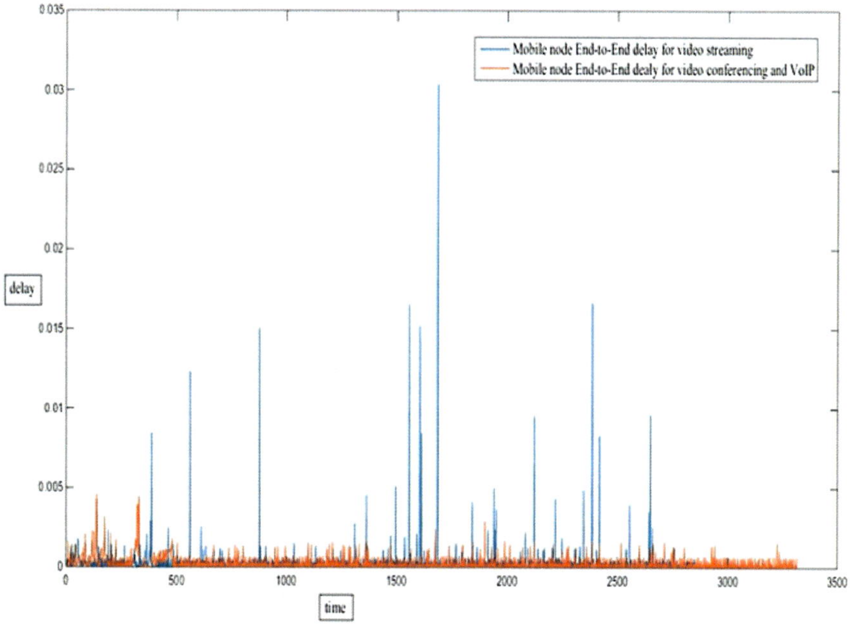

Fig. 9 The packet transmission end-to-end delay time for video conferencing, video streaming, and VoIP of the networks when data is streamed through the Mobile edge devices

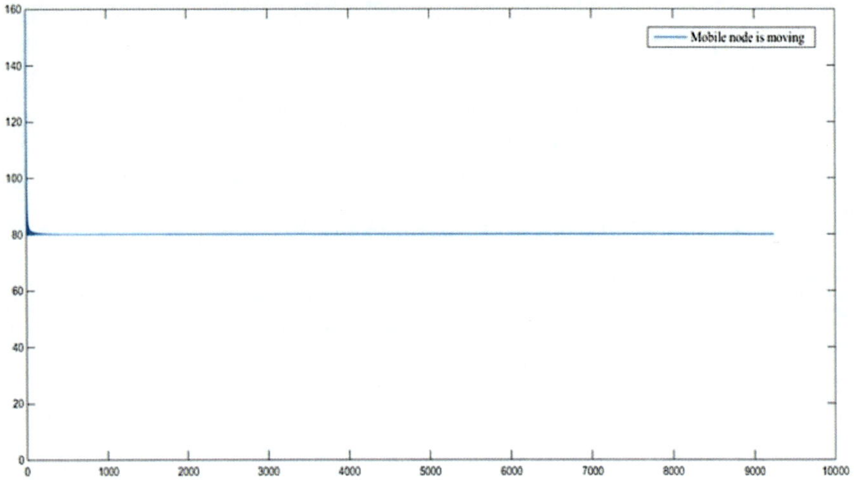

Fig.10 The average throughput value of the network when data is streamed through the Mobile edge Devices

Acknowledgements At this juncture, I would like to express my sincere gratitude to my guide, Sajal Saha, for his continuous support, motivation, and guidance throughout my research work. This research paper would not have been possible without his expertise and unflinching support. I also express my profound gratitude to all other respected teachers of the university for their insightful comments, advice, and co-operation.

References

1. Wilianto, Fitri, I.: Information technology service management with cloud computing approach to improve administration system and online learning performance. J. Commun. Inf. Technol. **9**(2), 51–57 (2015)
2. Tuli, S., Mahmud, R., Tuli, S., Buyya, R.: FogBus: a blockchain-based lightweight framework for edge and fog computing. J. Syst. Softw. 154:22–36 (2019). https://doi.org/10.1016/j.jss.2019.04.050
3. Shih, P.-J.: From the cloud to the edge: the technical characteristics and application scenarios of fog computing. In: Institute for Information Industry Market Intelligence and Consulting Institute, pp. 61–64 (2018). https://doi.org/10.5875/ausmt.v8i21836
4. Chen, M., Miao, Y., Humar, I.: OPNET IoT Simulation. Springer, Berlin. ISBN 978-981-32-9169-0 (2018). https://doi.org/10.1007/978-981-32-9170-6
5. Aamir, M., Mustafa Ali Zaidi, S.: QoS analysis of VoIP traffic for different codecs and frame counts per packet in the multimedia environment using OPNET. In: 15th International Multi-topic Conference (INMIC), pp. 1–7, IEEE. ISBN: 978-1-4673-2252-2 (2013). https://doi.org/10.1109/INMIC.2012.6511508
6. Parvez, I., Guvenc, I.: A survey on low latency towards 5G: RAN, core network and caching solutions. IEEE Commun. Surv. Tutor. **20**(4), 3098–3130 (2018)
7. Premsankar, G., Di. Francesco, M., Taleb, T.: Edge computing for the internet of things. IEEE Internet Things J. **5**(2), 1275–1284 (2018)

8. Panse, V.H., Neole, B. Ghate, M., Lad, B.: A survey on use cases and security issues in fog computing. Helix **8**(6), 4413–4417 (2018). https://doi.org/10.29042/2018
9. Mohorko, J., Fras, M., Cucej, Z.: Real video stream transmission over a simulated wireless link. In: IEEE, 2008 International Conference on Advanced Technologies for Communications, pp. 231–234 (2008). https://doi.org/10.1109/ATC.2008.476052
10. Aamir, M., Zaidi, M., Mansoor, H.: Performance analysis of DiffServ based quality of service in a multimedia wired network and VPN effect using OPNET. Int. J. Comput. Sci. Issues (IJCSI) **9**(3) (2012)
11. Memon, S.A., Hasan, S.R., Remon, N.A.: Evaluation of video streaming performance over peer-to-peer network. In: 2014 International Conference on Collaboration Technologies and Systems (CTS), pp. 413–420. IEEE (2014). https://doi.org/10.1109/CTS.2014.6867597
12. Haratcherev, L., Taal, J., Langendoen, K., Lagendijk, R., Sips, H.: Optimized video streaming over 802.11 by cross-layer signaling. IEEE Commun. Mag. **44**(1), 115–121 (2006). https://doi.org/10.1109/MCOM.2006.1580941
13. Musaddiq, A., Hashmi, U.S., Jawed, S.: Performance and cost evaluation of IEEE 802.11g and 802.3i protocols for network connectivity at a University Campus using OPNET Simulation. In: 2013 UK Sim 15th International Conference on Computer Modelling and Simulation, pp. 762–767. IEEE (2013). https://doi.org/10.1109/UKSim.2013.111
14. Sondi, P., Gantsou, D., Lecomte, S.: Performance evaluation of multimedia applications over an OLSR-based mobile ad hoc network using OPNET. In: 12th International Conference on Computer Modelling and Simulation, pp. 567–572. IEEE (2010). https://doi.org/10.1109/UKSIM2010.109
15. Khan, R., Mian, A.N.: Sustainable IoT sensing applications development through graph-based abstraction layer. J. Electron. **9**(564), 1–23 (2020). https://doi.org/10.3390/electronics9040564
16. Al-Ghadanfary, K.A., Al- Somaidai, M.B.: Simulation of some routing protocols in a client wireless mesh network for multimedia applications. In: International Conference of Electrical, Communication, Computer, Power and Control Engineering (ICECCPCE), pp.1–6. IEEE (2014). https://doi.org/10.1109/ICECCPCE.2013.6998743
17. Khan, A., Sun, L., Fajardo, I.J.O., Liberal, F.: Video quality prediction model for H.264 video for H.264 video over UMTS network application in mobile video streaming. In: International Conference on Communication (2010). https://ieexplore.ieeeorg/xpl/comhome/5497983/proceeding